Heavy Ion
Collisions
Cargèse 1984

NATO ASI Series

Advanced Science Institutes Series

A series presenting the results of activities sponsored by the NATO Science Committe which aims at the dissemination of advanced scientific and technological knowledge, with a view to strengthening links between scientific communities.

The series is published by an international board of publishers in conjunction with the NATO Scientific Affairs Division

A	**Life Sciences**	Plenum Publishing Corporation
B	**Physics**	New York and London
C	**Mathematical and Physical Sciences**	D. Reidel Publishing Company Dordrecht, Boston, and Lancaster
D	**Behavioral and Social Sciences**	Martinus Nijhoff Publishers
E	**Engineering and Materials Sciences**	The Hague, Boston, and Lancaster
F	**Computer and Systems Sciences**	Springer-Verlag
G	**Ecological Sciences**	Berlin, Heidelberg, New York, and Tokyo

Recent Volumes in this Series

Volume 126—Perspectives in Particles and Fields: *Cargèse 1983*
edited by Maurice Lévy, Jean-Louis Basdevant, David Speiser, Jacques Weyers, Maurice Jacob, and Raymond Gastmans

Volume 127—Phenomena Induced by Intermolecular Interactions
edited by G. Birnbaum

Volume 128—Techniques and Concepts of High-Energy Physics III
edited by Thomas Ferbel

Volume 129—Transport in Nonstoichiometric Compounds
edited by George Simkovich and Vladimir S. Stubican

Volume 130—Heavy Ion Collisions: *Cargèse* 1984
edited by Paul Bonche, Maurice Lévy, Philippe Quentin, and Dominique Vautherin

Volume 131—Physics of Plasma–Wall Interactions in Controlled Fusion
edited by D. E. Post and R. Behrisch

Volume 132—Physics of New Laser Sources
edited by Neal B. Abraham, F. T. Arecchi, Aram Mooradian, and Alberto Sona

Series B: Physics

Heavy Ion Collisions

Cargèse 1984

Edited by

Paul Bonche

CEN Saclay
Gif-sur-Yvette, France

Maurice Lévy

Université Pierre et Marie Curie
Paris, France

Philippe Quentin

Université de Bordeaux
Bordeaux, France

and

Dominique Vautherin

Institut de Physique Nucliare
Université de Paris XI
Orsay, France

Plenum Press
New York and London
Published in cooperation with NATO Scientific Affairs Division

Proceedings of a NATO Advanced Study Institute entitled
Heavy Ion Collisions: From Collective Motions to Quarks,
held September 2–15, 1984,
in Cargèse, Corsica, France

Library of Congress Cataloging in Publication Data

NATO Advanced Study Institute (1984: Cargèse, Corsica)
  Heavy ion collisions.

  (NATO ASI series. Series B, Physics; v. 130)
  "Proceedings of a NATO Advanced Study Institute entitled Heavy Ion Colli-
sions: From Collective Motions to Quarks, held September 2–15, 1984, in
Cargèse, Corsica, France"—T.p. verso.
  "Published in cooperation with NATO Scientific Affairs Division."
  Bibliography: p.
  Includes index.
  1. Heavy ion collisions—Congresses. I. Bonche, P. II. North Atlantic Treaty
Organization. Scientific Affair Division. III. Title. IV. Series.
QC794.6.C6N376   1984   539.7'54   85–24403
  ISBN-13: 978-1-4684-5017-0   e-ISBN-13: 978-1-4684-5015-6
  DOI: 10.1007/978-1-4684-5015-6

© 1986 Plenum Press, New York
Softcover reprint of the hardcover 1st edition   1986
A Division of Plenum Publishing Corporation
233 Spring Street, New York, N.Y. 10013

PREFACE

The 1984 Cargèse Advanced Study Institute was devoted to the study of nuclear heavy ion collisions at medium and ultrarelativistic energies. The origin of this meeting goes back to 1982 when the organizers met at the GANIL laboratory in Caen, France which had just started accelerating argon ions at 44 MeV per nucleon. We then realized that 1984 should be the appropriate time to review the first results obtained with such new kinds of facilities. The material contained in this volume, presenting many beautiful results on nuclei at high excitation, fully confirms this point.

Many stimulating exchanges between experts in rather different fields already took place during the school and we hope that this cross fertilization will lead to further developments.

About half of the present volume is also devoted to the field of relativistic heavy ion collisions, which is now expanding rapidly. As an illustration, let us recall that the construction of a 30 on 30 GeV per nucleon collider at Brookhaven has been recognized last year as one of the major priorities by the US Nuclear Science Advisory Committee.

We would like to express our gratitude to NATO for its generous financial support which made this institute possible.

We also wish to thank the Institut de Physique Nucléaire et de Physique des Particules (France), the Commissariat à l'énergie atomique (France) and The National Science Foundation (USA) for the attribution of travel grants.

Special thanks are due to our colleagues Gordon Baym (Urbana), S.E. Koonin (Caltec) and W.J. Swiatecki for valuable suggestions regarding the program of the school.

It is also a pleasure to express our gratitude to the Université de Nice for making available the facilities of the Institut d'Etudes Scientifiques de Cargèse. Last but not least we wish to thank Ms Marie-France Hanseler for her outstanding organizational work, and the Staff of the Cargèse Institute for making our stay in Cargèse a very enjoyable one.

M. Levy
P. Bonche
P. Quentin
D. Vautherin

CONTENTS

Collective variables and dissipation   1
   R. Balian

Energy dissipation in nucleus-nucleus collisions
    around 40 MeV per nucleon: New phenomena
    or transition trends ?   33
   M. Lefort, B. Borderie, D. Jacquet, and the
   Nuclear Chemistry Group

Angular momentum dynamics in damped nuclear reactions,   69
   J. Randrup* and T. Dossing

Nuclear structure at high spin   99
   B. Herskind

Chaos in nuclei or statistical mechanics of small systems :
    fluctuations   119
   H.A. Weidenmüller

Spectral fluctuations and chaotic motion   145
   O. Bohigas, M.J. Giannoni*, and C. Schmit

Time-dependent variational principle for the expectation
    value of an observable: Mean-field applications   165
   M. Vénéroni

Dynamics of the relativistic heavy ion collisions   209
   J. Cugnon

Skyrmions, dense matter and nuclear forces   283
   C.J. Pethick

Nucleus-nucleus collisions at high energies   305
   M.A. Faessler

Comments on multiplicity in proton nucleus collisions   361
    A.K. Kerman*, T. Matsui, and B. Svetitsky

An introduction to lattice QCD   367
    G. Martinelli

Index   407

Note : in the case of several authors, the name of the
       speaker is followed by an *.

COLLECTIVE VARIABLES AND DISSIPATION

Roger Balian

Service de Physique Théorique, CEA, Centre de Saclay

91191 Gif-sur-Yvette Cedex, France

ABSTRACT

   This course is an introduction to some basic concepts of non-equilibrium statistical mechanics. We emphasize in particular the relevant entropy relative to a given set of collective variables, the meaning of the projection method in the Liouville space, its use to establish the generalized transport equations for these variables, and the interpretation of dissipation in the framework of information theory.

TABLE

1   - Introduction
2   - The Liouvillean formulation of quantum mechanics
3   - Contracting the description
4   - Reduced densities
5   - Relevant entropy relative to a set of variables
6   - Metric structure of the spaces of states and observables
7   - Projection in the Liouville space
8   - Memory kernel
9   - Generalized transport equations
10   - Mean-field approximations
11   - Perturbation methods and short-memory approximations
12   - Conclusion

1. INTRODUCTION

During the last decade, phenomenological theories of heavy nuclei have gradually been superseded by more microscopic approaches. It has become feasible to understand and even to compute many dynamical processes, for instance to analyse collisions of heavy ions, by starting from a many-body model of interacting nucleons without any adjustable parameter. This type of studies in nuclear physics appears as a new branch of statistical mechanics, and it takes advantage of many concepts and techniques drawn from the study of other types of matter.

A specific feature of nuclear systems is their *finite size*, which brings in more complications than simplifications. On the one hand, properties associated with the thermodynamic limit of statistical mechanics do not hold here : the extensivity of nuclear matter can be considered only as an approximation. Surface effects are essential, as well as size effects like shell structures. From the dynamical viewpoint, the finite size of nuclear systems does not allow them to relax towards thermal equilibrium as is usual in statistical mechanics, and the analysis of dissipative processes requires special care, as we shall see. On the other hand, the number of degrees of freedom is still much too large to allow complete microscopic studies. We are therefore faced with a standard problem in non-equilibrium statistical mechanics : is it possible to describe adequately the system in terms of some small set of collective variables which would replace the untractable set of positions and momenta of all nucleons ?

We shall be mainly concerned in this course with the systematic construction of the equations of motion for a set of collective variables, taking the microscopic dynamics as a start. The unified framework and most of the techniques which we shall present have been introduced long ago in the context of non-equilibrium statistical mechanics, but they are not so widely known and applied as they deserve. The formalism is general enough to encompass many existing theories and approximations currently used in nuclear theory. In addition, the geometric interpretation which we shall give may help to visualize the steps involved in the derivation of the collective motion. Our scope will be merely formal, because we wish to emphasize the conceptual rather than the technical aspects. This should help the reader adapting the general theory to such or such specific problem.

Several important qualitative differences exist between the dynamics at the microscopic level and at the level of collective variables, and the origin of these differences will be enlightened by the general approach. As usual in statistical mechanics, new phenomena appear when the description is contracted, i.e., when the non-collective variables are eliminated : the physics changes with the level of description. Let us recall a few general features of the type of contraction performed here. On the one hand, the Schrödinger equation is linear ; however, isolating the collective degrees of freedom leads to *non-linearities* which may have important consequences. In contrast to linear equations, which provide nothing but superpositions of sinusoidal oscillations, a nonlinear dynamics may lead to solitons, as well as to chaotic behaviors (like in the Euler equations of hydrodynamics or in the Hénon-Heiles model). On the other hand, the collective dynamics may present a *classical* structure in spite of the underlying quantum nature of the problem. Actually, a nucleus is a quantum object, and the quantities characterizing its state are therefore statistical ; but they may follow classical equations of motion of the Ehrenfest type. Finally, *dissipation*, which is absent at the microscopic level, may take place in the collective equations of motion.

Our approach will introduce naturally most quantities and concepts of current use in the thermodynamics of irreversible processes, such as *local temperatures, chemical potentials, hydrodynamic velocities*, or *transport coefficients*. In particular, the whole course will emphasize *entropy*, both as a measure of the *missing information* and as a tool for eliminating the non-collective variables considered as irrelevant. For instance, we shall characterize the *dissipation rate* by using the concept of *relevant entropy* introduced in § 5. The more standard approach to this question in nuclear physics relies on energetic considerations : the energy is split, in a more or less natural way, into a "mechanical" part, associated with the collective degrees of freedom, and a "thermal" part accounting for the remainder, and the dissipation is defined as an energy flow from the former to the latter part. However, from the viewpoint of statistical mechanics and of thermodynamics, entropy is a more basic quantity than energy, since energy is only one among the conserved variables of the system. Having defined the relevant entropy as the missing information relative to the set of collective variables, we shall understand and measure the dissipation as the increase of this relevant entropy, an extension of the Carnot principle.

We shall make use of several techniques of statistical mechanics. We shall first introduce (§ 2) the *Liouvillean formulation* of quantum mechanics. More general than the standard formulation in Hilbert space, it encompasses useful formalisms such as the polarization representation for a spin or the Wigner representation. Working in the Liouville space is also essential for the *reduction* of the number of degrees of freedom retained in the dynamical description (§ 3 and 4). The concept of entropy will be used to endow this Liouville space with a natural *metric* (§ 6). We shall then interpret the elimination of the irrelevant variables and the construction of the trajectory of the collective variables as a *projection* (§ 7). The successive steps of the *projection method* (§ 7-9) lead to *exact* equations of motion for the collective variables, which may have the form of *transport* equations or of *balance* equations according to the context. The generalized mean-field approximation (§ 10) will help us to identify the terms in these equations which are responsible for dissipation. Finally, we shall briefly review (§ 11) various approximation methods which provide closed expressions for dissipation or for transport coefficients.

We shall try to avoid all technicalities. The interested reader will find more details in an extensive article in preparation [1]. Many articles and books [2] deal with the projection method and its applications. By commenting here the ideas which underlie this method, and by sketching how they should be worked out in practice, we would like to make this immense and fruitful literature more popular among nuclear (and hopefully other) physicists.

2. THE LIOUVILLEAN FORMULATION OF QUANTUM MECHANICS

Quantum mechanics associates with a physical system two types of objects, the *observables*, which characterize all possible experiments, and the *states*, from which the expectation value of any observable can be derived. Most textbooks introduce the states as wave functions or as kets $|\psi>$, elements of some Hilbert space on which the observables act. We shall take however as a starting point of our analysis the representation of states as *density operators* D (or as density matrices when a basis is chosen in the Hilbert space). In the special case of a pure ket $|\psi>$, the density operator is the projector $|\psi><\psi|$. More generally, any state is represented by a density operator which is diagonalized as

$$D = \sum_n |\psi_n> p_n <\psi_n| \qquad , \qquad (2.1)$$

where the numbers p_n are positive and have a unit sum. The expectation value of the observable A is given by

$$<A> = \text{Tr } AD \qquad (2.2)$$

when the system is in the state D.

Let us resume in this language the postulates of quantum mechanics. The set of all conceivable experiments is characterized by an algebra of Hermitean observables A. A state of the system is represented by a density operator D, playing the role of a probability distribution and providing the expectation value of any observable as (2.2). Density operators are restricted by their properties of Hermiticity, positivity, and normalization, which are equivalent to the requirements that $<A>$ be real, that $<A^2>$ be positive, and that the expectation value of the unit observable be 1, respectively. Finally, if the system is isolated and if its Hamiltonian

H is known, the states evolve according to the Liouville-von Neumann equation ($\hbar = 1$)

$$i \, \dot{D} = [H,D] \qquad . \qquad (2.3)$$

This formulation includes the standard formulation in terms of kets and bras as the special case when (2.1) reduces to a single term. It is however more general and somewhat simpler, both formally and conceptually. It is suited to quantum statistical mechanics, and it allows to deal with situations beyond the scope of the quantum mechanics of kets and bras, namely the consideration of subsystems of a quantum system, or the analysis of a measurement process, or the existence of a random part in the Hamiltonian.

We are now in position to introduce the Liouvillean formulation of quantum mechanics (including quantum statistics). It relies on the idea that a state is nothing but a *means for evaluating the expectation values* of all the observables of the system, and that the correspondence between the observables and their expectations is *linear*. A state will therefore be *defined* as such a linear correspondence, and the formalism of density operators will appear as a special representation of this general formulation.

More precisely, let us disregard the algebraic structure of the set of observables, and let us focus upon its vector space structure only. We choose in this set a complete basis of linearly independent observables Ω^μ. Then, any observable A is expressed as a linear combination of the observables of this basis, with scalar coefficients A_μ interpreted as the *coordinates of the observable* A:

$$A = \sum_\mu A_\mu \, \Omega^\mu \qquad . \qquad (2.4)$$

A state D is characterized by the knowledge of the expectation values

$$D^\mu = \langle \Omega^\mu \rangle \qquad (2.5)$$

of the observables Ω^μ of the basis, which are interpreted as the *coordinates of the state* D. The expectation value of any observable (2.4) is then given by

$$\langle A \rangle = \sum_\mu A_\mu \, D^\mu = (A;D) \qquad , \qquad (2.6)$$

which appears as the *scalar product of the observable* A *by the state* D. In the following, we shall omit the summation over repeated indices.

The *postulates of quantum mechanics* are easily expressed in this Liouvillean formulation. A system is characterized by the *algebra* of its observables, i.e., by the structure constants $C^{\mu\nu}_\rho$

entering the products

$$\Omega^\mu \, \Omega^\nu = C^{\mu\nu}_{\ \ \rho} \, \Omega^\rho \qquad . \qquad\qquad (2.7)$$

The *expectation value* of the observable A in the state D is the scalar product (2.6). The coordinates D^μ of a state are *constrained* by the following requirements :

   a) the expectation value of any *Hermitean* observable is *real* ;
   b) the expectation value of the *unit* observable is 1 ;
   c) the *fluctuation* $<A^2> - <A>^2$ of any Hermitean observable is *positive*.

Finally, for an isolated system with a known Hamiltonian, the coordinates D^μ *evolve* according to the (linear) Liouville equation

$$i \, \dot{D}^\mu = L^\mu_{\ \nu} \, D^\nu \qquad , \qquad\qquad (2.8)$$

equivalent to (2.3), where L is defined as the *Liouville operator*.

The representation of states as density matrices is recovered through a special choice of the basis of observables Ω^μ. Denoting by $|\alpha>$ a basis of the Hilbert space, it is easily seen that the dyadic operators

$$\Omega^\mu = |\alpha><\beta| \qquad , \qquad\qquad (2.9a)$$

labelled by the index $\mu \equiv (\alpha, \beta)$, are linearly independent and span the space of observables. The expansion (2.4) reads then

$$A = \sum_{\alpha, \beta} |\alpha><\alpha|A|\beta><\beta| \qquad , \qquad\qquad (2.9b)$$

and A_μ is the matrix element $<\alpha|A|\beta>$, while the coordinate D^μ of a state D, defined by (2.5), is equal for $\mu \equiv (\alpha, \beta)$ to

$$<\Omega^\mu> = \text{Tr} |\alpha><\beta|D = <\beta|D|\alpha> \qquad . \qquad\qquad (2.9c)$$

The scalar product (2.6), which reads here

$$<A> = A_\mu \, D^\mu = \sum_{\alpha, \beta} <\alpha|A|\beta><\beta|D|\alpha> = \text{Tr } AD , \qquad\qquad (2.9d)$$

is identified with (2.2). Finally, the identification of (2.3) with (2.8) is achieved by expressing the Liouville operator L in terms of the Hamiltonian H through

$$L^\mu_{\ \nu} = \delta_{\alpha\gamma} <\beta|H|\delta> - \delta_{\beta\delta} <\gamma|H|\alpha> \qquad , \qquad\qquad (2.10)$$

$$\mu \equiv (\alpha, \beta) \qquad , \qquad \nu \equiv (\gamma, \delta) \qquad .$$

While the Liouvillean formulation encompasses the usual representations of quantum mechanics in terms of matrices (2.9b) and

(2.9c) for observables and for states, it brings in an interesting new freedom. Indeed, considering the pair $\mu = (\alpha, \beta)$ as a single index allows *linear changes of the basis* Ω^μ which mix the right and left indices α and β of the matrices (usual changes of representation in Hilbert space act on the α and on the β separately). In such changes of basis, the coordinates A_μ behave *covariantly*, and the coordinates D^μ behave *contravariantly*.

Instead of starting from a Hilbert space representation and performing a change of basis, it is usually more expedient to build a Liouville representation *directly*, by starting from the choice of a convenient basis Ω^μ in the space of observables. The new representations thus introduced may present attractive features. Consider for instance a spin 1/2. The vector space of observables has 4 dimensions and it is spanned as in (2.9) by the basis

$$|+\rangle\langle+| \quad , \quad |+\rangle\langle-| \quad , \quad |-\rangle\langle+| \quad , \quad |-\rangle\langle-| \quad . \qquad (2.11a)$$

But it may alternatively be spanned by the basis

$$1 \; , \; \sigma_x \; , \; \sigma_y \; , \; \sigma_z \quad , \qquad (2.11b)$$

including the unit observable and the Pauli operators. The use of (2.11b) defines the *polarization representation*. In this representation, the 2×2 Hermitean density matrix is replaced by the set of 4 coordinates D^μ which are the averages of (2.11b), i.e., 1 and the average polarization vector $\langle\sigma\rangle$. The Hermiticity and positivity requirements on D are expressed by the fact that $\langle\sigma\rangle$ is a real vector, of length equal at most to 1. Thus, the physical quantities $\langle\sigma\rangle$ enter *directly* the formalism. If the magnetic moment $-\sigma/2$ (in suitable units) lies in a field **B**, the Hamiltonian is $H = -\mathbf{B}.\sigma/2$, and the Liouville equation (2.8) reads

$$\frac{d}{dt} \langle\sigma\rangle = \mathbf{B} \times \langle\sigma\rangle \quad , \qquad (2.11c)$$

exhibiting the Larmor precession better than the equation of motion of the density matrix.

Another example is provided by the *Wigner representation*, especially useful when quantum effects are weak. The Wigner transform $D_W(x,p)$ of the density operator is introduced traditionally by performing a Fourier transform with respect to $x'-x''$ (the quantity $2x = x' + x''$ being kept fixed) on the elements $\langle x'|D|x''\rangle$ of the density matrix in the x-representation. Such a construction enters the general Liouville framework : indeed, as already indicated, the various Liouville representations result from one another by a linear transform over the operators Ω^μ, and the Fourier transform on $x'-x''$ is such a linear transform. More directly, we can construct the Wigner representation as a special case of

Liouville representation by making the following choice of basis Ω^μ in the space of observables. Let the index $\mu \equiv (x,p)$ denote a point in phase space (we restrict to a one-dimensional system for simplicity). From the conjugate observables $\hat{x}$ and $\hat{p}$, we construct the observables

$$\Omega(x,p) \equiv h(2\pi)^{-2} \int d\alpha \; d\beta \; \exp[i\alpha(\hat{x}-x) + i\beta(\hat{p}-p)] \; , \quad (2.12a)$$

labelled by the index $\mu \equiv (x,p)$ and whose algebra is generated by $[\hat{x},\hat{p}] = i\hbar$. In the classical limit, (2.12a) would become $h \; \delta(\hat{x}-x) \; \delta(\hat{p}-p)$, and $\Omega(x,p)$ would be identified with the (classical) observable associated with the probability density at the point (x,p) of phase space (with the measure $dx \; dp/h$). The observables (2.12a) are thus a quantum generalization of these classical observables localized in phase space. One can show as an exercise that they constitute a basis in the space of observables, and that (2.4) reads here

$$A = h^{-1} \int dx \; dp \; A_w(x,p) \; \Omega(x,p) \; , \quad (2.12b)$$

where the coordinates A_μ are identified with the Wigner representation

$$A_w(x,p) = \text{Tr} \; \Omega(x,p) \; A \quad (2.12c)$$

of A. One can also verify that (2.5) is identified with

$$D_w(x,p) = \langle\Omega(x,p)\rangle = \text{Tr} \; \Omega(x,p) \; D \; , \quad (2.12d)$$

and the expression (2.6) of an expectation value as a scalar product is

$$\langle A\rangle = h^{-1} \int dx \; dp \; A_w(x,p) \; D_w(x,p) \; . \quad (2.12e)$$

This expression reminds classical statistical mechanics, D_w playing the role of a probability distribution ; however, the positivity of any $\langle A^2\rangle$ is not expressed by the positivity of the function $D_w(x,p)$.

The Liouvillean formulation includes *classical* statistical mechanics as the special case in which the algebra (2.7) of observables is *commutative*. In the phase space representation (which is the classical limit of the Wigner representation), D^μ is recovered to be the classical probability density in phase space, while (2.8) is replaced by the classical Liouville equation involving a Poisson bracket instead of a commutator. Special features of quantum mechanics, such as the superposition principle or the uncertainty relations, are hidden in the structure (2.7) of the operator algebra which interferes with the positivity constraint (c) on the coordinates D^μ of the states.

We have worked implicity in the Schrödinger picture, where
the states evolve according to (2.8) and the observables remain
fixed. In the equivalent Heisenberg picture, the states remain
fixed and the observables evolve according to the equation
$i \dot{A} = [A,H]$, which reads in the Liouville formulation

$$i \dot{A}_\mu = A_\nu \, L^\nu_{\ \mu} \quad . \tag{2.13}$$

The Liouville operator L may act therefore either with its lower
index on the states (eq.(2.8)) or with its upper index on the observables (eq.(2.13)). We shall introduce other such objects having
an upper and a lower index, called *superoperators*. They generate
linear transformations either over the space of states (with their
lower index) or over the space of observables (with their upper
index). They are handled like the operators acting on a Hilbert
space, but here the kets are replaced by density operators (states)
and the bras by observables.

The interest of the Liouvillean formulation lies in its flexibility. One can for instance decide to take such or such observable of interest (for instance the component σ_z of a spin) as one
of the operators of the basis Ω^μ (such as in the choice (2.11b));
the corresponding coordinate D^μ will then be just the expectation
$\langle\sigma_z\rangle$ of interest. This will be useful to split the set of observables into relevant ones and irrelevant ones.

In addition, although the Liouville equation (2.8) is equivalent to (2.3) for a choice (2.10) of the Liouville operator, its
form is general enough to account for *non-Hamiltonian evolutions*.
For instance, the depolarization of a spin may be described by a
relaxation equation

$$\frac{d}{dt} \langle\sigma\rangle = -\gamma \, \langle\sigma\rangle \quad , \tag{2.14}$$

which enters the Liouvillean framework (2.8), although no corresponding Hamiltonian exists : any evolution of the type (2.3) would
conserve the length of $\langle\sigma\rangle$. A dynamics of the type (2.14) can be
obtained either if the Hamiltonian contains a random part (several
evolutions are possible and we have to take an average over them),
or if the spin interacts weakly with other degrees of freedom. We
need thus to work in the Liouvillean formalism whenever we are
interested in the dynamics of some degrees of freedom, the other
ones being eliminated.

Notice finally that the Liouville equation (2.8) is formally
simpler than its counterpart (2.3) in Hilbert space. Its explicit
solution can be written in Liouville space as

$$D^\mu(t) = U^\mu_{\ \nu}(t,t_o)D^\nu(t_o) \quad , \quad U(t,t_o) = e^{-iL(t-t_o)} \quad , \tag{2.15a}$$

which is equivalent to the solution

$$D(t) = U(t,t_0)D(t_0)U(t_0,t) \quad , \quad U(t,t_0) = e^{-iH(t-t_0)} \quad (2.15b)$$

in Hilbert space, but which involves only one evolution superoperator U instead of the two evolution operators U of (2.15b). The form (2.15a) allows in particular to take the classical limit. It also affords perturbation expansions having no equivalent in (2.15b), since nothing prevents to introduce in (2.15a) unperturbed Liouville operators L_0 not reducible to a form (2.10), such as the relaxation Liouville operator (2.14).

3. CONTRACTING THE DESCRIPTION

The number of variables characterizing the state of a system is most often immense. These variables, the expectation values of all the observables, are represented in the Liouville formulation as the coordinates D^μ of a point in a many-dimensional space (if the Hilbert space associated with the system has n dimensions, we need n^2 real numbers to specify D). Both experimentally and theoretically, it is out of question in practice to describe the trajectory of this point. It is necessary to *contract the description*, i.e., to restrict to some limited number of variables, which should be chosen adequately by taking into account both experiment and known general characteristics of the dynamics. These variables are the expectation values of some selected observables, called according to the context *collective, simple, relevant, macroscopic, coarse-grained,...* . We shall most often use the term *relevant*.

In the Liouvillean formalism, the relevant observables ω^i are some linear combinations $\omega_\mu^i \Omega^\mu$ of the operators Ω^μ of the basis. Instead of characterizing the state by the whole set of coordinates $D^\mu = \langle \Omega^\mu \rangle$, we content ourselves with following only the small number of expectation values $\langle \omega^i \rangle$. We have thus to distinguish between a *microscopic state*, specified *completely* (albeit *statistically*) by the set of averages $\langle \Omega^\mu \rangle$, and a *macroscopic state*, in which *the description is incomplete* and limited to the knowledge at each time of the averages $\langle \omega^i \rangle$.

Geometrically, the relevant observables are represented by the points of a p-dimensional plane spanned by the set $\omega^1, \omega^2, .. \omega^P$ in the space of observables (fig.1). In the space of states, the knowledge of the expectations $\langle \omega^i \rangle = \omega_\mu^i D^\mu$ characterizes a plane Δ of codimension p (fig.2). All the points D of this plane are equivalent from the viewpoint of the relevant variables. A microscopic state is represented by a plane Δ passing through D, the dimensionality and direction of which depend on the choice of the relevant observables.

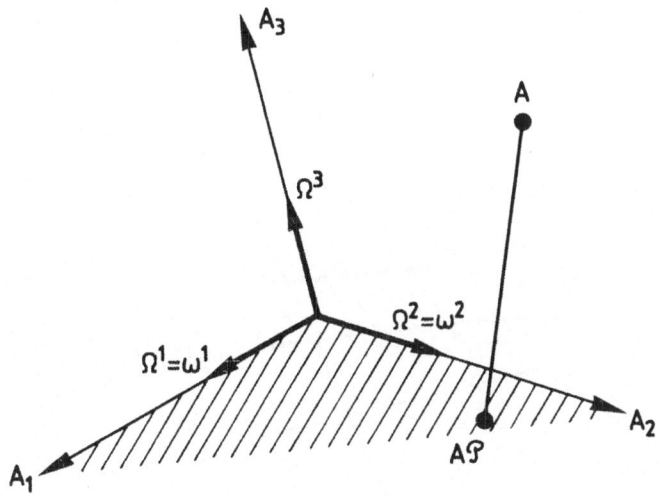

Fig. 1. *The space of observables*. An observable (2.4) is visualized as a point in the vector space spanned by the basis
$\Omega^1, \Omega^2, \Omega^3 \ldots$. The subspace of relevant observables is a
plane, spanned by the observables ω. We have chosen here
two relevant observables ω, identified with the first two
operators Ω. The projector P of § 7 performs an orthogonal
projection with respect to the metric referring to D_0.

The number and the nature of the relevant observables ω^i depend on the physical system, on the circumstances, and on the quality of the approximations which one is ready to accept. One should
include in this set the *conserved observables* (energy, momentum,
angular momentum, number of nucleons), since the constraints imposed on the dynamics by their constancy are obviously essential[1].
It is also natural to include the *slow* variables, which are the
most accessible experimentally and the most significant theoretically. In particular, the reduction to *hydrodynamic variables* is
suited to situations in which the system remains locally close to
equilibrium. In this case, the variables $\langle \omega^i \rangle$ characterizing the
macroscopic state are chosen to be the densities at each point of
conserved quantities (densities of energy, of momentum, of particles, of charge). If the system is well described in terms of broken invariances, one should select among the relevant variables

---

(1) In particular, the *unit observable* will always be included
in the relevant set and denoted by ω^0.

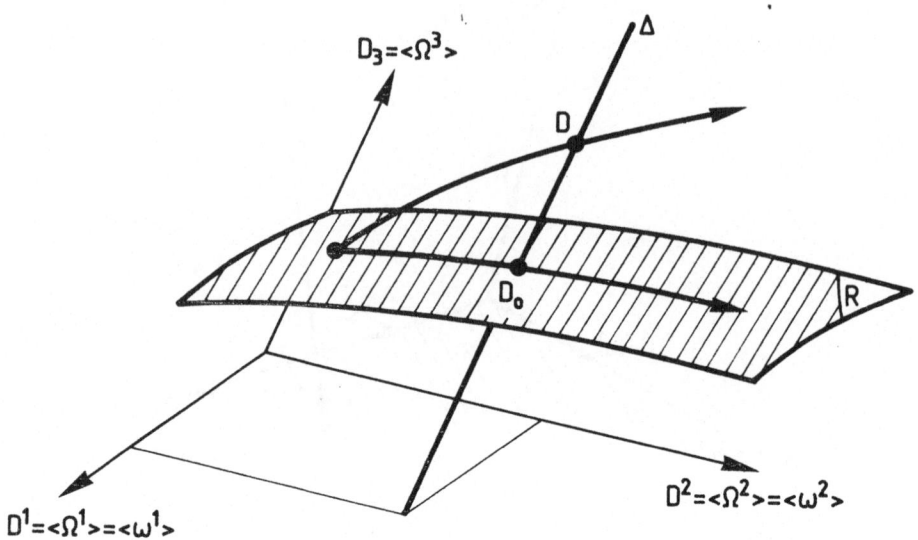

Fig. 2. *The space of states.* The knowledge of the variables $\langle\omega^i\rangle$
characterizes a plane Δ. A completely specified state is
represented by a point D, an incomplete description by the
plane Δ. Among the points D of Δ, which are equivalent
with regard to the collective variables $\langle\omega\rangle$, the point D_0
represents the reduced density associated with a minimum
amount of information (§ 4). The surface R of reduced sta-
tes is parametrized by the variables $\langle\omega\rangle$ or the multipliers
λ. With the metric of § 6, Δ and R are orthogonal. The trajectory of D is given by the Liouville equation, the resulting trajectory of D_0 by the generalized transport equation (8.4).

the *order parameters* (possibly the *local* order parameters). For
instance, if a system is ferromagnetic, the states close to thermal
equilibrium are characterized by the global or the local magnetization. Similarly, when pairing takes place, the dynamics should follow the motion of the pair $\langle c_\alpha c_\beta\rangle$.

If a system is described as a collection of weakly interacting particles or quasi-particles, it is natural to disregard
their correlations. The relevant observables are then the *single-
particle observables*. In classical mechanics, a basis ω for this
set is $\sum_j \delta^3(\hat{r}_j - r)\, \delta^3(\hat{p}_j - p)$, where $j = 1, \ldots N$ denotes the particles,
and the variables $\langle\omega\rangle$ are identified as the density $f(r,p)$ of particles in the one-body phase space. In quantum mechanics, taking

as a basis ω the set $c_\beta^\dagger c_\alpha$ provides as relevant variables the expectation values $\rho_{\alpha\beta} = \langle c_\beta^\dagger c_\alpha \rangle$, which are the quantum mechanical counterpart of the density $f(r,p)$. This correspondence is made more precise by taking as a basis in the single-particle space the observables (2.12a), generalized in an obvious fashion to 3 dimensions. The average of these observables is a function $f(r,p)$, which is the Wigner transform of $\rho_{\alpha\beta}$, and the quantum equivalent of the particle density in phase space.

Nuclear physics makes an extensive use of the idea of contracting the description, although with a more phenomenological scope than here. Besides the examples just given, incomplete descriptions involving a smaller number of *collective variables* are of current use. One would like for instance to consider only the energy, the number of nucleons, and some variables characterizing the *shape* of the nucleus considered as a liquid drop.

An important general problem in the *dynamics* of many-body systems consists in the establishment of the equations of motion for the relevant variables $\langle \omega^i \rangle$ only. The equations (2.8) *couple all the variable* $\langle \Omega^\mu \rangle = D^\mu(t)$. We wish to follow the motion of a subset containing only some of their linear combinations, $\langle \omega^i \rangle = \omega^i_\mu D^\mu(t)$, and to eliminate all other coordinates of D, extremely numerous and uninteresting. We shall sketch in the following the projection method, a general and systematic technique for performing this elimination once the relevant observables ω^i have been selected.

4. REDUCED DENSITIES

We have seen that the sole knowledge of the variables $\langle \omega^i \rangle$ is not sufficient to specify the microscopic state D of the system, and that it characterizes but a plane Δ. A question then arises : among all the microscopic states D compatible with the given values

$$\langle \omega^i \rangle = \omega^i_\mu D^\mu = \mathrm{Tr}\ \omega^i D \quad , \qquad (4.1)$$

which one is the *least biased* state D_0 that should be assigned to characterize the system? This question is analogous to a standard problem in statistics : which probability distribution should be assigned to a statistical ensemble of events on which only partial information is available? When nothing is known, it is reasonable to associate the same probability with each possible event. The knowledge of some expectation values provides constraints on the probabilities. The probability law to be used for making further predictions should be compatible with these constraints, but otherwise should be as random, as widely spread, as possible.

In order to decide which probability law is the least biased, it is necessary to be able to *measure the bias*, or the amount of randomness of a law, and to look for the maximum value of this randomness.

In statistical mechanics, the role of the probability law is played by the density operator (2.1), the randomness of which is usually estimated by means of the *statistical entropy*, defined as

$$S(D) = -\text{Tr } D \ln D = \sum_n p_n \ln p_n \quad . \tag{4.2}$$

This quantity vanishes when the state is specified as much as possible, i.e., for a pure state $D = |\psi\rangle\langle\psi|$. It is largest when all probabilities p_n are equal in (2.1), i.e., for a density matrix D proportional to the unit matrix. The form (4.2) is a consequence of the requirement of additivity : if a system is composed of two parts a and b, and if these two parts are uncorrelated, its density operator has the form of a tensor product $D = D_a \otimes D_b$, and its associated statistical entropy S(D) should be equal to the sum $S(D_a)$+ $S(D_b)$ of the entropies of the two parts. The statistical entropy measures the *dispersion* of the distribution D. It may also be interpreted as a *measure of disorder*, or as a *measure of the information which is missing* because the state D is known only in a probabilistic fashion.

It is therefore natural to select, as the microscopic probabilistic state representing the most adequately a system specified only by the variables (4.1), the density operator D_0 which renders the statistical entropy S(D) maximum subject to the constraints (4.1). This density operator D_0 is the least biased one in the sense of information theory, among the microscopic states D belonging to a plane Δ which represents the macroscopic state (4.1).

By introducing a Lagrange multiplier λ_i for each constraint, the research of a maximum entropy provides for D_0 the *generalized canonical form*

$$D_0 = \exp[-\lambda_i \, \omega^i] \quad , \tag{4.3}$$

where the values of the parameters λ_i are related to the expectation values $\langle\omega^i\rangle$ through

$$\langle\omega^i\rangle = -\frac{\partial}{\partial\lambda_i} \text{Tr } \exp[-\lambda_j \, \omega^j] \quad . \tag{4.4}$$

The normalization of D_0 is ensured by the constraint $\langle\omega^o\rangle = 1$, where ω^o is chosen as the unit observable. The density operator (4.3) is the one which corresponds to the *largest disorder* in the plane Δ. In the language of information theory, D_0 is the density operator in the family Δ which contains *no more information than the minimum required to account for the knowledge of the variables* $\langle\omega^i\rangle$.

Reduced densities of the form (4.3) are encountered in varied circumstances, depending on the choice of the relevant set of observables. When the observables ω^i are the energy, the particle number and unity, D_0 is the *grand canonical equilibrium* density operator. This equilibrium state is the most disordered one for given values of the constants of the motion. The corresponding multipliers λ_i are identified simply in terms of the temperature, the chemical potential and the grand potential. In the reduced *hydrodynamic* description, for which the variables $\langle\omega^i\rangle$ are the densities of energy, momentum and particles, D_0 is a state of *local equilibrium*, and the multipliers are interpreted in terms of a local temperature, a local velocity and a local chemical potential. In the quantum *one-body* reduction, D_0 is an *independent-particle* state. For a system composed of several parts, if the correlations between these parts are disregarded, the reduced density D_0 is a tensor product of density operators related to each part.

The reduced states D_0 of the form (4.3) are represented in the space of states (fig.2) by the points of a p-dimensional surface R. This surface is parametrized either by the variables $\langle\omega^i\rangle$ themselves, or equivalently by the multipliers λ_i which are related to the expectations $\langle\omega^i\rangle$ through (4.4), and which often have a *thermodynamic interpretation* as we just saw.

The above analysis applies in particular to the *initial state* of a system. Preparing the system at the time t_0 amounts to provide the values of some variables, and it is natural to include these variables in the relevant set $\langle\omega^i\rangle$. The density operator describing such a situation in a non-biased way has the form (4.3), in which Lagrange multipliers are introduced only for the quantities given at the time t_0. The other relevant quantities do not appear in this state $D_0(t_0)$. Equivalently their associated multipliers λ_i are set equal to zero. For instance a chemical potential vanishes if the associated particle number is not specified.

At a later time, in a Hamiltonian evolution, the state is given by (2.15). Except in special cases (for instance when all the observables ω^i are constants of the motion), $D(t)$ will not retain the form (4.3) although we have $D(t_0) = D_0(t_0)$ at the initial time. The knowledge of the microscopic state $D(t)$ then provides the values $\langle\omega^i\rangle$ of the relevant variables at the time t, and the associated macroscopic state is represented by a plane $\Delta(t)$ (fig.2). If we are only interested in the variables $\langle\omega^i\rangle$, and if we *disregard any other information* contained in $D(t)$, we are compelled to choose as a representative point in the plane $\Delta(t)$ the maximum entropy state $D_0(t)$ instead of $D(t)$. We thus associate with the microscopic state $D(t)$ a *reduced density* $D_0(t)$ defined by (4.3) and (4.4). With regard to the *useful information*, the densities $D(t)$ and $D_0(t)$ are equivalent, since they provide the same values for the variables $\langle \omega^i \rangle$. However, $D(t)$ involves a statistical bias with respect to the other (irrelevant) variables, which has been introduced by the dynamics. Contracting the description leads to follow two trajectories in the space of states (fig. 2). The Hamiltonian trajectory of $D(t)$, generated by the equation of motion (2.8), represents the evolution associated with a *complete statistical description*. The trajectory of the corresponding reduced density $D_o(t)$ represents the evolution associated with the considered *incomplete description*. Although the trajectory of $D(t)$ is easier to characterize formally, it contains lots of irrelevant information. The trajectory of $D_o(t)$ in the surface R of reduced states, or equivalently the motion of the relevant variables $\langle \omega^i \rangle$, or of the parameters λ_i, is the only dynamics of interest. We shall write in § 9 this dynamics, eliminating the irrelevant parts of the microscopic state $D(t)$.

5. RELEVANT ENTROPY RELATIVE TO A SET OF VARIABLES

The statistical entropy $S(D)$ defined by (4.2) is identified in the textbooks of statistical mechanics with the entropy of thermodynamics, at least when D describes an *equilibrium* situation. However, during the Hamiltonian evolution of an isolated system, the statistical entropy (4.2) remains constant in time. Indeed, (4.2) and (2.3) yield

$$\dot{S} = -\mathrm{Tr}\ \dot{D}(\ln D + 1) = i\ \mathrm{Tr}[H,D](\ln D + 1)$$
$$= i\ \mathrm{Tr}\ H[D,(\ln D + 1)] = 0 \quad . \tag{5.1}$$

This property seems to contradict the second principle of thermodynamics, and constitutes the *paradox of irreversibility*.

In order to solve this paradox, let us remind that the concept of state in thermodynamics is a macroscopic one. Consider for instance a situation in which two systems a and b, initially at thermal equilibrium but at different temperatures, evolve slowly towards an overall equilibrium, by means of heat exchanges through the wall which separates them. We assume this wall to be thin enough to retain no energy, but not much permeable to heat. At each time, the thermodynamic variables characterizing the state of the system are the energies E_a and E_b of each part. The thermodynamic entropy is the sum $S_a + S_b$ of the entropies of both parts, each one evaluated as if the subsystems were at equilibrium at the considered time. At the microscopic level, the energies E_a and E_b are the expectation values $\langle H_a \rangle$ and $\langle H_b \rangle$ of the Hamiltonians of the subsystems. The total Hamiltonian involves in addition a very small coupling term responsible for the energy (heat) exchanges between a and b. A contraction of the description in which the only relevant observables are taken to be H_a and H_b (and the unit observable) provides at each time a reduced density (4.3) of the form

$$D_o = \exp[-\beta_a H_a - \beta_b H_b - \zeta] \quad .$$

The multipliers β_a and β_b are identified with the inverse macroscopic temperatures of the subsystems a and b, and ζ is a normalization factor. Since D_o is factorized into two parts $D_o^a \otimes D_o^b$, its associated statistical entropy (4.2) is of the form $S_a + S_b$, where S_a and S_b are the equilibrium entropies of each part :

$$S_a = -\text{Tr } D_o^a \ln D_o^a \quad , \quad D_o^a = \exp[-\beta_a H_a - \zeta_a] \quad .$$

We recover therefore in the framework of statistical mechanics the properties postulated in thermodynamics and observed at the macroscopic scale, *provided we describe the microscopic state by the reduced density* $D_o(t)$, instead of the density $D(t)$ evolving along the Liouville-von Neumann equation (2.3). This replacement has no incidence upon the macroscopic variables $E_a(t)$ and $E_b(t)$. It is however necessary for a correct microscopic interpretation of the thermodynamic entropy and of the temperatures of the subsystems a and b (which in the state $D(t)$ are not quite at thermal equilibrium).

More generally, the thermodynamics of a system evolving in the vicinity of equilibrium requires that the system can be divided into volume elements small enough to be nearly homogeneous and nearly at equilibrium. From the viewpoint of statistical mechanics, this means that a *hydrodynamic contraction of the description* has been performed, the whole set of microscopic variables being replaced by the sole densities of conserved variables (energy, momentum, particle number). The correlations between volume elements have in particular been disregarded. Here again, whereas the statistical entropy S(D) associated with $D(t)$ remains constant in time along the Hamiltonian evolution (2.3) of $D(t)$, the *thermodynamic entropy* (which is defined as the sum of the entropies of each volume element) should be identified with the statistical entropy $S(D_o)$ associated with the *reduced density* $D_o(t)$ pertaining to the hydrodynamic contraction. Nothing prevents then this entropy from increasing.

The interpretation of statistical entropy as a missing information helps to understand better the connection between macroscopic thermodynamics and statistical mechanics. The full density operator D of a system at a given time gathers all conceivable predictions, and its associated entropy S(D) measures the amount of missing information corresponding to this (statistical) knowledge. Among all states equivalent to D with regard to the expectation values $\langle \omega^i \rangle$, the reduced density D_o has been defined by expressing that the statistical entropy $S(D_o)$ is largest. This quantity is interpreted as the *missing information associated with the sole knowledge of the expectations* $\langle \omega^i \rangle$. It is a function of the variables $\langle \omega^i \rangle$ (or equivalently of the multipliers λ_i), given by (4.2), (4.3) as

$$S(\langle \omega \rangle) = S(D_o) = \lambda_i \langle \omega^i \rangle \qquad , \qquad (5.2)$$

where the λ_i and the $\langle \omega^i \rangle$ are related by (4.4). Returning to the state D itself, we may also interpret $S(\langle \omega \rangle)$ as a *relevant entropy of the state* D *relative to the observables* $\langle \omega^i \rangle$. The difference $S(\langle \omega \rangle) - S(D)$, which is positive by construction, is interpreted as the *information lost when disregarding the irrelevant variables* other than $\langle \omega^i \rangle$.

The constancy in time of $S(D)$ expresses that *we lose no information* when following the Hamiltonian motion of *all* the variables $D^\mu = \langle \Omega^\mu \rangle$. However, we are interested in practice in a smaller number of relevant variables $\langle \omega^i \rangle$, and the corresponding missing information or *relevant entropy* $S(\langle \omega \rangle)$ *may rise*. We can thus interpret *dissipation* as a *leakage of information* from the relevant towards the irrelevant degrees of freedom, produced by their coupling. The total information is conserved, but part of the *relevant information* may be *lost* by flowing towards the irrelevant variables. Dissipation takes place because we are *discarding* some information, associated with degrees of freedom considered as *inaccessible*.

The entropy of non-equilibrium macroscopic thermodynamics, defined by dividing the system into small elements, has been identified as the relevant entropy relative to the hydrodynamic densities. However, the above analysis allows to face situations beyond the scope of thermodynamics. For instance, the interpretation of spin echo experiments in magnetic resonance or the understanding of Boltzmann's H-theorem rely on various relevant entropies, associated with different choices of the observables ω^i considered as relevant. The concepts of disorder and of dissipation present thus a *relative* and even *anthropomorphic* character, depending on the variables $\langle \omega^i \rangle$ on which we focus. Disorder (at a given time) is relative to the variables on which information is available, and is measured by the associated relevant entropy $S(\langle \omega \rangle)$. Dissipation during the evolution also depends on a selection of relevant variables, and is measured by the rate of decrease $dS(\langle \omega \rangle)/dt$ of the useful information relative to these variables $\langle \omega^i \rangle$.

6. METRIC STRUCTURE OF THE SPACES OF STATES AND OBSERVABLES

In sections 4 and 5, we have not taken full advantage of the Liouville formulation. Actually, the statistical entropy $S(D)$, which is a function of the coordinates D^μ of D, has a simple expression (4.2) only in the Hilbert space representation (2.9). Accordingly, the expression (4.3) of the reduced densities D_o also came out written in the Hilbert space representation of states as density matrices. Expressing quantities such as $S(D)$, D_o or $S(D_o)$ in arbitrary Liouville representations requires a rather complicate change of basis, and the reader may wonder why we have bothered to introduce such more general representations. The interest of the Liouvillean formulation will become apparent soon, when we shall need to introduce superoperators acting on the spaces of observables or of states.

The consideration of entropy in the Liouvillean framework allows us in addition to enrich the *geometric structure* of the spaces of states and of observables represented by figs. 1 and 2. In § 3, these spaces have been defined as two *dual vector spaces* : the only existing structure was the scalar product (2.6) of an element of one space by an element of the other. It was not possible to consider the scalar product between two elements of the same space ; there was no metric. In order to define in a natural fashion a *metric* in the space of states, we notice that, besides the scalars (2.6), there exists *another scalar*, i.e., another quantity which is invariant in a change of basis, namely the *entropy* $S(D)$. For an infinitesimal variation δD of D, the second differential of S which reads in the Hilbert space representation

$$d^2 S = - \, \mathrm{Tr} \; \delta \, \ln D \; \; \delta D \qquad\qquad (6.1)$$

is a quadratic form in δD. In an arbitrary Liouville representation, δD is represented by its coordinates δD^μ, and (6.1) has the form

$$d^2 S = - \, G_{\mu\nu} \; \delta D^\mu \; \delta D^\nu \qquad\qquad (6.2)$$

where $G_{\mu\nu}$ is some known function of the point D. The reader will check as an exercise, by evaluating G in the Hilbert space representation, that the matrix $G_{\mu\nu}$ is symmetric and positive (for Hermitean density matrices). This expresses the *concavity of entropy*. Therefore, (6.2) defines a *Riemannian metric*, with $G_{\mu\nu}$ and its inverse $G^{\mu\nu}$ as metric tensors.

The existence of a natural metric allows to define *distances between neighbouring states*, and *angles* between two directions in the space of states. The tensors $G_{\mu\nu}$ and $G^{\mu\nu}$ define local *canonical isomorphisms* from one space to the other. In particular, the coordinates of $\delta \ln D$ considered as an element of the space of observables are related to δD by

$$\delta \, \ln D_\mu = G_{\mu\nu} \; \delta D^\nu \quad . \qquad\qquad (6.3)$$

The metric tensor $G^{\mu\nu}$ also defines the *scalar product of two observables* $A^{(1)}$ and $A^{(2)}$ as

$$G^{\mu\nu} \; A^{(1)}_\mu \; A^{(2)}_\nu \quad . \qquad\qquad (6.4)$$

This quantity depends on the state through G.

We have noticed (fig.2) that a plane Δ representing a macroscopic state and passing through the microscopic state D intersects the surface R at the point D_O representing the reduced state associated with D. It is an easy exercise to show that the plane Δ and the surface R intersect at D_O *perpendicularly* with respect to the metric tensor $G_{\mu\nu}(D_O)$. Therefore, the correspondence between D and D_O defined in § 4 can be visualized as an *orthogonal projection of the microscopic state* D *onto the surface* R *of reduced states*. In the sense of the metric just defined, D_O appears not only as the point of the surface R of reduced states which has the same coordinates $<\omega^i>$ as D, but also as the point of R which is *the closest to* D.

7. PROJECTION IN THE LIOUVILLE SPACE

We are going to take advantage of this remark and deduce from the Liouville equation of motion (2.8) for D(t), namely

$$i \, \dot{D} = L \, D \quad , \tag{7.1}$$

the equation of motion for its associated reduced density D_O. The latter is defined by its form (4.3), and by the identity between the expectations $<\omega^i>$ as expressed by (4.4) and as evaluated with D. Having set

$$D = D_O + D_1 \quad , \tag{7.2}$$

we thus wish to eliminate D_1.

We have just seen that D_O is the orthogonal projection of D onto R. The mathematical object performing this projection, the *projector* $P^\mu_{\ \nu}$, is a *superoperator* (§ 2). The above geometric considerations suggest that P should be constructed from the metric tensor $G_{\mu\nu}(D_O)$ with respect to which the angle of $D_1 = D-D_O$ with the surface R is evaluated. Thus, the projector P *depends on the point* D_O itself. One can check as an exercise that the explicit expression of P is given as follows. Let us denote by ω^i_μ (as in § 3) the components of the relevant observables ω^i on the basis Ω^μ (with $\omega^i = \omega^i_\mu \Omega^\mu$), and let us consider the components D^μ_O of D_O as functions of the averages $<\omega^i>$ through (4.3) and (4.4). The projector P associated with the metric $G_{\mu\nu}$ at the point D_O is then

$$P^\mu_{\ \nu} = \frac{\partial D^\mu_O}{\partial <\omega^i>} \, \omega^i_\nu \equiv \sigma^\mu_i \, \omega^i_\nu \quad . \tag{7.3}$$

It is easy to verify that the superoperator defined by (7.3) is a projector, i.e., that it satisfies

$$P^2 = P \quad . \tag{7.4}$$

Indeed, the objects

$$\sigma_i = \frac{\partial D_0}{\partial <\omega^i>} = \frac{\partial D_0}{\partial \lambda_j} \frac{\partial \lambda_j}{\partial <\omega^i>}$$

span a subspace of the space of states which is the *tangent plane* to R at the point D_0. This subspace, the *relevant subspace of states* (which depends on D_0), can be considered as the dual of the relevant subspace of observables spanned by the set ω^i. The property

$$(\omega^k; \sigma_i) = \frac{\partial <\omega^k>}{\partial \lambda_j} \frac{\partial \lambda_j}{\partial <\omega^i>} = \delta_i^k \qquad (7.5)$$

is easily checked and it implies

$$P^\mu_{\ \rho} P^\rho_{\ \nu} = \sigma^\mu_k \omega^k_{\ \rho} \sigma^\rho_i \omega^i_{\ \nu} = \sigma^\mu_i \omega^i_{\ \nu} = P^\mu_{\ \nu} \qquad .$$

Because the unit observable ω^0 belongs to the relevant subset, we get from (4.3) $D_0 = -\partial D_0/\partial \lambda_0$. Hence D_0 is a linear combination of the elements σ_i, and we have

$$P D_0 = D_0 \qquad . \qquad (7.6)$$

Moreover, taking into account the identity between the expectation values $<\omega^i> = (\omega^i; D) = (\omega^i; D_0)$, we get from (7.3) $PD = PD_0 = \sigma_i <\omega^i>$, and therefore P realizes as anticipated the correspondence from D to D_0 :

$$P D = D_0 \qquad . \qquad (7.7)$$

The superoperator $P^\mu_{\ \nu}$ acts with its upper index on the observables, and AP is some linear combination of the relevant observables ω^i, depending on the state D_0, and given by

$$A P = (A; \sigma_i) \omega^i \qquad . \qquad (7.8)$$

In the sense of the metric $G^{\mu\nu}(D_0)$, it may be checked that AP is the *orthogonal projection of* A *on the relevant plane* (fig.1). In other words, AP is the *relevant observable closest to* A. In particular, the projector P leaves the relevant observables invariant:

$$\omega^i P = \omega^i \qquad . \qquad (7.9)$$

We are now in position to project the equations of motion (7.1) and get the dynamics of D_0 (or equivalently of the $<\omega^i>$, or of the λ_i). Let us introduce the projector Q complementary to P,

$$Q = I - P \qquad , \qquad (7.10)$$

where I is the unit superoperator. The equations (7.2),(7.6) and (7.7) read

$$D_o = P\,D \qquad , \qquad D_1 = Q\,D \quad . \qquad (7.11)$$

By taking the derivative of (7.11) and using (7.1), we get the set of coupled equations for D_o and D_1 :

$$i\,\dot{D}_o = P\,L\,P\,D_o + i\,\dot{P}\,D_o + P\,L\,Q\,D_1 \qquad , \qquad (7.12)$$

$$i\,\dot{D}_1 = Q\,L\,Q\,D_1 - i\,\dot{P}\,D_o + Q\,L\,P\,D_o \quad . \qquad (7.13)$$

The projectors P and Q depend on time self-consistently through D_o, which enters the definition (7.3) of σ_i. We made use in (7.12), (7.13) of the property

$$\dot{P} = Q\,\dot{P}\,P = -\dot{Q} \qquad , \qquad (7.14)$$

a consequence of (7.3),(7.4),(7.10).

8. MEMORY KERNEL

It remains to eliminate from (7.12) the complicated part D_1 of the state D in order to obtain the motion of the simple part D_o. This elimination can be performed formally by introducing the Green's function $W(t,t')$ associated with the equation (7.13), which is defined by

$$i\,\frac{\partial}{\partial t}\,W^\mu{}_\nu(t,t') = [Q\,L\,Q]^\mu{}_\rho\,W^\rho{}_\nu(t,t') \qquad , \qquad (8.1)$$

and by the initial condition at $t = t'$:

$$W^\mu{}_\nu(t,t') = Q^\mu{}_\nu(t') \qquad . \qquad (8.2)$$

The superoperator W plays the role of an *evolution superoperator, in the irrelevant space* characterized by the projector Q. It is similar with the *complete* evolution superoperator U entering (2.15a) and associated in the Liouville space to the equation of motion (7.1) : U would result from the replacement of Q by unity in (8.1) and (8.2). The superoperator W is also termed the *memory kernel*, because as we shall see it describes memory effects due to the coupling between the relevant and the irrelevant spaces.

It is easy to check that the formal solution of (7.13) can be expressed in terms of W as

$$D_1(t) = W(t,t_o)\,D_1(t_o) + \int_{t_o}^{t} dt'\,W(t,t')[-\dot{P}-iQ\,L\,P]D_o(t') , (8.3)$$

for a given initial value $D_1(t_o)$. We have indicated in § 4 that D should take the reduced form (4.3) at the initial time. Hence, $D_1(t_o)$ vanishes, and only the second term of (8.3) exists.

By inserting (8.3) into (7.12), we get an *integro-differential equation of motion for* D_o :

$$i \, \dot{D}_o = (PLP + i\dot{P})D_o - PLQ \int_{t_o}^{t} dt' \; W(t,t') (\dot{P} + iQLP) \, \dot{D}_o(t') \; . \quad (8.4)$$

The dynamics of the variables $\langle \omega^i \rangle$ is obtained by taking the scalar product of (8.4) with ω^i, which provides

$$i \, \frac{d}{dt} \langle \omega^i \rangle = \langle \omega^i \, L \, P \rangle + (\omega^i \, L \, ; D_1) \; , \quad (8.5)$$

where D_1 should be replaced by (8.3).

The first term of (8.4) describes an evolution which would be generated by the part PLP of the Liouville operator acting within the relevant space. The second term, involving $\dot{P}$, arises from the motion of the relevant space (spanned by the moving frame σ). The remainder is a memory term. Its factor $\dot{P} + i \, QLP = Q(\dot{P} + iL)P$ couples $D_o(t')$ to the irrelevant subspace at any time t' between t_o and t. The kernel $W(t,t')$ then describes a truncated evolution in this irrelevant subspace. The last coupling factor PLQ leads back to the relevant space at the time t.

Notice that the last factor in (8.4) is also equal to

$$(\dot{P} + i \, QLP)D_o = Q(\dot{D}_o + i \, L \, D_o) \; , \quad (8.6)$$

where we made use of (7.6). It is therefore small, and retarded effects are weak, if D_o follows a trajectory close to the Hamiltonian flow $\dot{D} + i \, L \, D = 0$.

9. GENERALIZED TRANSPORT EQUATIONS

The equations (8.4), or (8.5), or the equations of motion for the variables λ_i (deduced from (8.5) and (8.3) through use of (4.4)), take varied forms, depending on the nature of the relevant observables ω^i. Since we are restricting ourselves to the conceptual aspects, we refer to the literature[1,2], which is extremely rich, for the applications to such or such problem of statistical mechanics or to such or such many-body system, as well as for the discussion of approximation techniques. We shall only illustrate the flexibility of the method by alluding to a few examples.

In the hydrodynamic contraction, the time-derivatives $\langle \dot{\omega} \rangle$ represent the variations of the densities at a given point of each conserved quantity. The right side of (8.5) is identified with the divergence of the *flux* associated with the *transport of energy, matter or momentum* causing the variation of the local density. For charged particles, (8.5) describes similarly the macroscopic *charge* transport. The *transport coefficients* are obtained in such cases by expressing the right side in terms of the variables λ, which are related to the local temperature, the local velocity or the local electrochemical potential, and by working out the expressions so as to exhibit the *gradients* of such quantities. In the linear regime, it is sufficient to stop the gradient expansion to lowest order. The ratio between flux of conserved variables and gradients of variables λ then defines the *thermal or electrical conductivity*, the *viscosity*, the *diffusion coefficient*, and any other transport coefficient.

When the observable ω^i is a *projector* on a region of Hilbert space characterized by some property, its expectation $<\omega^i>$ is the *probability* that this property be satisfied. The equation of motion (8.5) for $<\omega^i>$ is then interpreted as a *balance equation*, of the type of the *Pauli master equations*.

The formalism is also suited to the study of open systems, interacting with sources or *external baths* through a coupling which produces exchanges of energy (or of particles, or of any other conserved quantity). The observables ω^i suited to account for such situations are the collection of all observables of the open system, plus the Hamiltonian (or the other conserved observable considered) for the union of the system with the reservoirs. The equations (8.5) are then the *equations of motion of the open system*. For instance, if the system is a spin which may exchange energy with other degrees of freedom behaving as a thermal bath, these equations are directly adapted to the study of magnetic resonance. The coupling between the spin and its surroundings is responsible for the *relaxation*. Similarly, if the system is an atom, and if its surroundings are the vacuum with which it may exchange photons, the formalism is suited to the study of *emission* and *absorption* of electromagnetic radiation, and in particular of *line shapes*.

Similarly, for a Brownian classical particle, heavier than the particles of the fluid in which it is moving and with which it exchanges energy through collisions, a possible choice of relevant observables is $\delta^3(\hat{r}-r)\ \delta^3(\hat{p}-p)$ together with the overall Hamiltonian of the particle and the bath. The averages of these observables represent the probability $f(r,p)$ that the particle lies at the point (r,p) of phase space, and the overall energy. The equation of motion (8.5), (8.3) is a generalized *Fokker-Planck equation* for the evolution of the probability $f(r,p)$.

It is also possible to derive generalized *Langevin equations* by applying the projection method to the equation of motion (2.13) in the Heisenberg picture. An observable A, which depends now on time, is split into its relevant part A^P given by (7.8) and its irrelevant part A^Q. The time-derivative of A appears then as the sum of 3 terms, an *instantaneous* term in the relevant space, a

24

random force evolving in the irrelevant space with the kernel W, and a retarded *friction term* arising from the coupling between both spaces. The last two terms are related by a generalized *fluctuation-dissipation theorem*, a relationship which may be useful when writing Langevin equations on a semi-phenomenological basis[1,2].

Notice that, in agreement with the remarks made in the introduction, the elimination of the irrelevant variables has led to the generalized transport equations (8.5) which present important qualitative differences with the original microscopic Liouville-von Neumann equation. The transport equations are *non-linear*, through the dependence of P and W on the state D_0. They involve *retardation effects* governed by the memory kernel W. They present a *classical structure* : while the quantum mechanical character was obvious on the matrix form of the Liouville equation (2.3), the transport equations (8.5) do not have a different structure for classical and for quantum systems. Finally, as discussed in § 5, the relevant entropy $S(<\omega^i>)$ may increase, in contrast to the microscopic entropy $S(D)$. *Dissipation* has been introduced when going from the microscopic detailed description to the macroscopic description involving only the relevant variables.

10. MEAN-FIELD APPROXIMATIONS

The generalized transport equations (8.4) or (8.5) are *exact*. They contain however the memory kernel W defined by (8.1),(8.2), which is very complicated, because the irrelevant space in which QLQ and W act has an extremely large number of dimensions. It is therefore always necessary to resort to *approximations* or to *phenomenological* assumptions.

The simplest approximation consists in neglecting the second term of (8.5) which describes the coupling between relevant and irrelevant spaces. This *decoupling* leads to the approximate equations for the relevant variables :

$$\frac{d}{dt} <\omega^i> = -i <\omega^i LP> = -i(\omega^i L;D_0) \quad . \tag{10.1}$$

The observable $-i \, \omega^i L = -i[\omega^i, H]$ is interpreted as a *velocity* or a *force* associated with ω^i. The right side of (10.1) accounts only for its projection on the space of relevant observables (fig.1), which is interpreted as an average, *non-fluctuating part* of the total force $-i \, [\omega^i, H]$. The equations (10.1) are non-linear differential equations involving only the relevant variables $<\omega^i>$. The *retardation* terms which described the coupling with the irrelevant space have been dropped, and the retained *instantaneous* terms behave as a mean field.

The *dissipation rate* of the relevant entropy (5.2) is equal to

$$\frac{d}{dt} S(<\omega>) = -\frac{d}{dt} \text{ Tr } D_o \ln D_o = -\text{Tr } \dot{D}_o \ln D_o$$

$$= \lambda_i \frac{d}{dt} <\omega^i> \quad . \tag{10.2}$$

The approximation (10.1) provides, together with (4.3),

$$\frac{d}{dt} S(<\omega>) = i(\ln D_o \ L \ ;D_o) = i \text{ Tr } [\ln D_o,H]D_o$$

$$= i \text{ Tr } H[D_o,\ln D_o] = 0 \quad . \tag{10.3}$$

A mean-field evolution (10.1) produces therefore *no dissipation*. Dissipation arises from the coupling between relevant and irrelevant degrees of freedom. It should be remembered that this coupling involves the memory kernel : hence, *dissipation* is the result of *retardation* effects.

The mean-field approximations are *self-consistent* and *non-linear*, although instantaneous. These properties appear as a result of the elimination of the irrelevant part D_1 of the density operator, and of the approximation naturally induced by the projection method.

As an exercice, the reader may recover the *time-dependent Hartree-Fock equations* in the present framework by selecting as relevant observables ω^i the single-particle operators $c_\beta^\dagger c_\alpha$ (and the unit observable). The reduced densities D_o defined by (4.3) are then the independent particle densities $\exp[-\Sigma_{\alpha\beta}M_{\alpha\beta}c_\alpha^\dagger c_\beta -m]$. The expectation values of the relevant observables $\rho_{\alpha\beta} =<c_\beta^\dagger c_\alpha>$ are the Wick contractions with respect to D_o (and unity). By expressing an observable A as a sum of Wick normal products with respect to D_o, it can be shown that the projector P associated with D_o, when applied to A, keeps unchanged the normal products involving either 0 or 1 pair of non-contracted creation and annihilation operator, while suppressing the normal products involving more non-contracted operators. The mean-field approximation (10.1) is then identified with the time-dependent Hartree-Fock approximation.

A similar construction provides mean-field equations for any set of relevant observables ω^i chosen beforehand. In order to work out the formalism, the only technical requirement is the possibility of expressing explicitly the variables λ_i in terms of the $<\omega^i>$ by means of (4.4), i.e., we need to evaluate the generalized partition function

$$\text{Tr exp}[-\lambda_i \ \omega^i] \quad .$$

Once this is done, the construction of the projector P is straightforward and the approximate equations follow.

The mean-field equations associated with the hydrodynamic

contraction of the description are dissipationless transport equations equivalent to the set of the *Euler equations* and *conservation equations*. If the observables ω^i describe occupation numbers of fermions in single-particle states, *dissipationless balance equations* are obtained. If these observables ω^i are chosen as collective variables characterizing for instance the deformations of a nucleus together with the conjugate momenta, the mean-field equations will describe a *dissipationless collective motion*, having a classical structure although quantum mechanics is properly taken into account.

11. PERTURBATION METHODS AND SHORT-MEMORY APPROXIMATIONS

The dissipative term of the exact transport equations (8.4) or (8.5) is difficult to evaluate for two reasons. On the one hand, it involves the memory kernel W defined by the equations (8.1), (8.2), which are *practically impossible to solve*. The evolution superoperator W is even more complicate than the full evolution superoperator U, since it depends self-consistently on the projector Q. On the other hand, even if W were known, the dissipative term of the transport equation would be difficult to handle in view of its *retarded* character.

The first difficulty is usually overcome either by *simulating* W with more or less crude approximations presenting the expected behaviour, or in more elaborate treatments by resorting to various types of *perturbation* methods[1,2], that we restrict ourselves to list here. If the equations (8.1)(8.2) can be explicitly solved for an unperturbed Liouville operator L_0 not too different from L, expansions in terms of a *coupling $L-L_0$* may be used. In regimes close to equilibrium, it is natural to take as an expansion parameter the *deviation from equilibrium*. This allows to make use of a simpler unperturbed metric and a simpler unperturbed projector, referring to the equilibrium state rather than to the time-dependent reduced state D_0. Similarly, in the regimes considered in non-equilibrium macroscopic thermodynamics, the small quantity is the deviation from *local* equilibrium. In such hydrodynamic regimes, the natural expansion parameters are the *affinities*, defined as the gradients of the intensive variables λ_i (temperature, chemical potential, local velocity), which vanish in the state of global equilibrium. To lowest order, this expansion provides again the dissipationless hydrodynamic equations, previously obtained as mean-field equations in the projection method context. To first order, hydrodynamics in the linear regime is deduced from statistical mechanics, and explicit expressions for the transport coefficients are obtained. An example of such a treatment, for a classical gas, is the Chapman-Enskog solution of the Boltzmann equation.

The second difficulty, the retarded character of the dissipation, is usually by-passed through a suitable choice of the relevant observables ω^i. It is required that the characteristic *time-scales* associated with the dynamics of the variables $\langle\omega^i\rangle$ are *much larger* than the time-scales associated with the motion of the irrelevant variables. This requirement is natural : collective or macroscopic variables $\langle\omega^i\rangle$ should evolve more slowly than the other variables.

If this condition is fulfilled, the kernel $W(t,t')$ which describes the motion in the irrelevant space has very short characteristic time-scales. In addition, since it involves a very large number of degrees of freedom, it is expected to have a rather erratic behaviour as $t-t'$ increases. Thus, in the integral of (8.4), $W(t,t')$ oscillates rapidly with $t-t'$, while the last factor given by (8.6) is a slowly varying function of t'. The contributions from large values of $t-t'$ interfere destructively, and we may replace t' by t in the factor (8.6) of (8.4). Moreover, in the same short-memory approximation, the time-dependence of $Q(t)$ may be disregarded in the equations (8.1)(8.2) defining W, which are then solved formally as

$$W(t,t') \simeq Q(t) \, \exp[-i\,QLQ\,(t-t')] \quad . \qquad (11.1)$$

Integration over t' provides a factor $[i\,QLQ + 0]^{-1}$, the inverse of the Liouville operator in the irrelevant space (the term $+ 0$ is introduced to account for the vanishingly small real eigenvalues of QLQ). Insertion in (8.4) finally yields

$$i\,\dot{D}_0 = (PLP + i\dot{P})D_0 + i\,PLQ\,[QLQ - i0]^{-1}(\dot{P} + i\,QLP)D_0 \; .$$
$$(11.2)$$

The exact integro-differential operator (8.4) has thus been replaced in the short memory approximation by a differential equation *involving only one time t*. Besides the mean-field term, (11.2) contains a term which comes from the coupling with the irrelevant variables. The form of this approximate term will be adequate if the separation of the variables into a relevant and an irrelevant set implies also a clear-cut separation in the time-scales.

It can be shown[1,2] that, in rather wide circumstances, the second part of (11.2) produces *dissipation*, i.e., an *increase of the relevant entropy* relative to the variables $\langle\omega^i\rangle$. This fact is understood by reminding that the irrelevant variables have short time-scales and are extremely numerous. When the coupling transfers some amount of information from the relevant to the irrelevant set, this information remains trapped : it gets lost in experimentally inaccessible degrees of freedom, from which it cannot usually return in a finite time. This produces an *irreversible transfer of disorder* from the complicated towards the simple variables.

Dissipation, which is in principle a *retarded* phenomenon, appears as *instantaneous* in the considered approximation. In particular, in the hydrodynamic regime, the only slow variables are the densities of conserved quantities, and hence the short-memory approximation is justified. This is why hydrodynamics, heat or charge conduction, in situations to which we are accustomed at the macroscopic scale, can be described in terms of instantaneous transport coefficients.

An interesting exercise consists in applying the above formalism to the classical Brownian motion of a particle in a fluid[1,2]. Various equations, usually written on the basis of phenomenological arguments, will thus be recovered by using the short-memory approximation in the generalized transport equations after some choice of relevant observables has been performed. We have already indicated (§ 9) which choice leads to the *Fokker-Planck* equation. If the set of observables ω^i is restricted to $\delta^3(\hat{r}-r)$ and to the overall energy, (11.2) provides for the density of Brownian particles in ordinary space a *diffusion equation*. If the choice is further restricted to the momentum $\hat{p}$ of the Brownian particle and to the overall energy, (11.2) describes the slowing down of the Brownian particle by a *friction* force proportional to its velocity.

For less simple problems, an explicit evaluation of the dissipative term of (11.2) may still remain impossible in practice, and more or less coarse approximations are further required. For instance, by neglecting the structure of the Liouville operator in the irrelevant space, (11.1) may be simulated as

$$W(t,t') \simeq Q(t) \, e^{-(t-t')/\tau} \quad , \tag{11.3}$$

where τ is a unique short time characteristic of the evolution in the irrelevant space. By working out (11.2), we obtain then the *relaxation time approximation* for the transport equations,

$$<\dot{\omega}^i> = -i <[\omega^i,H]> - \tau <[[\omega^i,H]Q,H]> \quad , \tag{11.4}$$

where the averages are meant on D_o. We could imagine accounting better the structure of QLQ by introducing several relaxation times, or *form factors*, or by combining (11.2) with a perturbation expansion. For instance, a dissipative term can be introduced in this fashion in the time-dependent Hartree-Fock equation.

These examples illustrate the usefulness of the exact generalized transport equations (8.5). Although these equations as they stand cannot be dealt with explicitly, various approximation schemes may be worked out on them, discussed, and hopefully justified. The projection method is thus a *guide to a safe phenomenology*. It helps to analyze better the approximations leading to the dynamics of a reduced set of variables, to keep some control on these approximations, and to derive improved new approximations.

12. CONCLUSION

The method which we have sketched constitutes a general, powerful and systematic approach to the dynamics of many-body systems. It relies crucially upon a choice of variables considered *as the most representative*, either because their characteristic *time-scale is the largest* or because they are *weakly coupled* to the rest. This possibility of *choice* brings in a great flexibility, while the existence of a *general theoretical framework* allows the unification of various approximations and a better understanding of their meaning.

However, not only the technical tools, reduced density, projector, memory kernel, depend on the choice of variables considered as relevant, but the *concepts themselves are relative* to this choice. For instance, the temperature of a system off-equilibrium is defined naturally as the inverse of the Lagrange multiplier β associated with energy, but the value of this multiplier depends on the number and on the nature of the other observables ω^l. The entropy is also, as we have seen, a relative quantity : we have distinguished the detailed entropy $S(D)$, associated with the whole algebra of observables and remaining constant, from the relevant entropy relative to the hydrodynamic densities, which is larger and which is identified with the thermodynamical entropy. For a gas, Boltzmann has also introduced another relevant entropy, intermediate between the latter entropies, and relative to the single-particle observables. This entropy disregards the information contained in the correlations between the particles of the gas, and the H-theorem of entropy increase applies to it.

Thus, *dissipation depends on the level of description*. The larger the number of degrees of freedom studied, the weaker the dissipation is. Some choices may even give rise to a negative dissipation : in a spin echo experiment, the relevant entropy relative to the average magnetization of the sample increases during the relaxation process, but it decreases after the impulse which rotates the spins by π. Such a phenomenon is the signature of an *inadequate description* : it indicates the existence of *hidden variables*, able to restore some information which they have stored up. Such variables must be taken into account for understanding the observed dynamics.

REFERENCES

1. R. Balian, Y. Alhassid et H. Reinhardt, Physics Reports, in preparation. This article treats with more details most of the themes presented here, and includes a bibliography.
2. The projection method underlies many articles, quoted in the books of F. Haake, Statistical treatment of open systems by generalized master equations, Springer Tracts in Modern Physics, vol. 66 (Berlin, 1973) and of H. Grabert, Projection

operator techniques in non equilibrium statistical mechanics,
Springer tracts,vol. 95 (Berlin, 1982), as well as in the
review article of O. Penrose, Rep. on Progress in Phys. 42
(1979) 1937. It has been introduced by S. Nakajima, Progr.
Theor.Phys. 20 (1958) 948, by R. Zwanzig, J.Chem.Phys. 33
(1960) 1338 ; Physica 30 (1964) 1109, and by H. Mori and
K. Kawasaki, Progr.Theor.Phys. 27 (1962) 529. Its metric interpretation is given in ref. 1.

ENERGY DISSIPATION IN NUCLEUS-NUCLEUS COLLISIONS AROUND 40 MeV PER NUCLEON: NEW PHENOMENA OR TRANSITION TRENDS ?

Marc Lefort, Bernard Borderie, Dominique Jacquet
and the Nuclear Chemistry Group

Institut de Physique Nucléaire
91406 - Orsay, France

ABSTRACT

We have tried to present some of the new aspects which appeared recently in nucleus-nucleus reactions at collision velocities in the range of 0.2 c to 0.3 c. From a set of experimental data obtained at CERN and GANIL by the Orsay Nuclear Chemistry Group, we have tentatively extracted some preliminary conclusions about the possibility of very large energy deposits occuring in more or less central collisions. We have shown that, when nuclear matter is heated at excitation energies greater than 3 to 4 MeV per nucleon, quite a number of new aspects appear in the de-excitation process. Also, for intermediate impact parameters, a new phenomenon, predicted by Bondorf and by Campi et al., might have been observed : "multifragmentation". We should emphazise that what is presented here results from a collective work and is an attempt to show how physicists try to progress from the raw materials of their experiments towards some more elaborated considerations.

INTRODUCTION

After a couple of years during which nucleus-nucleus collision: have been studied with new beams in the range 20-80 MeV per nucleon, one may distinguish three main classes of phenomena, depending on impact parameters as sketched in fig. 1.

i) The first one corresponds to fragments with masses lighter than the projectile mass and velocities close to the beam velocity. A typical example is shown in figure 2, extracted from the study[1] of the reaction ^{40}Ar + Ni at 44 MeV.A. The process looks like what

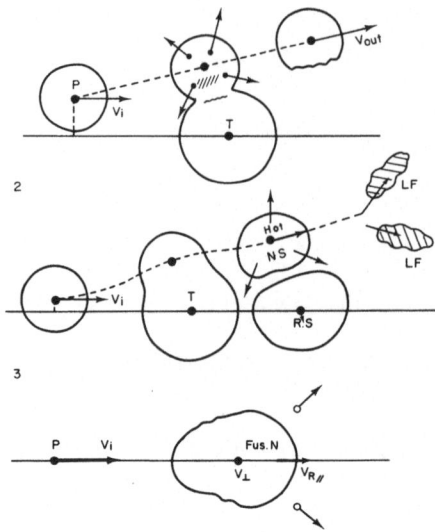

Fig. 1. Schematic presentation of three classes of violent nuclear
collisions depending on the impact parameters.

was described at much higher energies under the name of "fragmentation". Since the collision is rather grazing and fast and the entrance velocity has not been very much degraded, the energy dissipation is small and the momentum width reflects essentially the
Fermi distribution of the projectile nucleons. In that respect, one
observes a transition from the high energy behaviour towards the
well known low energy "deep inelastic" process, since a low momentum tail grows in amount with the decreasing size of the ejectile
and also exhibits a larger yield at lower bombarding energies.

ii) The persistance of this collective behaviour should be
emphasized as one of the characteristic of a second class of collisions which corresponds to lower ℓ-waves. The more the nuclei
overlap, the stronger is the damping and the lighter are the ejectiles, as shown on the contour yields versus $v_\perp$ and $v_{//}$ (fig. 3).
With smaller impact parameters other new aspects appear : light fragments are emitted with lower energies and wider angular distributions. Their origin is not yet clearly attributed, but there are a
number of recent data which show that at least a large fraction is
issued from a zone of overlap where the participant nucleons are
heated locally at a high temperature. This thermal source travels at
an intermediate velocity (slightly lower than $v_{beam}/2$). Due to the
lower impact parameters, more target nucleons are involved in the
collision and the small energy dissipation of the grazing collision
is not any more possible. The fast ejectile production is replaced
by an intermediate piece of hot nuclear matter which is teared-out

34

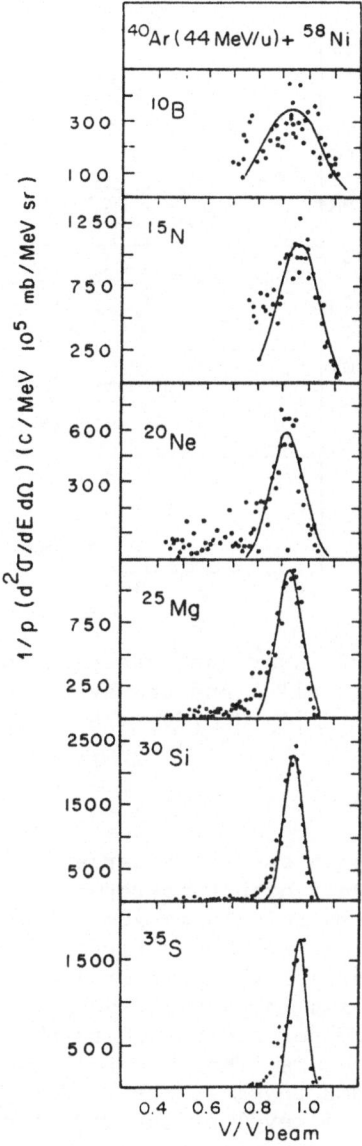

Fig. 2. Velocities of fragments emitted in the forward direction (at 3°) in the bombardment of Ni by [40]Ar at 44 MeV.A. (ref.[1]).

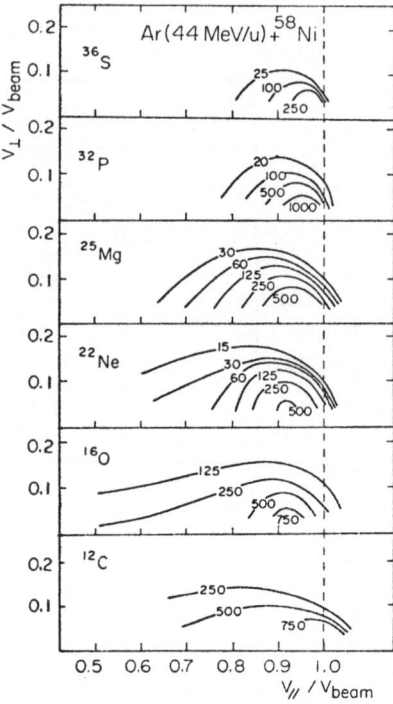

Fig. 3. Contour plot of invariant cross-sections for different ejectiles versus parallel and transverse velocities normalized
on the beam velocity. $d^3\sigma/dv^3$ are expressed in $(10^5$ mb/MeV sr).
$(\text{MeV/c})^{-1}$.

off the target, whereas a spectator target residue is left nearly at
rest. Because of its very high temperature, the hot piece breaks
down into fragments very quickly in some kind of multifragmentation.
However, this is not the well known fireball model of higher energies since the whole projectile participates and there is no projectile fragment. This class is perhaps the most typical of the velocity region around 40 MeV per nucleon where the coexistence of one and
two-body dissipation occurs. The dynamical evolution may be described
by solving, as in Plasma physics, the Landau-Vlasov equation[2] for
nucleons above the Fermi level of the target nucleus (mean field
and collision terms present simultaneously). For the moment, a few
data are available about these fragments which could be considered
as "ashes" from the heated zone, and we need more informations on
their yields, the dependence on Z and A, the angular and energy distributions.

iii) The third class corresponds to cases where large linear
momentum transfers have occured. We should notice that for the range

of bombarding energies which are concerned, even small impact parameters may correspond to rather high angular momenta. Typically, an
argon projectile accelerated at 30 MeV/u, with a linear momentum of
9.38 GeV/c, induces an angular momentum of 45 $\hbar$ for an impact parameter of only one Fermi. At the same velocity, krypton ions induce
100 $\hbar$, whereas 60 MeV/u carbon ions deliver 45 $\hbar$ for an impact parameter of 2.5 Fermi. Therefore, even central collisions may share
enough rotational energy in order to decrease the fission barrier
substantially.

Our purpose is to discuss briefly the second and third classes,
where energy deposits are large enough so that high temperatures
could be reached in nuclear matter. We shall begin by central collisions corresponding with low impact parameters and large linear
momentum transfers. They lead to very excited nuclear systems the
decay of which can be studied. But we have first to be able to trigger as specifically as possible those collisions which we call central ones.

2. COLLECTION OF RESIDUAL NUCLEI WITH LARGE RECOILING VELOCITIES

In principle, it is easy to calculate what should be the recoil velocity after complete fusion. The evaporation of many nucleons from the highly excited nucleus should not change the average
velocity, but would spread it over a width which depends on which
sort of particles are emitted. There exists codes which simulate the
effects of particle emission[3] (Linda). Also, the angular distribution indicates if there has been some perpendicular momentum transfer, so that the system has undergone only an incomplete fusion.
There has been two kinds of work in this line.

a) Off line experiments measuring activation products in various collectors corresponding to a set of ranges. One can deduce
the set of recoil energies for radioactive residual nuclei which are
identified by X and γ rays decay schemes after irradiation.

b) Mass and velocity determinations at various angles with
time of flights measurements between channel plates and ΔE, E detectors, or PPAC and MWPC (parallel plate avalanche detector, multiwire proportional chambers).

One can have a first glance indicating that in violent collisions, there still exists cases where one may approach more or less
complete fusion. This is shown for example very simply by measurements[4] of the ranges of various residual nuclei as they are collected in the forward direction from the reaction $^{14}N + ^{124}Sn$ at 420
MeV. It is possible to transform the range distribution into an histogram of the recoil kinetic energies. Finally a comparison is made
with respect to the cases where full momentum is transferred. As an
example, figure 4 shows that, although the average recoil energy (or

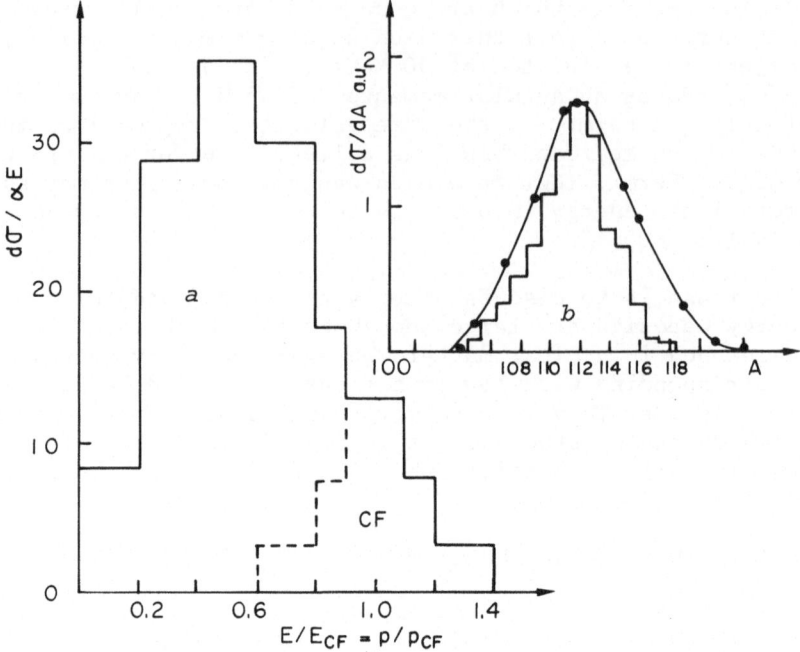

Fig. 4. a) Histogram of the various recoil energies deduced from various ranges measured in collectors for all the products issued from the reaction ^{14}N + ^{124}Sn at 30 MeV.A. A tentative estimate of the fraction where full momentum transfer has occured is shown.
b) A comparison between the mass distribution for the residues recoiling with full velocity and the distribution calculated with the evaporation code EVA, assuming the corresponding excitation energy after complete fusion (^{14}N+^{124}Sn). (ref.[4]).

linear momentum) is around 0.7 the full complete fusion energy , there is a substantial percentage of cases (35 %) where the residual nuclei are recoiling with the full velocity. Furthermore, these residues, being characterized by their γ decay, correspond to masses much lighter than the compound nucleus. There is indeed a good agreement[4] with the observed loss of A units and the calculated results on the number of particles evaporated by a compound nucleus sharing the corresponding excitation energy (fig. 4b). Heavier projectiles are now available and some results begin to appear for higher momentum transfers. Whereas full linear momentum transfer is of the order of 4.7 GeV/c with neon beams at 30 MeV/n, one can reach 9 GeV/c with argon projectiles and nearly 20 GeV/u with Kr at the same velocity.

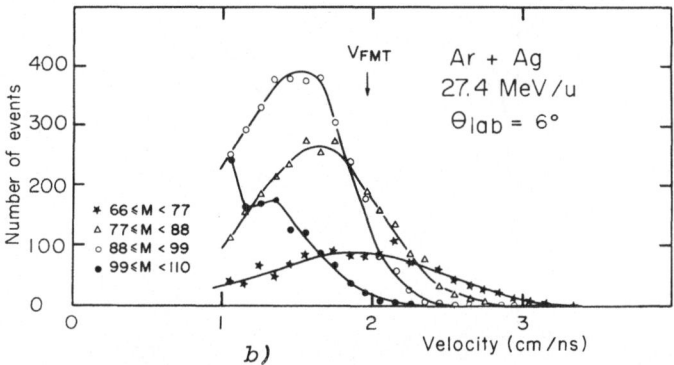

Fig. 5. Velocity distributions for various sets of masses for recoiling nuclei at θ = 6° from the collision ^{40}Ar + Ag at
27 MeV.A (ref.[5]).

More details can be extracted from velocity measurements of the
residual nuclei emitted in the forward direction as this is shown in
figure 5 extracted from a series of results obtained at GANIL[5] on
the system Ar + Ag at 27 MeV.A. A group of masses between A = 66
and A = 100 are identified in the reaction products with velocities
corresponding to momentum transfers around 70 % of the full momentum, and the cross section exceeds probably 10 per cent of σ_R. In
the mass range between 66 and 77, the measured velocity is equal to
the complete transfer. Since the corresponding excitation energy is
of the order of 800 MeV, the compound nucleus would have to endure
a temperature approaching 7 MeV. The amount of mass ΔM expected from
evaporation could be estimated around 60 units assuming roughly that
14 MeV per mass unit are necessary in the process. But the observed
residues correspond to more than 75 units. Moreover, the very wide
angular distribution of theses products seems to indicate that not
only nucleons but alpha particles and heavier fragments are emitted
from such hot nuclei. This is a rough indication that interesting
decay features might appear in those systems and differ from the
usual statistical model formalism (fig. 6).

As a matter fact, recent measurements[6] on the light fragments
(Z = 3 to Z = 8) emission at various angles indicate three different
origins that we shall discuss more thoroughly later on.Let us consider for the moment the presentation of contour lines for the inva-
riant cross sections 1/pc . $d^2\sigma/d\Omega dE$ versus $v_\perp$ and $v_{//}$. It shows the
slowest source travelling at a velocity corresponding to a full mo-
mentum transfer (fig. 7). Although the cross sections for that part
are not very large, they are in agreement with the prediction of an
equilibrium statistical process and a decay width for a given Z
number given by Moretto et al.[7].

$$\Gamma_Z = T_Z (E/(E-B_Z))^2 \exp (2\sqrt{a(E-B_Z)} - 2\sqrt{aE}) \qquad (1)$$

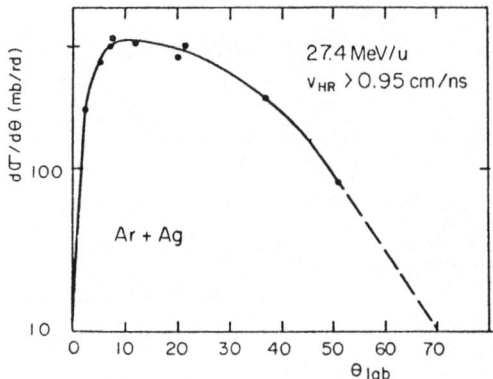

Fig. 6. Angular distributions, $d\sigma/d\theta$, for all products recoiling
with a velocity larger than half the full recoil velocity
in ^{40}Ar + Ag collisions at 27 MeV.A. (ref.[5])

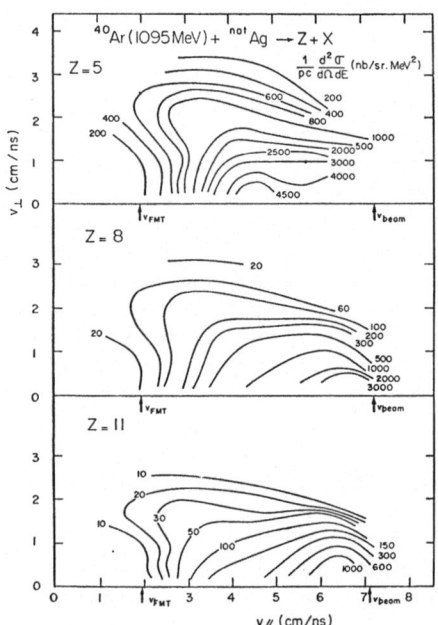

Fig. 7. Contour plot of invariant cross sections for fragments
Z = 5, Z = 8 and Z = 11 versus parallel and transverse
velocities in collisions ^{40}Ar + Ag at 27 MeV.A. The beam
velocitiy and the full momentum transfer velocity (FMT) are
indicated by arrows (ref.[6]).

which is proportional to $(E-B_Z)$ exp $(E-B_Z/T)$, where B_Z is the emission barrier for fragment Z and T_Z is the temperature evaluated using

$$E^* - B_Z = a\ T^2_Z \qquad\qquad\qquad (2)$$

With an excitation energy of the order of 600 MeV, a barrier around 50 MeV for Z = 5 and a level density parameter a = A/8, T is equal to 6 MeV and the expectation yield is of the order of 1 mb per steradian, whereas for Z = 8 it should be around 100 μbarn per sr.

There has been quite a number of attempts to measure the ratio of fusion-like reactions to the total reaction cross section, σ_F/σ_R, in an energy range between 10 and 40 MeV.A. This ratio has been plotted by Viola[8] versus the effective beam momentum $p^{eff} = \sqrt{2M(E-B)}$, where B is the coulomb barrier. It is not yet fully established that the incoming linear momentum is the appropriate parameter. One could as well try to look for the velocity dependence. The behaviour of σ_F/σ_R on p^{eff} appears to decrease linearly when p increases from 1 to 3 GeV/c, for various projectiles, He, carbon, oxygen, neon (fig. 8). From that line, one would expect that the probability for complete fusion vanishes around 4 GeV/c. The new results obtained with ^{20}Ne at 30 MeV/n (4.7 GeV/c) and ^{40}Ar at 27 MeV/n (9 GeV/c) indicate that the extrapolation should not be followed so far.

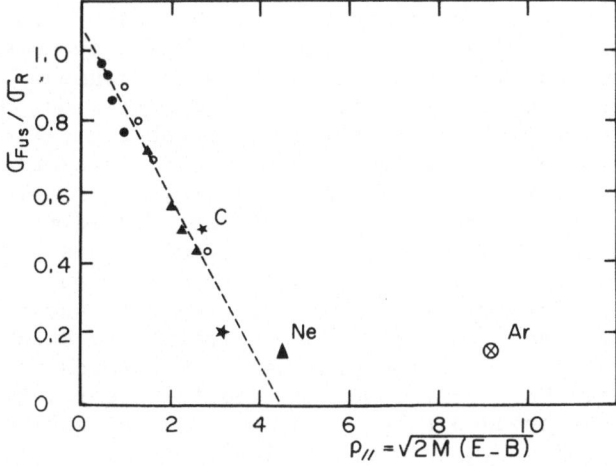

Fig. 8. Part of the total reaction cross section going into full momentum transfer, σ_{fus}/σ_R, versus the effective linear momentum, $p^{eff} = \sqrt{2M(E-B)}$ as plotted by Viola (ref.[8]).

3. SELECTION OF LARGE LINEAR MOMENTUM TRANSFERS WITH THE FISSION FRAGMENT ANGULAR CORRELATION

The fission fragment angular correlation has been used over twenty years in order to separate complete linear momentum transfer and low linear momentum transfer. The technique was suggested by Halpern and Nicholson[9] and developed by Sikkeland and Viola[10].

The amount of linear momentum transferred from a projectile to a target-like nucleus can be derived from the folding angle distribution of the correlated fission fragments emitted by the resulting nucleus. Whereas a correlation at 180° corresponds to a so called cold fission where no momentum has been transmitted to the fissioning nucleus, in the full momentum transfer the velocity vector of the recoiling compound nucleus produces in the laboratory system a closing of the correlation angle in the forward direction. The relation between the angle $(\theta_1 + \theta_2) = \theta_{corr}$ and the recoiling velocity v_{rec} is given by :

$$v_{rec} = \bar{v}_{fiss} \frac{\sin \theta_{cor} + \mu \sin (\theta_1 - \theta_2) \sin \overline{\theta_1}}{\sin \theta_1 \sin \theta_2 + 2 \mu \cos \overline{\theta_1}} \tag{3}$$

where $\bar{v}_{fiss}$ is the velocity of fission fragments in the C. of m. of the fissioning nucleus, μ is $m_1 - m_2/m_1 + m_2$, θ_1 and θ_2 are the laboratory angles for fragments 1 and 2 respectively and $\overline{\theta_1}$ the center of mass angle for fragment 1.

We assume that there is a negligible amount of transverse momentum. In order to derive the connection between momentum and angle, one needs the knowledge of the fission fragment kinetic energy for the particular nucleus which undergoes fission. One can use the systematics of Viola[11] for the most probable kinetic energy in symmetric fission events v_{fiss}. One should notice that the momentum transfer scale is derived from the correlation angle scale in the hypothesis of a symmetric mass division. Considering, in the mass distribution, those cases far from symmetry has the effect of extending towards smaller θ_{corr} values the corresponding scale for angles. Also particle emission from the fragments and the opening of the detectors tends to broaden the correlation. Below 10 MeV per nucleon, the central collision component corresponds to complete fusion mechanism where full momentum transfer occurs. Peripheral collisions either produce fission when the target nucleus has a low fission barrier, but θ_{AB} is close to 180°, or cannot yield fission phenomena for lighter targets like gold or Pb. But, for higher energies, the central collision peak, unstead of being located at the full linear momentum transfer correlation angle, shifts towards larger angles, i.e., incomplete momentum transfer. Therefore one has possibility to appreciate how much of the initial momentum has been put into the system, and to derive the amount of energy deposit, providing the knowledge of the mass. This approach has been initia-

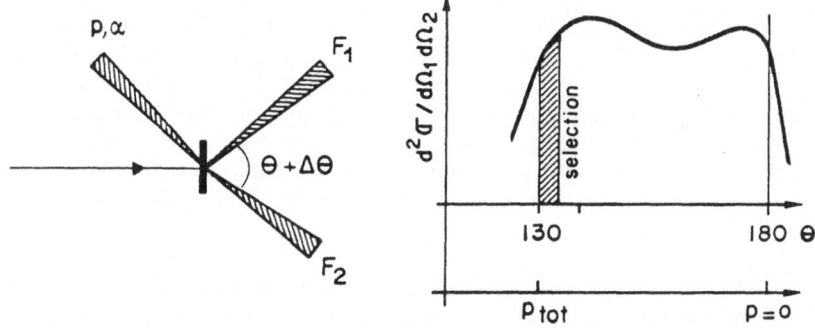

Fig. 9a. Schematic illustration of the correspondence between angular correlation of the fission fragments and linear momentum transfer.

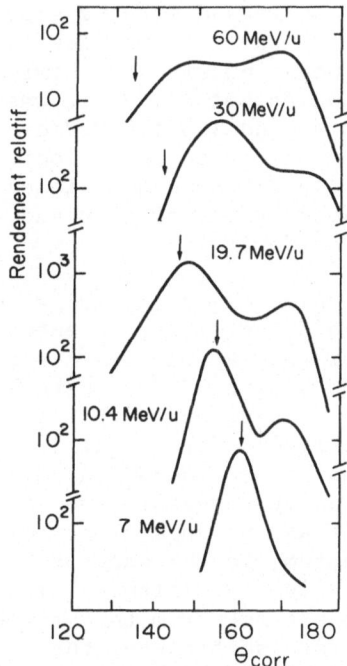

Fig. 9b. Typical evolution of the yields at various fission fragment angular correlations when the carbon projectile enerby increases from 7 to 60 MeV per nucleon.

ted by Viola et al.[12], in the range 10-20 MeV/u, and by the Orsay group working at ALICE, CERN and GANIL[13] in the range 15-84 MeV/u.

A typical presentation of various angular correlations depending on the incident energies is given in figure 9. Therefore, one may select a narrow range of angular correlations and, by that mean, trigger a particular set of linear momentum transfers for which light particle emission is observed in coincidence.

Actually, a complete linear momentum transfer corresponds to a distribution amongst the correlation angles, from which the centroid angle θ^0_{F1F2} is obtained with a symmetric mass division and the most probable kinetic energy release in fission. The dispersion around that value is due i) to the distribution of asymmetric fissions around symmetry ; ii) to neutron evaporation. Therefore, all the events observed at a particular correlation angle could not be strictly attributed to a precise linear momentum transfer. The method suffers a lack of precision in that sense. However, quite interesting conclusions have been derived from a systematic study as a function of the incident momentum. Above 15 MeV/n, at least for carbon ions, it is clear that, on the average, complete momentum transfer becomes less probable and we have shown[13,14], in 1981, that the centroid of the linear momentum levels off around 2 GeV/c, as shown in figure 10. Comparing with the limitation observed for α projectiles by Saint Laurent et al.[15], it was suggested an universal value per nucleon around 160-180 MeV/c. This corresponds to a projectile velocity comparable to the velocity of sound in nuclear matter. Raha and Weiner[16] have speculated that perhaps soliton propagation occurs above this velocity and therefore may limit longitudinal momentum transfer.

Starting with this value of 180 MeV/cn, we could tentatively predict[14] that the centroid for linear momentum transfer <LMT> would be at 2.8 GeV/c for ^{16}O, at 3.6 GeV/c for ^{20}Ne and 7.2 GeV/c for ^{40}Ar. As a matter fact, since 1982, there has been experimental results which show some consistency with this estimate, since a value around 2.6 was observed[17] for ^{16}O at 19.9 MeV/n, the centroid for linear momentum transfer was found[18] at 3.6 for ^{20}Ne at 30 MeV/n, as shown in figure 11, and very recently we have observed[19] in the analysis of the fission fragment angular correlation a peak around 75 per cent of full momentum, in the case of 27 MeV/n Argon beam on ^{238}U, i.e. at 7.0 GeV/c (fig. 12). Although it might not be very accurate, we have tried to extract from that representation the contribution to the total cross section for the complete fusion (around full momentum transfer), for large energy deposits (between full MT and 0.6 FMT) and for what we attribute to more peripheral reactions which contribute finally to a fission process. Depending on the kind of integration over all angles that we use, the values should be taken with a large uncertainty both for the cross section and for the limiting ℓ wave : ℓ_c.

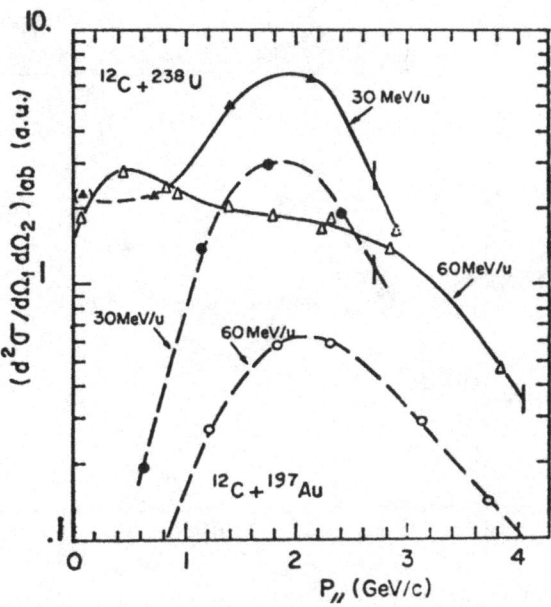

Fig. 10. Yields for various fission fragment angular correlations
and for the corresponding linear momentum transfer in the
collisions C + Au and C + U at 30 and 60 MeV per nucleon
(ref.[14]).

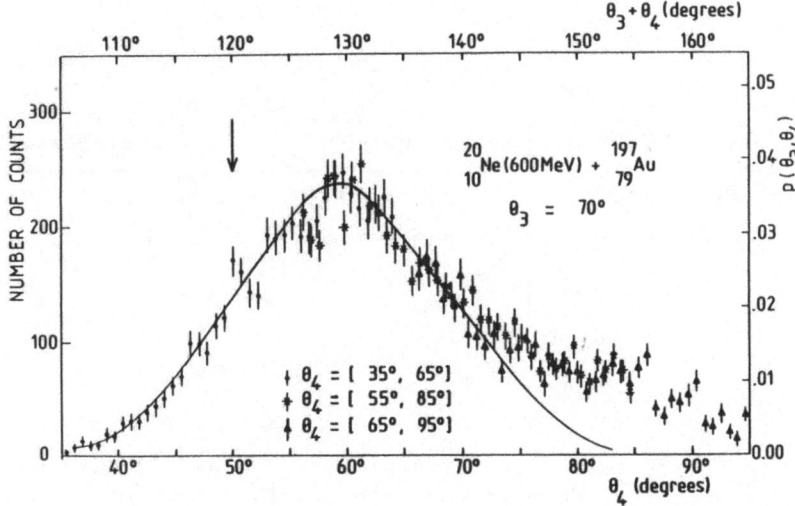

Fig. 11. Angular correlations for the fission fragments in the
reaction ^{20}Ne + ^{197}Au at 30 MeV.A (ref.[18]).

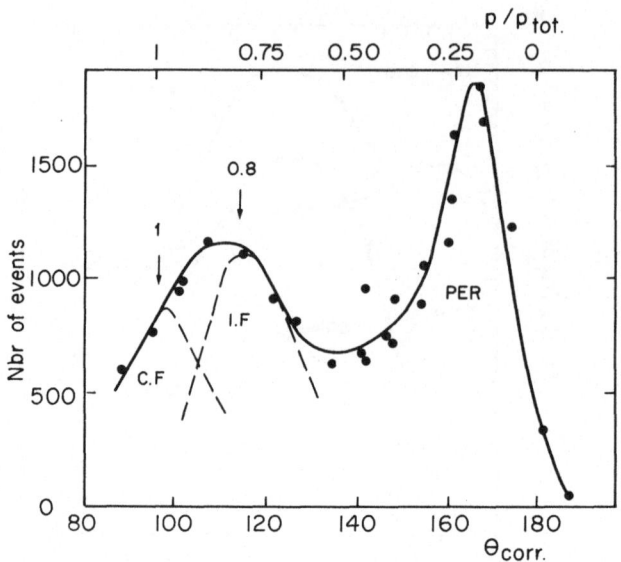

Fig. 12. Distribution of the yield for various linear momentum transfers in the collision $^{40}Ar + ^{238}U$ at 27 MeV/n. A very rough estimation is made for the amount of events corresponding to FMT (1), to incomplete momentum transfer (around 80 %) and to more peripheral reactions leading to fission.

TABLE I

	Complete Fusion	Large Energy Deposits	Peripheral Fission
σ_{mb}	250 ± 100	1250 ± 250	2000 ± 250
ℓ_c in $\hbar$	125 ± 25	280 ± 40	$\ell > 250$
b_{max} in fm	2.8 ± 0.4	6 ± 1	b > 6 fm

We ignore if this very naive picture can be extended to more massive projectiles. However, one can try to derive what amount of energy can be deposited in the resulting nucleus. We make the assumption that the ratio $\overline{E}_{tr}/\overline{E}_{CF}$, of the energy deposit to the full complete fusion energy is proportional to the ratio $<LMT>/p_i$ of the linear momentum transfer to the incoming momentum, more precisely :

$$\overline{E}_{tr} = \overline{E}_{CF} \; \frac{<LMT>}{p_i} \cdot \frac{A_T + A_i}{A_T + A_{tr}} \qquad (4)$$

where A_T and A_i are the target and projectile mass number respectively and $A_{tr}/A_i = <LMT>/p_i$ is the fraction of the incoming projectile mass. This implies that the missing linear momentum is due to a lower mass of the incoming ion, and not to a decrease of the incident nucleon velocity. We shall see later that this seems to be the case. Therefore, when $<LMT>$ reaches a limit, $\overline{E}_{tr}$ still increases with $\overline{E}_{CF}$ and with the bombarding energy. So, the following table (Table II) has been calculated for a medium mass target and an uranium target[20], for a series of projectiles at 30 MeV/u.

We have mentioned the excitation energy and the nuclear temperature, assuming $E^* = A/8 \, T^2$, for those nuclei corresponding to the most probable momentum transfer as derived from 180 MeV/c per nucleon.

TABLE II   "Hot Nuclei"

Target A = 100 $<p>/A$ = 180 MeV/c A^{-1}

E = 30 MeV.A $<p>/p_{tot}$ = 0.77 $T = \sqrt{\dfrac{8E^*}{A}}$

Ion	$\overline{E}$	E_{rec}	P_{tot}	$<p_{//}>$	$<E^*>_{res}$	$<T>_{res}$	T_{CF}
^{12}C	331	39	2.8	2.16	254	4.22	4.8
^{20}Ne	500	100	4.69	3.6	376	5.09	5.8
^{40}Ar	857	343	9.38	7.2	650	6.3	6.8
^{84}Kr	1369	1151	19.69	15.1	1118	7.4	7.55
^{84}Kr (40)	1826	1534	22.74	17.5	1433	8.5	8.56

Target A = 238 $<p>/A$ = 180 MeV/c A^{-1}

Ion	$\overline{E}$	E_{rec}	P_{tot}	$<p_{//}>$	$<E^*>_{res}$	$<T>_{res}$	T_{CF}
^{12}C(30)	345	15	2.8	2.16	264	2.9	3.6
^{12}C(60)	685	35	3.96	2.16	370	3.5	4.7
^{20}Ne(30)	554	46	4.69	3.6	413	3.55	4.14
^{40}Ar(30)	1028	172	9.38	7.2	779	4.8	5.43
^{84}Kr(30)	1863	657	19.69	15.1	1520	6.43	6.8
^{84}Kr(40)	2483	877	22.74	15.1	1948	7.3	7.95

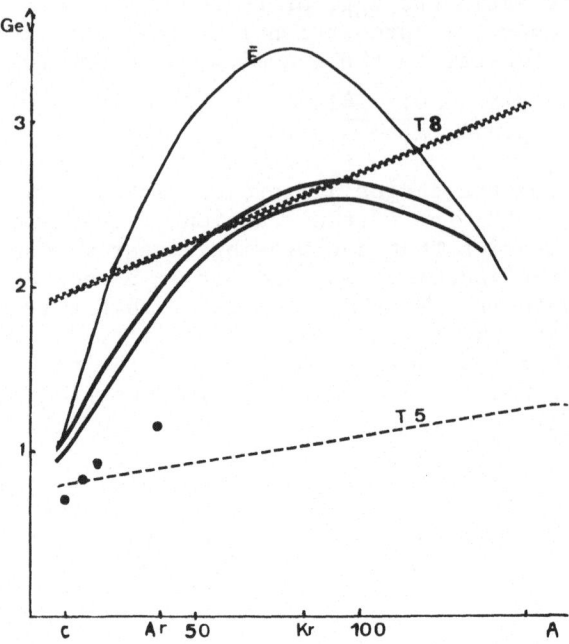

Fig. 13. A speculative prospect of the possible energy deposits in a heavy target by various ions at different bombarding energies. The curves labelled T = 5 MeV and T = 8 MeV show the energy deposit necessary for reaching those temperatures assuming $E^* = A/8 \, T^2$. Points indicate results which have been obtained at 30 MeV.A.

In addition, we have put the temperature which would correspond to complete fusion, i.e. full momentum transfer. Since there are now some data indicating that a fraction of the cross section still corresponds to these events, it seems worthwhile to notice that in principle, even quite heavy nuclei can be heated with krypton beams at excitation energies approaching 2 GeV which corresponds to temperatures near 8 MeV. For the moment, we know that a temperature of 5 MeV has been experimentally measured, but the most crucial experiments with krypton beams have not yet been performed. According to these considerations, figure 13 shows maximum available energies for collisions of various ions from Ganil with uranium targets.

As a matter of fact, a limiting value might exist around a temperature of 5 MeV, if one follows the conclusions given by Campi et al.[21], regarding the de-excitation mode of hot nuclei. These authors have noticed that the number of nucleon-nucleon collisions increases with the excitation energy, and then, the mean free path decreases down to some 1.5 fm for very high energies. The system basculates into a multifragmentation regime which has also been predicted by Bondorf[22] and the number of fragments tends to be equal to the

number of N-N collisions. Assuming an average number of N-N collisions,<n>, proportional to the number of nucleons, A, so that <n>/A
could be taken around 0.04 for the critical mean free path below
which fast fragmentation occurs, and an effective average energy
deposit, E_O, of 80 MeV at each collision, a total excitation energy per nucleon $<n>E_O/A$ around 3.2 MeV is reached, above which the
slow evaporation process becomes much less probable than fast fragmentation. This corresponds roughly to T = 5 MeV if one takes
$E* = A/8 \ T^2$.

With this scenario in mind, it might be worthwhile to have a
look at the results obtained in ref.[18,19] and [23] shown in figure 14
in a schematic presentation and at the numbers in Table III. Let us
first consider the yield at 18 per cent of the full momentum transfer and the corresponding excitation energy assuming that the excited system has accepted 18 % of the projectile mass. In the case of
Au target, the fission yield is very low whatever is the projectile
except in the very specific case of Ar + Au at 44 MeV.A. This can
be understood because the excitation energy is probably too low for
favoring fission against neutron evaporation for nuclei in the re-

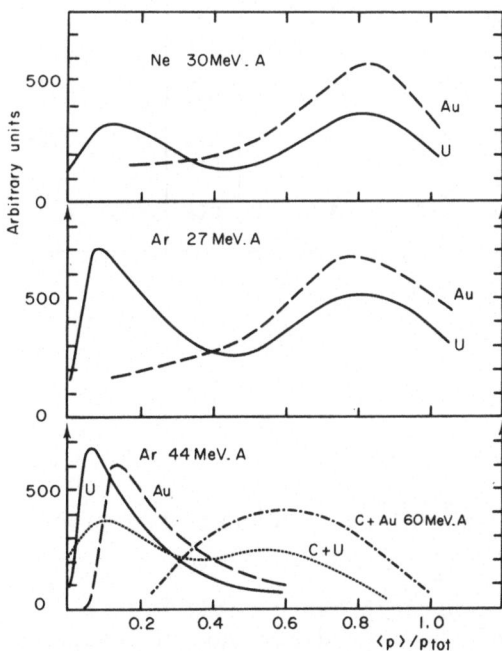

Fig. 14. General behaviour of the momentum distribution in fissioning nuclei resulting from various collisions, ^{20}Ne + Au,
Ne + U, at 30 MeV/n, ^{40}Ar + Au and ^{40}Ar + U at 27 MeV/n,
Ar + Au and Ar + Th at 44 MeV/n. Results obtained in ref.
18,19 and 22. This figure has been proposed by D. Jacquet.

gion ^{200}Hg - ^{206}Pb. On the contrary, a high excitation energy above 300 MeV (and probably a high angular momentum) is reached with the Argon ions at 44 MeV/n and fission occurs.

Let us turn now towards the high linear momentum transfers, at 70 % of FMT. Fission is observed for all cases (even with fast ^{12}C at 60 MeV/n) except[23] for the Ar + Au and Ar + U at 44 MeV/n. The corresponding excitation energies are below 3.2 MeV.A, except for the last case where the excitation energy around 1000 MeV corresponds to a temperature greater than T = 6 MeV, above the multifragmentation limit.

TABLE III

Excitation energies and nuclear temperatures for two partial momentum transfers p/p_{tot} = 0.18 and 0.7.

| Beam | E/n | p/p_{tot} = 0.18 | | | | p/p_{tot} = 0.7 | | | |
		E* MeV	E*/A	T= $\sqrt{8E*/A}$	Exp	E* MeV	E*/A	T= $\sqrt{8E*/A}$	Exp
^{12}C	60 MeV/n	120	0.6	2.2	no Fis	480	2.28	4.3	Fis
^{20}Ne	30 MeV/n	108	0.53	2	no Fis	390	1.78	3.8	Fis
^{40}Ar	27 MeV/n	190	0.9	2.8	no Fis	660	2.86	4.9	Fis
^{40}Ar	44 MeV/u	316	1.5	3.5	Fis	1090	4.73	6.2	no Fis

4. STUDY OF THE DE-EXCITATION OF NUCLEI RESULTING FROM VIOLENT CENTRAL COLLISIONS

A rather crude approach has already been touched. It consists in the comparison of the mass distribution resulting from an evaporation code and of the experimental measurements of the evaporation residues with excitation energies of the order of 500 or 600 MeV. The evaporation chain has a length of more than 50 nucleons. Furthermore, it is difficult to know what amount of angular momentum should be introduced.

A more precise way of investigation is to select a narrow range of linear momentum transfers (by the angular correlation of the fission fragments), and to look for light charged particles emitted in the backward direction, as it is sketched in figure 9. Such a procedure has been tempted in the case of ^{12}C on uranium and gold at 60 MeV per nucleon[24], and more recently at GANIL with ^{40}Ar on uranium at 27 MeV/n. A schematic presentation of the experimental array of detectors is shown in figure 15. From these experiments, one may stress a number of points.

i) A large fraction of ^{1}H and ^{4}He are emitted prior to fission. At forward angles the energy spectra extend up to more than 120 MeV as shown[19] in figure 16 in the case of Ar + U at 27 MeV/n. But the backward spectra (115 and 160°) exhibit a typical evaporation shape. Furthermore, they are exactly the same when the fission fragment is detected in coincidence at various angles. This indicates quite clearly that there is no kinematical effect of the velocity vector of the fragment. A typical demonstration appears in figure 17 extracted from a study[24] of carbon induced fission on thorium at 60 MeV/n.

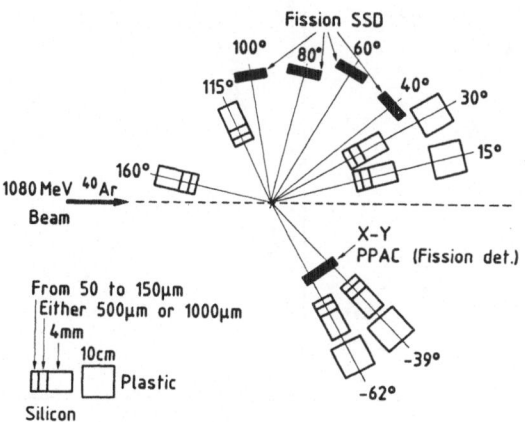

Fig. 15. Experimental array of detectors for fission fragments and light charged particles.

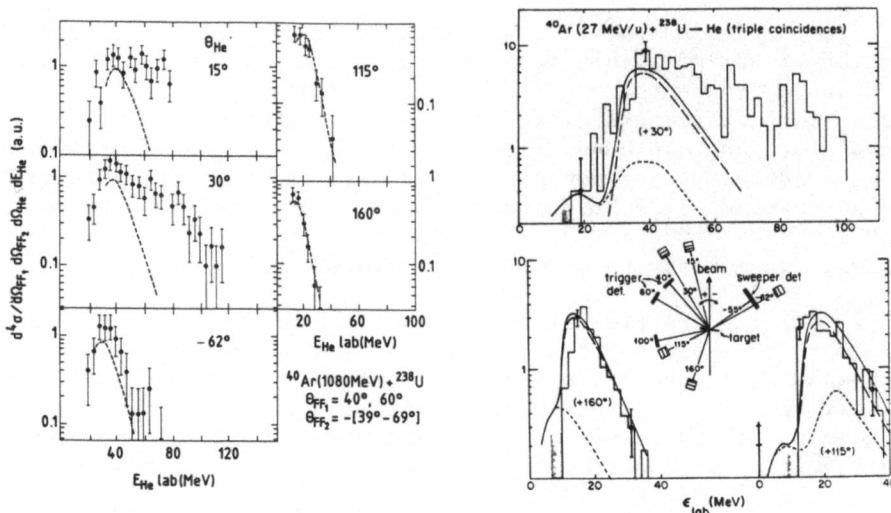

Fig. 16. Energy spectra of alpha particles measured at different
angles (from 15° to 160°) in coincidence with fission fragments in two given angular correlations (ref.[19]). On the
right, smooth curves are from simulations from fully accelerated fragments (......), from the composite (------), and
their sum (———).

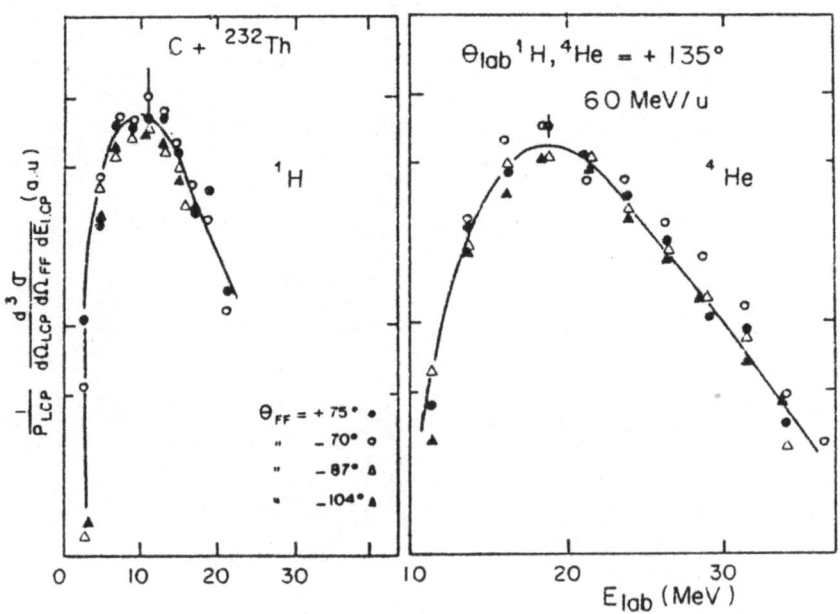

Fig. 17. Energy spectra for ^{1}H and $^4_2$He measured in the laboratory
at θ = 135° in coincidence with a fission fragment detected
at various angles (^{12}C + Th at 60 MeV/n).

Furthermore, the maxima of the spectra are located about at 10 MeV for protons and 20 MeV for α particles and when comparing the spectra relative to Au and Th targets, the difference of Z values is reflected by a shift in the maxima of about 1 MeV for ^{1}H and 2 MeV for ^{2}He.

ii) The light charged particle emission is found isotropic when a strong selection is applied to the fission fragment angular correlation so that only LMT values in the vicinity of full linear momentum are considered. This is obtained at $\theta_{F_1F_2}$ = 135° for ^{12}C + ^{238}U, 128.5° for ^{12}C + ^{197}Au at 60 MeV/u and 90° for Ar + U at 27 MeV/n.

iii) From the location of the ridge of the invariant cross sections for ^{4}He presented in the plane defined by $v_\perp$ and $v_{//}$, an average source velocity was deduced arount 0.018 c for ^{12}C + Au which represents 90 % of the center of mass velocity. A similar result has been observed[19] in the work on (Ar + U), when large momentum transfers greater than 70 % of the full momentum are selected. Fig. 18 shows that the source for alpha particles travels at a velocity equal to 0.035 c, nearly equal to the recoiling composite nucleus velocity. For small momentum transfers, the lower part of the figure indicates that several sources appear, one of them being around half the beam velocity. We shall come back later on this aspect.

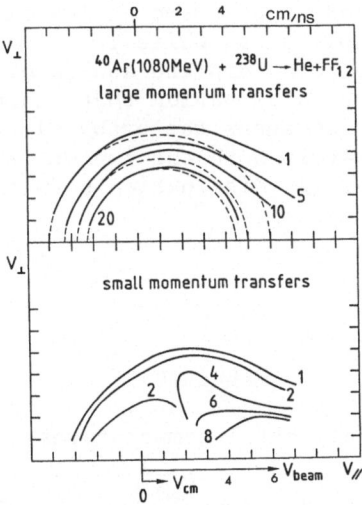

Fig. 18. Contour plot of invariant cross sections for alpha particles versus parallel and transverse velocities for two cases of momentum transfers. In the case of large momentum the circles are centered around a source moving with the center of mass full velocity (v_{cm}) (ref.[19]).

With the preceding observations in mind, one may have some confidence in studying the particle emission as a typical decay characteristics of the highly excited nucleus.

The temperatures derived from the slope of the tails are compared in Table IV with the expectation from a fully thermalized compound nucleus decay. Also the experimental maxima of the spectra in the reference frame can be compared with B + T, where B is the coulomb barrier extracted from compilation data for Z = 79 + 6 (Au), Z = 90 + 6 (Th), Z = 92 + 18 (Ar + U).

As a matter fact, such results are very surprising. It is indeed not expected at all that a compound nucleus Z = 96 (curium) or Z = 98 (californium) or Z = 110, would not first decay by fission. Even at lower energies, the Stony Brook group has observed since a several years that protons and alpha particles are emitted prior fission with a multiplicity of the order of 0.1 in the reaction Ar + U at 8 MeV per nucleon[25].

An interesting feature of the light particle production can be shown by plotting the differential multiplicity versus folding angle or momentum transfer. Forward-directed ^{4}He and H production (Fig. 16 and 19) arises from the whole range of momentum transfers, but with a distinct preference for the peripheral collisions at $\theta_{FF} \gtrsim 140°$. (This preference appears even before substracting the evaporative components which are largest for $\theta_{FF} \lesssim 140°$). By contrast, the ^{4}He production at backward angles is decidedly disfavored in the peripheral collisions while it is generated with comparable multiplicities in both complete and incomplete fusion. Evidently the massive energy dissipation that characterizes fusion-like reactions strongly enhances the probability of this evaporation-like emission. However, both the impact parameter (or entrance channel spin) and the energy dissipation seem to influence the

TABLE IV

Experimental temperatures

	T_{exp}	$T = \sqrt{\dfrac{8E}{A}}$	$E_{max} - T_{exp}$	E_B
C + Au	4.7 ± 0.5	5	16.3 ± 0.5	18.5
C + Th	4.8 ± 0.5	4.7	18.2 ± 0.5	20
Ar+ U	5.0 ± 1.0	5.5	20 ± 2	22.5±2

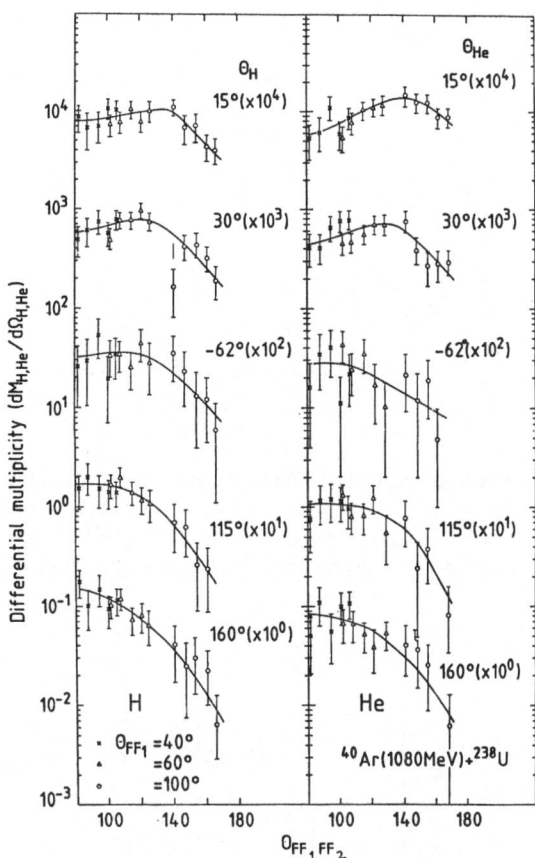

Fig. 19. Differential multiplicities for protons and alpha particles
measured at various angles as a function of the angular
correlation (or momentum transfer) of the fission fragments
emitted in Ar + U at 27 MeV/n (ref.[19]).

abundance of the forward-peaked ^{4}He with dominance achieved by
large spin. For ^{40}Ar + U, at a bombarding energy of 1080 MeV, where
some 900 MeV of excitation energy can be reached in the composite
system, more than 1 ^{4}He and 2 ^{1}H per event are evaporated (fig. 20).
More precisely some Monte-Carlo calculations have been made[26] in
order to simulate three classes of emitted particles :

i) the evaporation from a compound system yields an alpha particle
   multiplicity of 0.088 ± 0.01 at 334 MeV and a multiplicity of
   1.4 at 1080 MeV ;

ii) evaporation from the fission fragments would fit the experimental spectra with a multiplicity of 0.036 ± 0.012 at 334 MeV and
   between 0.0 and 0.4 at 1080 MeV ;

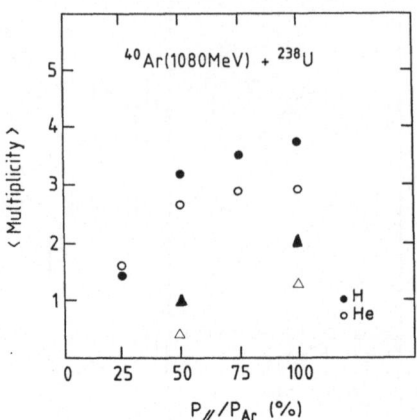

Fig. 20. Total average multiplicity for protons (full points) and
alpha particles (open circles). Triangles indicate the
average multiplicity for protons (full) and alpha (open)
corresponding to evaporation spectra measured at backward
angles.

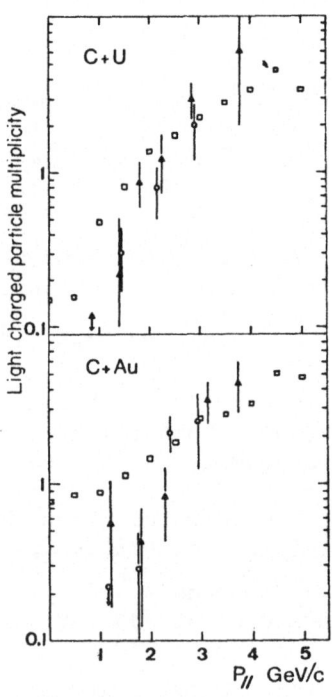

Fig. 21. The multiplicity of evaporated LCP emitted isotropically
versus the LMT in ^{12}C + ^{238}U at 3 energies (ref.[24]).

iii) the pre-equilibrium fast emission would correspond to 0.071 ± 0.018 at 334 MeV and 1.6 at 1080 MeV.

Also, for ^{12}C + ^{197}Au as well as ^{12}C + ^{238}U, the total number of evaporated light charged particles per event (p, d, t, ^{4}He) is of the order of 4 at the highest momentum transfer, around 5 GeV/c (fig. 21).

The last observation we want to mention is relative to the mass of the fragments. From the kinetic energy and the correlation angle measurements, one can deduce the masses of the fragments. The main observation is that those fragments which are associated with large momentum transfers, i.e. with the smallest angles $\theta_{F_1 F_2}$, have much lower masses than those corresponding to rather small LMT. This is typically shown in figure 22. Now we have shown that we can derive in equation 4 an excitation energy corresponding to a given LMT and therefore calculate the total mass ΔA of particles emitted in order to exhaust this energy. Each nucleon dissipates $(S_n + 2T)$ where S_n is the binding energy and T decreases at each step of the evaporation chain. The amount of mass ΔA expected from evaporation of ν nucleons could be compared to the measured mass deficit $m_D = A - 2A_F$, where A_F the measured mass number of the final product and A the mass of the fissioning nucleus. To make the comparison more precise in the case of the uranium target, the number of neutrons arising from cold fission was added to ΔA. The data points shown[27] in figure 23 lie in the vicinity of the diagonal, indicating a strong correlation. Now if we take the same amount of linear momentum transfer but different incoming energies, the values of ΔA as well as m_D increase with increasing energies. This means that a given amount of LMT is more efficient at 84 MeV/n than at 30 MeV/n. It would not be true if the transfer process was dominated by nucleonic interactions with all the nucleons of the projectile, since the defect of linear momentum would correspond to a limitation of the average nucleon velocity at the last stage of interaction. Therefore, a given <LMT> is due to an incomplete fusion occuring at larger impact parameters, the velocity is kept equal to the projectile velocity and the excitation energy is related to the <LMT> and to the complete fusion energy by equation 4, which can be written

$$E^* = \frac{A_T V_i}{2(A_T + A_{fus})} \quad <LMT> \tag{5}$$

where A_{fus} is the fraction of the projectile mass which is agglomerated to the target. It is obvious that for a constant value of <LMT>, E^* increases with the incoming velocity V_i. This is exactly what is found in figure 23 where E^* is transformed into a number of emitted nucleons, and the results is compared to the experimental mass deficit.

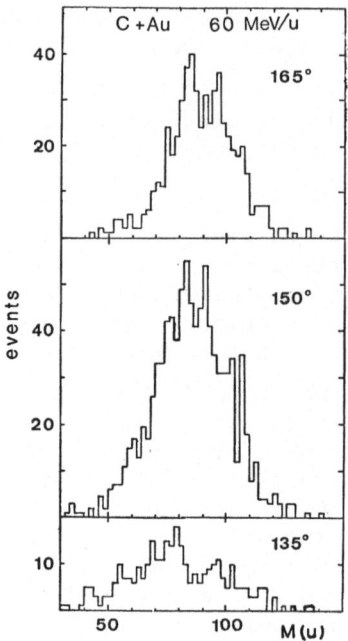

Fig. 22. Mass distribution of the trigger fragment for three correlation angles obtained in the fission of C + Au at 60 MeV/n.
The maximum is shifted down at A = 80 for the full linear
momentum transfer corresponding to θ = 135° (ref.[29]).

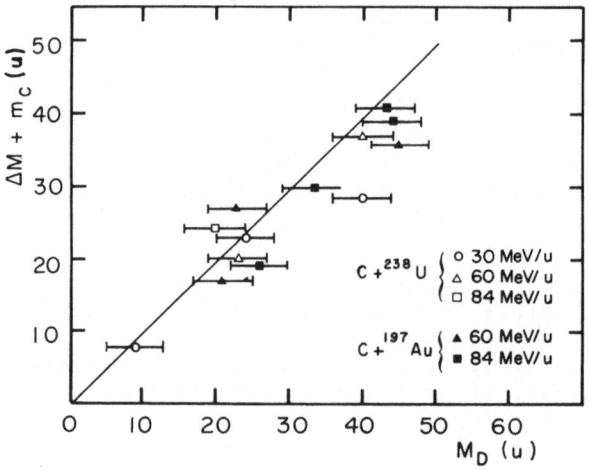

Fig. 23. A comparison of the calculated number of emitted nucleons
ΔA and the measured mass defect $m_D = A - 2A_F$ deduced from the
experimental average mass of the fragments (ref.[27]).

5. VERY HOT INTERMEDIATE NUCLEAR SYSTEMS. DEEP INELASTIC "MULTI-FRAGMENTATION"

5.a. Fragments issued from a moving source

In the introduction, we mentioned three classes of collisions and we pointed out that hot species could be created for intermediate impact parameters. Although there are not yet many complete data concerning this aspect, it seems worthwhile to endeavour to understand the very particular behaviour of light fragment emission. Let us come back to figure 7 where contour lines of the invariant cross section $1/pc \ d^2\sigma/d\Omega dE$ are plotted versus the velocity components $v_\perp$ and $v_\parallel$. Clearly appears a high velocity component, as circles centered at a velocity slightly lower than the beam velocity which can be attributed to the first class of phenomena quoted in the introduction. What is more surprising is the shape of the contour lines around a second source of lower velocity. It resembles the emission from the fireball as observed at higher energies[30]. As this has been made by several authors for high energy collisions[30,31,32], one may try to extract from the results the picture of the emission from a moving source and parametrize with a maxwellian expression in the c.m. system of this source :

$$W(E) \ \alpha \ E^{1/2} \ \exp \ (- \frac{E}{K})\tag{6}$$

The energy spectra as they were measured in figure 24 would fit the

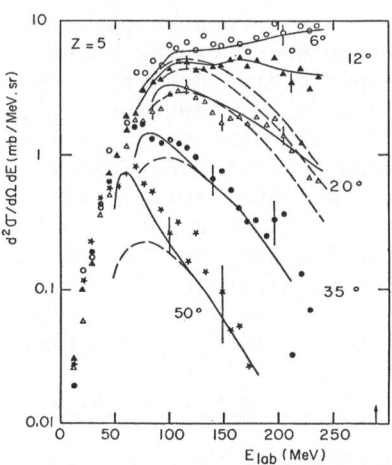

Fig. 24. Energy spectra for all fragments Z = 5 emitted at 5 angles (6 to 50°) from the collisions ^{40}Ar + Ag at 27 MeV.A. Dash lines = emission from an intermediate moving source and maxwellian shape. Full line = sum of the three sources.

maxwellian with a parameter K = 15 MeV, and the reconstruction in the laboratory system

$$W(E) \; \alpha \; (E - Z \, E_c)^{1/2} \; \exp - (\frac{E - Z E_c + E_s - 2 E_s^{1/2} \; (E - Z E_c)^{1/2} \; \cos \theta}{K}) \quad (7)$$

requires a kinetic energy for the source E_s corresponding to half the beam velocity, and a repulsion coulomb term E_c equal to 7 MeV.

The interesting remarks that could be done about these parameters are the following :

i) the term in $E^{1/2}$ indicates a volume dependence and not the surface dependance expected for an evaporation process ;

ii) the coulomb repulsion in the laboratory system, $E_c Z$, (35 MeV for Z = 5) is very large and affects very much the contour lines. It indicates that the fragment is repelled by a coulomb effect from a highly charged partner with a Z value of the order of 30 or 40 if it was spherical ;

iii) the parameter K is often presented as a temperature because the source is considered as a hot Fermi gas and it is related to the source velocity by

$$K = (2 m_o \; V_s \; \varepsilon_F / \pi^2)^{1/2} \quad (8)$$

for an equal number of target and projectile nucleons (ε_F = Fermi energy, m_o = nucleon mass).

As an example figure 25 shows a comparison between the invariant cross section pattern which can be calculated for boron fragment emission and the experimental data already presented in figure 7. It exhibits clearly that a component contributes to the contour lines, originated from a source having a parallel velocity at 3.6 cm per ns and a transverse velocity very close to zero. Coming back to the energy spectra in the laboratory system, the choice of the K value is rather critical for fitting not only the high energy slope but also the angular distribution. The hot moving source component is represented by the dashed lines in figure 24. The underestimates at the low energy part of the spectra arises from the fragment emission by the complete fusion nucleus we have discussed earlier.

5.b. A possible scenario for fragment emission by a hot moving source

At relativistic energies, the fireball model[30] introduced the concept of a very hot participant region, defined by the geometrical overlap between target and projectile. For the asymmetric system Ar + Ag studied here, such geometrical considerations would imply a larger number of participant nucleons from the target than from the projectile ; the velocity of that zone would therefore be small-

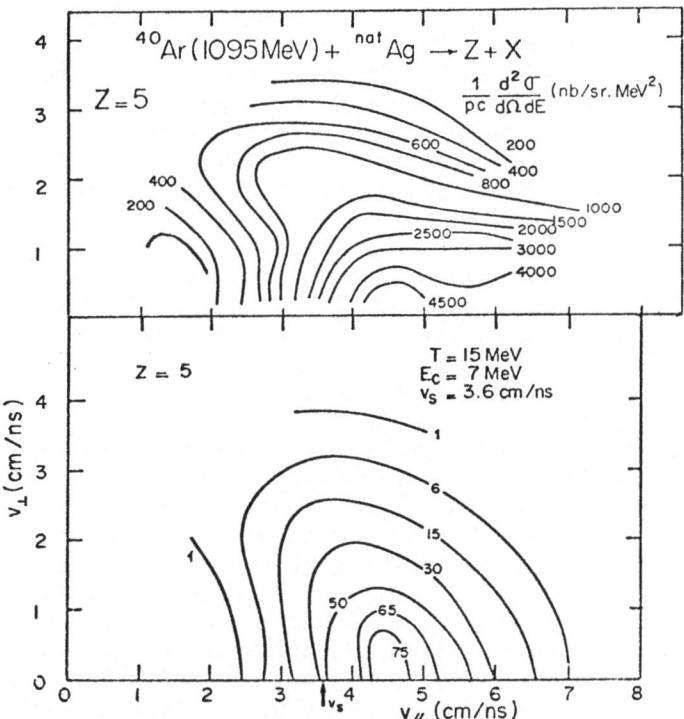

Fig. 25. Lower part contour plots of invariant cross sections for
Z = 5 calculated from the intermediate velocity source
V_s = 3.6 cm per ns, with the mentioned parameters in the
maxwellian expression (ref.[6]). Upper part : experimental
contour plot see fig. 7.

er than the observed source velocity equal to $v_{beam}/2$. The experimental observation of a source moving with $v = v_{beam}/2$, would suggest
that in the energy range below 100 MeV.A, intermediate impact parameters produce an agglomeration of the projectile and of part of the
target, forming a hot remnant without spectator projectile fragment[6].
The velocity value equal to $1/2 \, v_{beam}$ would be due to a number of
nucleons from the target equal to the whole Ar projectile nucleon
number. The corresponding impact parameter would be arount 3-5 fm.
This ball of hot nuclear matter made of 70-90 nucleons shares an
excitation energy of the order of 500 MeV. Then, in a very short
time, before leaving the target zone, this hot source emits fragments
in a multifragmentation process due to its temperature a bit higher
than 7 MeV, instead of decaying by evaporation.

Because the phenomena are produced at intermediate impact parameters and the so called multifragmentation process is originated
from a large energy dissipation in a hot remnant, the name of "deep

inelastic multifragmentation" is suggested. Finally it is perhaps worthwhile to notice the analogy between the contour lines of figure 7 for Z = 5 and the pattern obtained for alpha particles differential cross-sections in those Ar + U collisions at 27 MeV/n which were selected at intermediate linear momentum transfers (figure 18). In that region, the fission yield was found to be rather low, between the two humps at large p and at very small p. A possible explanation is that, in this range of impact parameters around 5 fm, the intermediate hot ball is produced which disappears in a multifragmentation decay and fission is inhibited, since what is left as a target spectator is much smaller and nearly cold.

6. THERMAL PROPERTIES OF "HOT" NUCLEI

The experiments which have been briefly reported in this article, and hopefully the data which will appear after a few month at GANIL and in other places, might bring a series of interesting data useful for a better thermodynamical description of nuclear matter. A first example is the lowering of the coulomb barrier for charged particle evaporation from highly excited nuclei, as expected since the density tails tend to become more diffuse when increassing T. Quite a few years ago, calculations have been made in the frame of thermal Hartree-Fock approximation by Fleckner et al[33] and predicts a T = 5 MeV a lowering of 0.7 MeV for α particles emitted by lead. In many experiments, the low part of energy spectra is not well detected, and it seems worthwhile to work more on this aspect. Although the precision is not yet good enough, a comparison between our experimental data and the compilation shown in Table II indicates a trend for lower experimental values than coulomb barriers calculated at low energies.

Another aspect is the competition between fission and particle evaporation in very excited nuclei. The standard analysis based on the "statistical model" for heavy compound nuclei where the neutron binding energy S_n is larger than B_f, predicts a fission width much greater than Γ_n and therefore neutrons should be emitted after fission by the fragments and, moreover light charged particles should correspond to very scarce events. The experimental observation of the opposite, i.e. a rather large multiplicity of LCP prior to fission, has stimulated[34] a reconsideration of the formulation of the fission process, in the line of Kramers' approach[35]. The dynamics introduces a transient time for established a stationary current across the fission barrier so that other decay modes than fission compete favourably with the fission channel[36]. The transient behaviour affects more steps of the decay chain when the excitation energy increases so that the neutron multiplicity prior fission which is small at E* around 50 MeV is enhanced considerably at 200 MeV (see figure 26). Although for the moment the emission of charged particles has not be calculated, it should also follow the same trend.

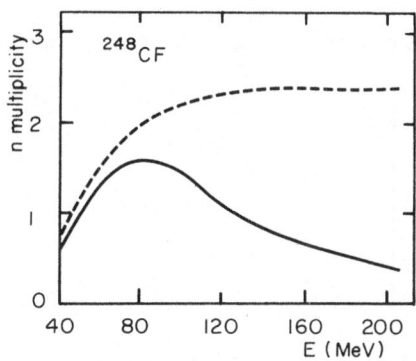

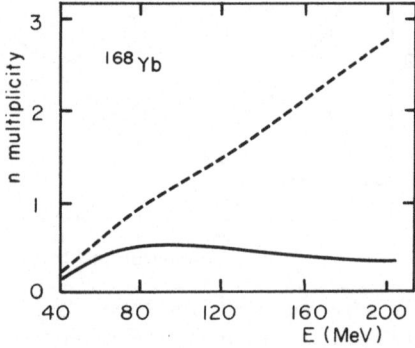

Fig. 26. The neutron multiplicity before fission versus the excitation energy calculated for two systems ^{248}Cf and ^{168}Yb (ref.[31]).

But an even more basic question may arise at very high excitation energies. If we want to apply any kind of statistical model for describing the decay, the hot nucleus itself should live the decay of time necessary for reaching thermalization. An estimate of the relaxation time can be made, with the knowledge of the nucleon mean free path and of the nuclear size. The detailed balance principle allows to calculate the lifetime[37]. Using the estimated of the decay probability for the neutron emission, one gets for the lifetime :

$$\theta = \frac{1}{W} = \frac{\pi^2 h^3}{2\mu} \frac{\rho(E^*_i)}{\int_0^{E^*_i - S_\nu} \varepsilon\sigma_{inv}\rho(E^*_f)d\varepsilon}$$

where μ is the neutron reduced mass, $\rho(E^*_i)$ is the level density at

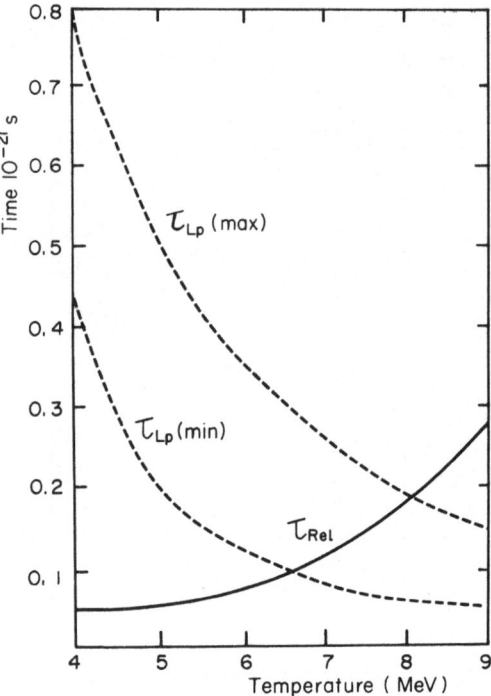

Fig. 27. Comparison of the relaxation time τ_{rel} and the life time
with particle evaporation τ_{Lp} for a heavy nucleus excited
as a function of temperature. The two curves $\tau_{Lp}(max)$ and
$\tau_{Lp}(min)$ correspond to two extreme cases, depending on the
level density parameters and on the inclusion of a large
angular momentum equal to 100 $\hbar$.

the excitation energy of the initial nucleus, σ_{inv} the inverse cross
section and $\rho(E_f^*)$ the level density of the excited final nucleus
after emission of a neutron with a binding energy S_ν. Using the usual relation $\rho(E^*) = \exp 2 \sqrt{A/8 \, E^*}$ and the nuclear temperature T,
one can derive a simple expression

$$\theta = 2.2 \; 10^{-21} \; \frac{A^{1/3}}{E_1^* - S_\nu} \; \exp \; (\frac{S_\nu}{T})$$

In the region of temperatures between 5 to 8 MeV, it is rather difficult to decide what is the longest time, as seen in figure 27 obtained with different parameters. However, the experimental energy
spectra shown above indicate a behaviour compatible with a thermalized composite where the evaporation treatment is still valid, as
far as T is smaller than 5 MeV.

Also, in a Fermi gas model the entropy should be proportional to the temperature since the ratio $S/2T = E^*/T^2$ is equal to the level density parameter of Bethe's formula[38]. New attempts have been made including the contribution of unbound states in Hartree-Fock calculations[39]. For temperatures above 4 MeV the mean field becomes shallower and the nuclear radius increases as well as the density of the single particle spectrum ; this should have consequences on the barrier for particle emission.

Furthermore, those calculations come to the conclusion that a boiling off occurs for a nucleus heated at a critical temperature somewhere between 8 and 12 MeV. It seems difficult to overcome an energy deposit much greater than 4 GeV in heavy well defined composite nuclei, and therefore one may reach only the shore of this region of critical temperature, as shown in figure 13.

Another scenario has also been proposed by Bondorf[22] and by Campi et al.[21], who propose that very hot pieces of nuclear matter, instead of being de-excited slowly along a statistical process, may break into light fragments. The critical excitation energy could stay around 3 MeV per nucleon, corresponding to some 5 MeV for the nuclear temperature, in such a way that the usual statistical decay is too slow as compared to the multifragmentation process.

All these considerations show that it is quite important to forecast a number of attempts in order to produce such hot nuclei. In that sense, peripheral collisions of class 2 may furnish a thermal source of high temperature as sketched in figure 1. And the observed light fragments originated from a source moving at half the beam velocity might indicate that we call a deep inelastic multifragmentation. This differs from the "fireball" observed at much higher energies where gas and liquid phases are expected to coexist in a system with nuclear density below normal[40]. In our case, the projectile is not flying apart as a travelling spectator, but is incorporated into a fraction of the target of similar magnitude and the all system at the first stage of its recoil explodes into light fragments.

To conclude, it seems worthwhile to stress that both central and peripheral collisions can provide informations on the behaviour of nuclear matter in extreme conditions. The central collisions provides data for real hot nuclei at temperatures between 3 and 8 MeV, whereas more peripheral collisions may be responsible for a new breaking of nuclear matter into small pieces when the local excitation energy exceeds 3 MeV/n.

ACKNOWLEDGMENTS

Belonging to the Nuclear Chemistry group, René Bimbot, Bernard Borderie, Claude Cabot, Joël Galin, Daniel Gardès, Henri Gauvin,

Daniel Guerreau, Dominique Jacquet, Marc Lefort, François Monnet, Marie-France Rivet and Xavier Tarrago have been involved at different stages of the published and unpublished data which were discussed in these lectures. Also we would like to thank Professor J. M. Alexander, Professor A. Kermann, Dr. X. Campi and D. Vautherin for very stimulating discussions.

REFERENCES

1. V. Borrel, D. Guerreau, J. Galin, B. Gatty, D. Jacquet, X. Tarrago, Z. Phys. A314, 191 (1983).
2. P. Grangé, H.A. Weidenmüller, G. Wolschin, Ann. of Phys. 136, 190 (1981).
3. Linda Code, E. Duek, L. Kowalski and J.M. Alexander, IPNO-DRE-82-20, Orsay (1982).
4. B. de Goncourt, Thèse Orsay (1984).
5. B. Borderie, M.F. Rivet, C. Cabot, D. Fabris, D. Gardès, H. Gauvin, F. Hanappe and J. Péter, Z. Phys. A316, 243 (1984).
6. B. Borderie et al., submitted to Z. Phys.A318 (1984), IPNO-DRE-84-28, Orsay (1984).
7. L.G. Sobotka, M.L. Padgett, G.J. Wozniak, G. Guarino, A.J. Pacheco, L.G. Moretto, Y. Chan, R.G. Stockstad, I. Tserruya and S. Wald, Phys. Rev. Lett. 51, 2187 (1983).
8. V.E. Viola Jr, Nukleonika (1983), preprint INC-40007-17 Indiana.
9. W.J. Nicholson and I. Halpern, Phys. Rev. 116, 175 (1959).
10. T. Sikkeland, E.L. Haines and V.E. Viola Jr, Phys. Rev. 125, 1350 (1962).
11. V.E. Viola Jr, Nucl. Data Tables 1, 39 (1966).
12. V.E. Viola, R.G. Clark, W.G. Meyer, A.M. Zebelman, R.G. Sextro, Nucl. Phys. A261, 174 (1976).
13. M. Lefort, Bad Honnef Meeting 1981, Lecture Notes in Physics 158 (1982).
14. J. Galin, H. Oeschler, S. Song, B. Borderie, M.F. Rivet, I. Forest, D. Gardès, B. Gatty, H. Guillemot, M. Lefort, B. Tamain and X. Tarrago, Phys. Rev. Lett. 48, 1787 (1982).
15. F. Saint-Laurent, M. Conjeaud, R. Dayras, S. Harar, H. Oeschler and C. Volant, Phys. Lett. 110B, 372 (1982).
16. S. Raha and R.M. Weiner, Phys. Rev. Lett. 50, 407 (1983).
17. Y. Chan, M. Murphy, R.G. Stockstadt, I. Tserruya, S. Wold, A. Budzanowski, Phys. Rev. C27, 447 (1983).
18. G. Nebbia, E. Tomasi, C. Ngô, X.S. Chen, G. La Rana, S. Leray, Z. Phys. A311, 247 (1983).
19. D. Jacquet, E. Duek, J.M. Alexander, B. Borderie, J. Galin, D. Gardès, D. Guerreau, M. Lefort, F. Monnet, M.F. Rivet and X. Tarrago, submitted to Phys. Rev. Lett.
20. M. Lefort, Nucl. Phys. A409, 141c (1983).
21. X. Campi, J. Debois and E. Lipparini, Int. Conf. on Theoretical Approaches to H.I. Mechanisms, Paris, Mai 1984.
22. J.P. Bondorf, Nucl. Phys. A387c, 25 (1982).

23. E.C. Pollaco, M. Conjeaud, S. Harar, C. Volant, Y. Cassagnou,
    R. Dayras, R. Legrain, M.S. Nguyen, H. Oeschler, F. Saint-
    Laurent, DPhN/Saclay, 2124 (1984).
24. S. Song, M.F. Rivet, R. Bimbot, B. Borderie, I. Forest, J. Galin,
    D. Gardès, B. Gatty, M. Lefort, H. Oeschler, B. Tamain,
    X. Tarrago, Phys. Rev. 130B, 14 (1983).
25. M.F. Rivet, D. Logan, J.M. Alexander, D. Guerreau, E. Duek,
    M.S. Zisman and M. Kaplan, Phys. Rev. C25, 2417 (1982).
26. D. Guerreau, personnal communication.
27. D. Jacquet, M.F. Rivet, R. Bimbot, B. Borderie, J. Galin,
    D. Gardès, B. Gatty, D. Guerreau, L. Kowalski, M. Lefort,
    X. Tarrago, IPNO-DRE-84-19, Orsay (1984).
28. J.R. Nix, Nucl. Phys. A130, 241 (1969).
29. M.F. Rivet et al.; Winter meeting. Bormio (1983)
30. J. Gosset, H.H. Gutbrod, W.G. Meyer, A.M. Poskanzer,
    A. Sandoval, R. Stock and G.D. Westfall, Phys. Rev. C16,
    629 (1977).
31. A.S. Goldhaber, Phys. Rev. C17, 2243 (1978).
32. C.K. Gelbke, Nucl. Phys. A400, 473c (1983).
33. J. Fleckner, G. Saner, U. Mosel, Phys. Lett. 65B, 316 (1976).
34. P. Grangé, Li Jun-Qing, H.A. Weidenmüller, Phys. Rev. C27,
    2063 (1983).
35. K.A. Kramers, Physica VII, 4, 284 (1940).
36. S. Hassani and P. Grangé, Phys. Lett. 137B, 281 (1984).
37. T. Ericson, Advances in Physics, 9, 425 (1960).
38. H.A. Bethe, Phys. Rev. 50, 332 (1936).
39. P. Bonche, S. Levit, D. Vautherin, Nucl. Phys. A (to appear).
40. P. Siemens and J. Kapusta, Phys. Rev. Lett. 43, 1486 (1979).

ANGULAR MOMENTUM DYNAMICS IN DAMPED NUCLEAR REACTIONS[*]

Jørgen Randrup

NORDITA, Blegdamsvej 17, DK-2100 Copenhagen Ø, Denmark
and
Nuclear Science Division, Lawrence Berkeley Laboratory
University of California, Berkeley, CA 94720, USA

Thomas Døssing

Niels Bohr Institute, Blegdamsvej 17,
DK-2100 Copenhagen Ø, Denmark

Damped nuclear reactions are studied in order to improve our understanding of macroscopic nuclear dynamics at moderate excitation. Towards this end, it is important to study angular-momentum related observables, as they carry considerable information about the reaction dynamics: While the correlated mass and charge distribution, which has received much attention in the past, can be characterized by two mean values and three second moments, the correlated fragment spin-spin distribution requires two mean values and *thirteen* non-trivial covariances.

The first of the present two lectures discusses the dynamical accumulation of angular momentum in a damped nuclear reaction, as caused by the stochastic exchange of nucleons. The second lecture focusses on the description of the correlated angular distribution of fission fragments resulting from damped reaction products with correlated spins.

---

[*] This work was supported by the Director, Office of Energy Research, Division of Nuclear Physics of the Office of High Energy and Nuclear Physics of the U.S. Department of Energy under Contract DE-AC03-76SF00098.

LECTURE 1: Accumulation of Angular Momentum

This lecture reports briefly on some selected results of a recent study of the dynamical evolution of angular momentum in damped nuclear reactions.[1]

1.1. Introduction

The study is carried out within the framework of the nucleon exchange transport model,[2] in which the dissipation of the macroscopic variables is caused by the inelastic interactions of individual nucleons with the time-dependent mean field. In the case of a binary system, as is temporarily created during a damped reaction, this one-body mechanism appears as a "window" dissipation caused by the transfer of nucleons between the two reaction partners, in addition to a "wall" friction caused by the reflection of nucleons from the changing potential in the interaction zone between the two nucleides.

The reacting system is idealized as two spherical nucleides A and B. The relative orbital angular momentum is $\vec{L} = \vec{R} \times \vec{P}$, where $\vec{R}$ is the relative position and $\vec{P}$ is the relative momentum. The associated moment of inertia is $\mathfrak{J}_R = \mu R^2$. The angular momenta, or spins, of the individual nucleides are $\vec{S}^A$ and $\vec{S}^B$, and $\mathfrak{J}_A$ and $\mathfrak{J}_B$ are the associated moments of inertia. Specific details about this model can be found in Appendix A of ref.3.

1.2. Equations of Motion

In ref.2, the mobility tensors relating to the two fragment spins $\vec{S}^A$ and $\vec{S}^B$ were found to be

$$\overset{\leftrightarrow}{M}{}^{AA} = mN\left(a^2 \overset{\leftrightarrow}{T} + c^2_{ave} \overset{\leftrightarrow}{I} \right)$$

$$\overset{\leftrightarrow}{M}{}^{AB} = mN\left(ab \overset{\leftrightarrow}{T} - c^2_{ave} \overset{\leftrightarrow}{I} \right) = \overset{\leftrightarrow}{M}{}^{BA} \qquad (1.1)$$

$$\overset{\leftrightarrow}{M}{}^{BB} = mN\left(b^2 \overset{\leftrightarrow}{T} - c^2_{ave} \overset{\leftrightarrow}{I} \right)$$

Here $\overset{\leftrightarrow}{I}$ is the identity tensor and $\overset{\leftrightarrow}{T} = \overset{\leftrightarrow}{I} - \hat{R}\hat{R}$ projects onto the plane perpendicular to the dinuclear axis R. The distances to the "window" plane from the two nuclear centers are denoted by a and b, with $a + b = R$, while c_{ave} is the average off-axis displacement of the transferred nucleons. The nucleon mass is denoted by m, and N is the overall form factor governing the rate of nucleon transfer between the two nucleides A and B.

In addition to the fragment spins $\vec{S}^A$ and $\vec{S}^B$, it is also necessary to consider the evolution of the orbital angular momentum $\vec{L}$.
This is because we wish to use a coordinate system whose direction
fluctuates with respect to an external inertial (and hence the components of the total angular momentum $\vec{J}$ will fluctuate). It is nota-
tionally convenient to denote any of the angular-momentum labels A,
B, L by the letters F, G, ... so that $\vec{S}^F = \vec{S}^A, \vec{S}^B, \vec{L}$ for F = A, B, L,
respectively. The mobility tensor relating to the orbital angular
momentum can then be obtained by exploiting the conservation of the
total angular momentum $\vec{J} = \vec{S}^A + \vec{S}^B + \vec{L}$.

In terms of the mobility coefficients the spin transport coefficients are given as follows. The diffusion coefficients are
simply the corresponding mobility coefficients multiplied by the
"effective temperature" τ^*: $D^{FG} = M^{FG}\tau^*$. The drift coefficients are
obtained by multiplying the mobility tensor with the corresponding
generalized forces, i.e. minus the rotational frequencies $\vec{\omega}^F = \vec{S}^F/\mathfrak{I}_F$:
$\vec{V}^F = - \sum_G \overleftrightarrow{M}^{FG}\cdot\vec{\omega}^G = - \sum_G \overleftrightarrow{M}^{FG}\cdot\vec{S}^G/\mathfrak{I}_G$. Here and in the following the sum
over the labels G extend over G = A, B, L.

In order to take full account of the so-called tilting mode, it
is necessary to employ a "body-aligned" orthonormal reference system.
Specifically, we define the coordinate system xyz: $\hat{z} = \hat{R}$, $\hat{y} = \hat{L}$, $\hat{x} = \hat{y} \times \hat{z}$.
The choice of $\hat{z} = \hat{R}$ ensures that the mobility tensors $\overleftrightarrow{M}^{FG}$ are diagonal
in the spatial indices. Since $\hat{y} = \hat{L}$, the orbital angular momentum $\vec{L}$
has only components in the y-direction. We need then consider the
temporal evolution of S_x^A, S_y^A, S_z^A, S_x^B, S_y^B, S_z^B, L_y. In a standard collision experiment, all are initially zero except for L_y which equals
the total angular momentum J. It follows from the symmetry of the
problem that the mean values $\langle S_x^F \rangle$ and $\langle S_z^F \rangle$ and also the covariances
σ_{xy}^{FG} and σ_{yz}^{FG} will remain zero throughout the reaction.

The equations of motion for the non-vanishing mean values and
covariances are given by

$$\dot{S}_y^F = \sum_G \left(M_t^{FG} S_y^G + \frac{1}{L_y} \sigma_{xx}^{FG} M_t^{GL} \right)/\mathfrak{I}_G + \frac{\tau^*}{L_y} \left(2M_t^{FL} - \frac{S_y^F}{L_y} M_t^{LL} \right)$$

$$\dot{\sigma}_{xx}^{FH} = 2\tau^* M_t^{FH} - \sum_G \left(\sigma_{xx}^{FG} M_t^{GH} + M_t^{FG} \sigma_{xx}^{GH} \right)/\mathfrak{I}_G - \omega_R \left(\sigma_{xz}^{FH} + \sigma_{zx}^{FH} \right)$$

$$- \frac{S_y^F}{L_y} \left(2\tau^* M_t^{LH} - \sum_G M_t^{LG} \sigma_{xx}^{GH}/\mathfrak{I}_G \right) - \left(2\tau^* M_t^{FL} - \sum_G \sigma_{xx}^{FG} M_t^{GL}/\mathfrak{I}_G \right)\frac{S_y^H}{L_y}$$

$$+ 2\tau^* \frac{S_y^F}{L_y} M_t^{LL} \frac{S_y^H}{L_y}$$

$$\dot\sigma_{yy}^{FH} = 2\tau * M_t^{FH} - \sum_G \left(\sigma_{yy}^{FG} M_t^{GH} + M_t^{FG} \sigma_{yy}^{GH} \right) / \Im_G$$

$$\dot\sigma_{zz}^{FH} = 2\tau * M_n^{FH} - \sum_G \left(\sigma_{zz}^{FG} M_n^{GH} + M_n^{FG} \sigma_{zz}^{GH} \right) / \Im_G - \omega_R \left(\sigma_{xz}^{FH} + \sigma_{zx}^{FH} \right)$$

$$\dot\sigma_{xz}^{FH} = - \sum_G \left(\sigma_{xz}^{FG} M_n^{GH} + M_t^{FG} \sigma_{xz}^{GH} \right) / \Im_G - \omega_R \left(\sigma_{xx}^{FH} - \sigma_{zz}^{FH} \right) + 2\tau * \frac{S_y^F}{L_y} M_t^{LG} \sigma_{xz}^{GH} / \Im_G$$

$$(1.2)$$

Here we have omitted the bracket around the mean values of S_y^F for notational simplicity, since confusion can hardly arise.

In the above equation for the mean value the first term is the drift coefficient. In the equations for the covariances, the first term is the diffusion coefficient (which vanishes for the non-diagonal components) while the subsequent term represents the restoring term acting to saturate the growth of σ. The terms containing ω_R arise from the orbital rotation which continually mixes the in-plane components. The remaining terms contain $\langle L_y \rangle$ in the denominator and arise from the transformation to the fluctuating coordinate system aligned with $\vec{L}$. These terms are derived under the standard assumption that all spin dispersions are small in comparison with $\langle L_y \rangle$. While this is only well satisfied for larger impact parameters, the equations do remain well-behaved for more central collisions (which contribute only a small part of the reaction cross section) and even for head-on reactions the solutions are correct to within 25%. Although these latter terms are of the corrective type, they are essential in ensuring the proper long-time behavior of the solutions, namely an approach towards statistical equilibrium.

In the preceding we have referred the spin moments to a co-ordinate system defined in terms of the instantaneous values of $\hat{R}$ and $\hat{L}$. However, the direction of $\hat{L}$ can not be determined in a collision experiment, so it is necessary to transform the results to a coordinate system which can be externally defined. In a collision experiment two directions are readily determined: the beam direction $\hat{t}$ and the asymptotic dinuclear direction $\hat{R}(\infty)$. In terms of these two directions we define the following external coordinate system XYZ: $\hat{Z} = \hat{R}$, $\hat{Y} = \hat{R} \times \hat{t}$, $\hat{X} = \hat{Y} \times \hat{Z}$. Since the internal and the external coordinate systems have the same z-axis, the two are related by a rotation around the z-axis, $\mathcal{R}_z(\zeta)$. Since the distribution of the angle ζ between the directions L and Y is determined by the in-plane spin variances, the transformation from xyz to XYZ can be made.

The resulting spin distribution corresponds to a definite impact parameter, given by the specified value of the total angular momentum J. The model also yields equations of motion for the kinetic energy loss and its covariance with the spin variables. Therefore, it is possible to obtain an "observable" spin distribution gated by energy loss rather than impact parameter.

1.3. Equilibrium

In the preceding we have outlined how the dynamical evolution of the dinuclear spins can be calculated. The results of such calculations can best be understood in terms of the appropriate equilibrium solutions and the associated relaxation times.

In analogy with the treatment of the two-particle problem, we introduce the following spins and associated moments of inertia,

$$\vec{S} = \vec{S}^A + \vec{S}^B \quad , \qquad \mathfrak{I}_+ = \mathfrak{I}_A + \mathfrak{I}_B$$

$$\vec{S}^- = \mathfrak{I}_- \left(\frac{\vec{S}^A}{\mathfrak{I}_A} - \frac{\vec{S}^B}{\mathfrak{I}_B} \right) \quad , \qquad \mathfrak{I}_- = \frac{\mathfrak{I}_A \mathfrak{I}_B}{\mathfrak{I}_A + \mathfrak{I}_B} \tag{1.3}$$

They are analogous to the total and relative motion, respectively. For a given total angular momentum J, and under the standard assumption that the variances are small compared to $\langle L_y \rangle^2$, it is straightforward (albeit tedious) to demonstrate that the dynamical spin equations (1.2) have a unique stationary solution given by

$$\langle L_y \rangle = \frac{\mathfrak{I}_R}{\mathfrak{I}_0} J \quad , \qquad \overleftrightarrow{\sigma}^{LL} = \tau^* \, \mathfrak{I}_+ \, \frac{\mathfrak{I}_R}{\mathfrak{I}_0} \, \hat{y}\hat{y}$$

$$\langle S_y^+ \rangle = \frac{\mathfrak{I}_+}{\mathfrak{I}_0} J - \tau^* \mathfrak{I}_+ \frac{\mathfrak{I}_0}{\mathfrak{I}_R} \frac{1}{J} \quad , \quad \overleftrightarrow{\sigma}^{++} = \tau^* \, \mathfrak{I}_+ \frac{\mathfrak{I}_0}{\mathfrak{I}_R} (\hat{x}\hat{x} + \hat{z}\hat{z}) + \tau^* \, \mathfrak{I}_+ \frac{\mathfrak{I}_R}{\mathfrak{I}_0} \hat{y}\hat{y}$$

$$\langle S_y^- \rangle = 0 \quad , \qquad \overleftrightarrow{\sigma}^{--} = \tau^* \, \mathfrak{I}_- \, \overleftrightarrow{I} \tag{1.4}$$

where we have included terms to the first order in the effective temperature τ^*. During the reaction, the moments of the spin distribution will at each instant evolve towards these equilibrium values, which in turn vary in time due to the time dependence of the relative moment of inertia $\mathfrak{I}_R$ and the effective temperature τ^*. Below we shall first discuss the stationary solution in terms of a statistical model, and next we shall discuss the time scales for the approach towards equilibrium.

The part of the macroscopic hamiltonian $\mathcal{H}$ containing the angular-momentum variables in the disphere is

$$\mathcal{H}_{rot} = \frac{\vec{S}^{A2}}{2\mathfrak{I}_A} + \frac{\vec{S}^{B2}}{2\mathfrak{I}_B} + \frac{\vec{L}^2}{2\mathfrak{I}_R} \qquad (1.5)$$

For a given value of $\vec{J} = \vec{S}^A + \vec{S}^B + \vec{L}$, the lowest-energy mode of rotational motion in the disphere is a rigid rotation with each of the three angular momenta given by $\vec{S}^F = \mathfrak{I}_F \vec{J}/\mathfrak{I}_0$ where $\mathfrak{I}_0 = \mathfrak{I}_A + \mathfrak{I}_B + \mathfrak{I}_R$. Relative to this yrast mode of motion, intrinsic rotational excitations are possible. These excitations carry no net angular momentum and can be classified in two groups according to whether the two spheres turn in the same or in the opposite sense, i.e. a purely positive mode has $\vec{S}^- = \vec{0}$ and a purely negative mode has $\vec{S}^+ = \vec{0}$, where $\vec{S}^+$ and $\vec{S}^-$ are given in eq. (1.3).

We first consider the problem using the coordinate system $x'y'z'$ defined by $\hat{z}' = \hat{R}$, $\hat{y}' = \hat{I}$, $\hat{x}' = \hat{y}' \times \hat{z}'$, where $\vec{I} = \vec{J} - \vec{J} \cdot \hat{R}\hat{R}$ is the projection of the total angular momentum $\vec{J}$ on the plane perpendicular to $\vec{R}$. In order to bring the rotational hamiltonian (1.5) on normal form we introduce the following auxiliary spin variable $\vec{s} = \vec{S}^+ - (\mathfrak{I}_+/\mathfrak{I}_0)J_{y'}\hat{y}'$. This transformation has unit jacobian since $\hat{y}'$ is independent of $\vec{S}^+$ and we obtain

$$\mathcal{H}_{rot} = \frac{1}{2\mathfrak{I}_A}\left(\frac{\mathfrak{I}_A}{\mathfrak{I}_+}\vec{S}^+ + \vec{S}^-\right)^2 + \frac{1}{2\mathfrak{I}_B}\left(\frac{\mathfrak{I}_B}{\mathfrak{I}_+}\vec{S}^+ - \vec{S}^-\right)^2 + \frac{1}{2\mathfrak{I}_R}(\vec{J} - \vec{S}^+)^2$$

$$= \frac{J^2}{2\mathfrak{I}_0} + \frac{1}{2\mathfrak{I}_+}\frac{\mathfrak{I}_0}{\mathfrak{I}_R}\left(s_{x'}^2 + s_{y'}^2\right) + \frac{1}{2\mathfrak{I}_+}\frac{\mathfrak{I}_R}{\mathfrak{I}_0}s_{z'}^2 + \frac{\vec{S}^{-2}}{2\mathfrak{I}_-} \qquad (1.6)$$

Here the first term represents the yrast energy associated with a rigid rotation while the additional terms arise from the six normal modes of intrinsic rotational excitations of the disphere. The first of these terms is the energy of the two degenerate *wriggling* modes, where the two spheres rotate in the same sense around an axis perpendicular to $\hat{R}$. The next term is associated with the *tilting* mode arising when $\vec{J}$ has a component along the dinuclear axis $\hat{R}$; the two spheres thus turn in the same sense around $\hat{R}$. These three are the positive modes. The last term arises from the three degenerate negative modes: the *twisting* mode, where the two spheres rotate oppositely around $\vec{R}$, and the two *bending* modes, where the spheres turn oppositely around an axis perpendicular to $\hat{R}$.

Assume now that the rotational modes are weakly coupled to the remainder of the system, which is considered as a heat reservoir with the temperature τ. When $\tau \ll J^2/2\mathfrak{I}_0$, the six normal rotational modes are approximately harmonic. It is then possible to show that the ensuing thermal equilibrium distribution is characterized by

$$\langle S_{y'}^{F} \rangle = \frac{\mathfrak{I}_F}{\mathfrak{I}_0} J - \frac{\mathfrak{I}_F \mathfrak{I}_+}{\mathfrak{I}_R} \frac{\tau}{2J}$$

$$\sigma_{y'y'}^{FG} = \sigma_{x'x'}^{FG} = \left(\frac{\mathfrak{I}_R}{\mathfrak{I}_0} \mathfrak{I}_F \mathfrak{I}_G + \varepsilon_{FG} \mathfrak{I}_A \mathfrak{I}_B \right) \frac{\tau}{\mathfrak{I}_A + \mathfrak{I}_B}$$

$$\sigma_{z'z'}^{FG} = \left(\frac{\mathfrak{I}_0}{\mathfrak{I}_R} \mathfrak{I}_F \mathfrak{I}_G + \varepsilon_{FG} \mathfrak{I}_A \mathfrak{I}_B \right) \frac{\tau}{\mathfrak{I}_A + \mathfrak{I}_B} \tag{1.7}$$

(The symbol ε_{FG} is one when $F = G$ and minus one otherwise.) This result is in accordance with the analysis by Moretto.[4]

The above result was expressed in the I-aligned coordinate system x'y'z'. A transformation to our standard "body-fixed" L-aligned system xyz yields the following equilibrium distribution

$$\langle S_y^{F} \rangle = \frac{\mathfrak{I}_F}{\mathfrak{I}_0} J - \frac{\mathfrak{I}_F \mathfrak{I}_0}{\mathfrak{I}_R} \frac{\tau}{J}$$

$$\sigma_{yy}^{FG} = \left(\frac{\mathfrak{I}_R}{\mathfrak{I}_0} \mathfrak{I}_F \mathfrak{I}_G + \varepsilon_{FG} \mathfrak{I}_A \mathfrak{I}_B \right) \frac{\tau}{\mathfrak{I}_A + \mathfrak{I}_B}$$

$$\sigma_{zz}^{FG} = \sigma_{xx}^{FG} = \left(\frac{\mathfrak{I}_0}{\mathfrak{I}_R} \mathfrak{I}_F \mathfrak{I}_G + \varepsilon_{FG} \mathfrak{I}_A \mathfrak{I}_B \right) \frac{\tau}{\mathfrak{I}_A + \mathfrak{I}_B} \tag{1.8}$$

This result is identical to the stationary solution (1.4) of the dynamical equations (1.2).

In the variances in (1.7) and (1.8) the first terms arise from the positive modes (wriggling and tilting) while the second terms arise from the isotropic negative modes (bending and twisting). The most pronounced effect of the transformation from x'y'z' to xyz is the increase in the in-plane wriggling variance σ_{xx}^{++} by the factor $(\mathfrak{I}_0/\mathfrak{I}_R)^2 \cong 2$ so that the isotropy in the plane perpendicular to $\vec{R}$ is replaced by isotropy in the plane perpendicular to $\vec{L}$. A different normal form of the rotational hamiltonian (1.5) for an asymmetrical disphere has been introduced by Schmitt and Pacheco.[5] This leads to different definitions of the wriggling and bending modes, but the result expressed in the original variables, eq.(1.7), is of course the same.

1.4. Evolution in a Symmetric Disphere

In sect. 1.3 we introduced the spins $\vec{S}^+$ and $\vec{S}^-$; they are particularly convenient variables when the two spheres are equal. In the symmetric case, where $a = b$ and $\mathfrak{I}_A = \mathfrak{I}_B$, the mixed mobility tensor $\overleftrightarrow{M}^{+-}$ vanishes so that the dynamical equations for $\vec{S}^-$ decouple from the rest; furthermore, the mobility tensor $\overleftrightarrow{M}^{--}$ is isotropic.

Typical time scales for the approach to equilibrium can be obtained by dividing the asymptotic values by the respective initial time derivatives. This yields for the transversal spin components σ_{xx}^{++} and σ_{yy}^{++} the time scales $(\mathfrak{I}_0/\mathfrak{I}_R)t_{++}$ and $(\mathfrak{I}_R/\mathfrak{I}_0)t_{++}$, repectively, where

$$t_{++} = \frac{\tau^* \mathfrak{I}_+}{2\tau^* M^{LL}} = \frac{\mathfrak{I}_+}{2mNR^2} \quad , \tag{1.9}$$

while for the components of $\overleftrightarrow{\sigma}^{--}$ we find

$$t_{--} = \frac{\tau^* \mathfrak{I}_-}{2\tau^* M^{--}} = \frac{\mathfrak{I}_-}{2mNc_{ave}^2} \quad . \tag{1.10}$$

Thus, $t_{++}/t_{--} = (c_{ave}^2/R^2) \ll 1$.

Solving the equations more rigorously for the idealized case of constant coefficients, the time development of the variances are governed by these relaxation times, for example:

$$\overleftrightarrow{\sigma}^{--} = \mathfrak{I}_- \tau^* \left[1 - e^{-t/t_{--}} \right] \overleftrightarrow{I} \quad . \tag{1.11}$$

The normal variance σ_{zz}^{++} does not receive contributions directly through the transfer process, but only indirectly by the orbital rotation of σ_{xx}^{++} via σ_{xz}^{++}. Solving the equations for these variances, the relaxation time for σ_{xx}^{++} is t_{++}, which we already discussed, while for σ_{zz}^{++} the typical time scale is

$$t_{+z} \cong \left(4\omega_R^2 \frac{L_y}{J_y} t_{++} \right)^{-1} \quad , \tag{1.12}$$

which is usually fairly long.

The time scales for the evolution of mean values are approximately twice the ones relevant for the variances:

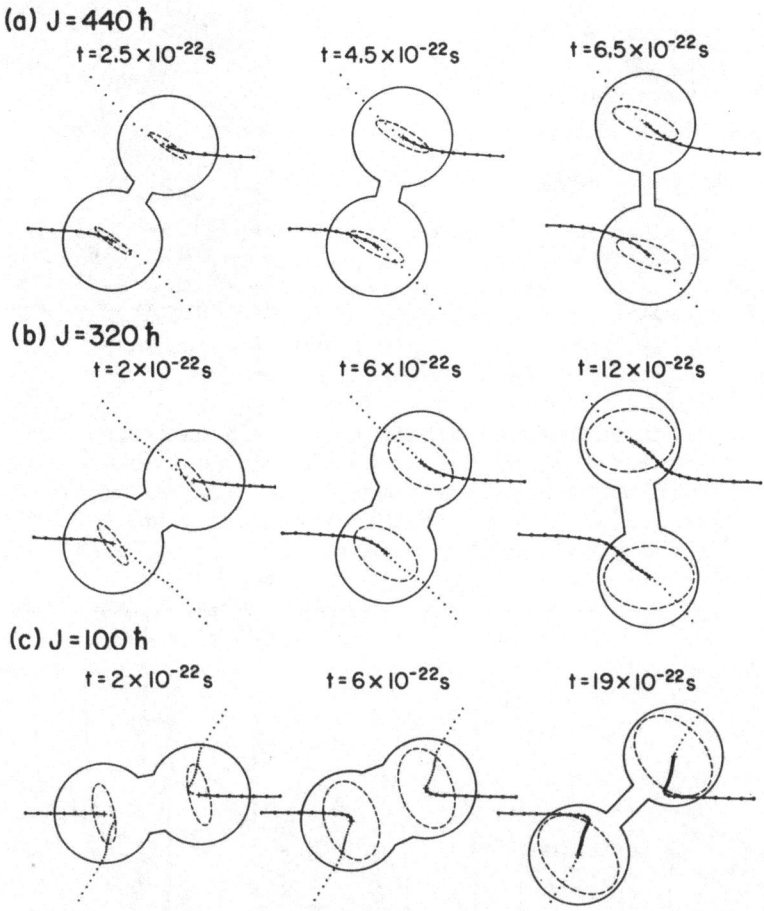

(a) J = 440 ℏ
t = 2.5 × 10⁻²²s   t = 4.5 × 10⁻²²s   t = 6.5 × 10⁻²²s

(b) J = 320 ℏ
t = 2 × 10⁻²²s   t = 6 × 10⁻²²s   t = 12 × 10⁻²²s

(c) J = 100 ℏ
t = 2 × 10⁻²²s   t = 6 × 10⁻²²s   t = 19 × 10⁻²²s

Fig.1.1. For three different values of the total angular momentum,
the dinuclear complex produced in the reaction 1400 MeV ^{165}Ho + ^{165}Ho
is shown at three different points in time: shortly after the neck
has opened, at the time of closest approach, and right before the
neck collapses. (The actual times indicated are measured from the
time of the nuclei approach to a surface separation of s = 4 fm.) The
dots indicate past and future locations of the nuclear centers at
intervals of 10^{-22} sec. The dashed ellipses indicate the one-sigma
contours of the in-plane distribution of the nuclear angular momenta
$\vec{S}^A$ and $\vec{S}^B$ scaled so that one fm corresponds to two ℏ (the nuclear
radii are 6.3 fm).

$$S_y^+ = \frac{\mathfrak{I}_+}{\mathfrak{I}_0} \left[1 - e^{-\frac{\mathfrak{I}_0}{\mathfrak{I}_R} \frac{t}{2t_{++}}} \right], \qquad S_y^- = 0 \ . \tag{1.13}$$

1.5. Illustrative Results

In the preceeding section we have discussed the characteristic features of the spin evolution with an emphasis on the qualitative aspects. We now wish to illustrate the theory quantitatively by making applications to one reaction of actual experimental interest, namely 1400 MeV ^{165}Ho + ^{165}Ho. A pictorial impression of the evolution of the dinuclear geometry can be gained from fig.1.1.

We now consider in some detail the calculated dynamical evolution of the angular momenta during the reaction phase. First, we consider the various relaxation times introduced in sect.1.4. They are shown in fig. 1.2 as functions of time, for a number of different values of the total angular momentum J. We note that throughout

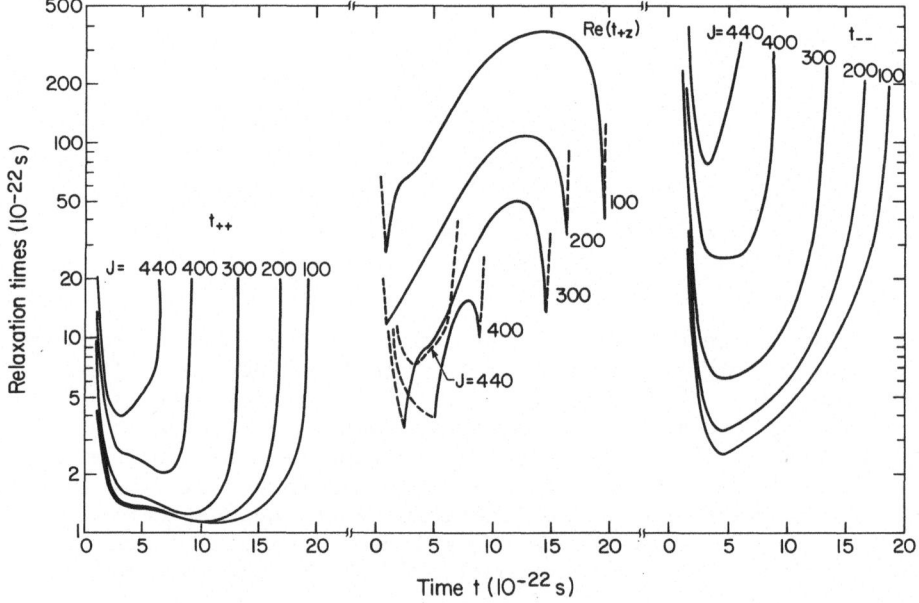

Fig.1.2. Calculated local relaxation times for the reaction 1400 MeV ^{165}Ho + ^{165}Ho for various values of the total angular momentum J. The relaxation times for the two positive perpendicular modes (wriggling) are denoted t_{++}, while that for the positive longitudinal mode (tilting) is denoted t_{+z}. The relaxation time for the three negative modes (bending and twisting) is denoted t_{--}.

the reaction phase the relaxation times t_{++} associated with the two wriggling modes are considerably shorter than t_{--} associated with the negative modes, as already expected since $c_{ave}^2 \ll R^2$. The relaxation time for the tilting mode is fairly long but has an opposite behavior, both as a function of time and in its dependence on J. By comparing the relaxation times with the reaction times it is possible to obtain an expectation for how far the various modes will evolve towards equilibrium. Thus, for not too large impact parameters, we expect the wriggling modes to achieve nearly complete relaxation, contrary to the negative modes for which this is at most expected for the smallest impact parameters. The tilting mode is generally expected to acquire little excitation.

The calculated dynamical evolution of the mean fragment spin projection is shown in fig.1.3, for three selected J-values. For the highest value, $J = 440 \, \hbar$, the reaction is over before the equilibrium mean value can be reached. For the intermediate value, $J = 320 \, \hbar$, the equilibrium value is nearly achieved around the time of closest approach. This equilibrium mean spin decreases as the two fragments recede' and the relative moment of inertia grows. Therefore, the mean spin exhibits a maximum as a function of time. The same is true at the most central reaction, $J = 100 \, \hbar$, but here the equilibrium values are of course smaller.

The calculated spin covariances are displayed in fig.1.4 as functions of time. The figure has three parts. The first shows the dynamical evolution during the reaction phase. It is clearly seen how σ_{xx}^{FG} and σ_{yy}^{FG} increase rapidly at early times; this is a reflection of the fast wriggling relaxation time (see fig.1.2). The local bumps in σ_{xx}^{FG} and σ_{yy}^{FG} around the time of closest approach ($t \approx 3 \cdot 10^{-22}$ s) are caused by a minimum in the effective temperature τ^*. [The effective temperature is initially nearly proportional to the relative nuclear velocity and hence at first it decreases. Later on, when the relative motion has subsided, τ^* is close to the intrinsic temperature τ which increases in time. Thus τ^* exhibits a minimum which occurs approximately at the turning point of the relative motion.] The evolution of σ_{zz}^{FG} is considerably slower, as expected from fig. 1.2. Most of σ_{zz}^{FG} is associated with the negative twisting mode as evidenced by the fact that the covariance σ_{zz}^{AB} is negative, but, as the difference between σ_{zz}^{AA} and σ_{zz}^{AB} indicates, there is also a fair amount of tilting. The second part of the figure shows, on a condensed time scale, the rotation of the covariances along the exit Coulomb trajectory. Finally, the third part shows the result of transforming to the external coordinate system XYZ. This transformation is seen to have a substantial effect of the x-components; in fact σ_{xx}^{AB} becomes negative.

The equiprobability contours of the fragment spin distribution are ellipsoids whose common shape and orientation are determined by the appropriate covariances. In order to give a visual impression of the spin evolution we have included in fig.1.1 contours of the spin distribution projected onto the xz-plane. One notes how the fairly peripheral collision (J = 440 ℏ) inhibits the build-up of negative spin modes so the distribution is very elongated. Furthermore, the

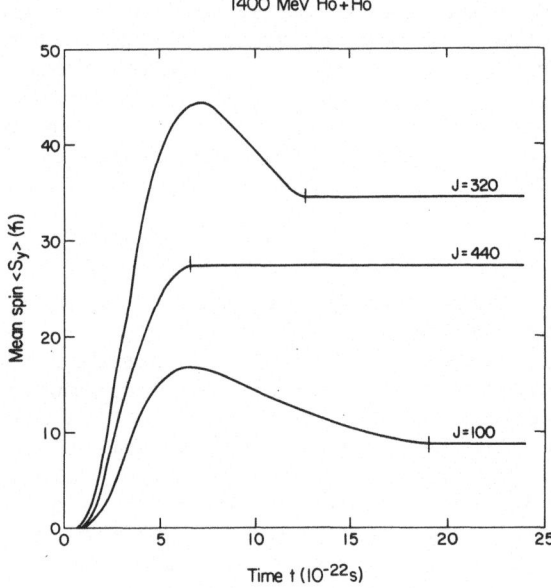

1400 MeV Ho+Ho

Fig.1.3. Calculated time evolution of the mean fragment spin $\langle S_y \rangle$ in the reaction 1400 MeV ^{165}Ho + ^{165}Ho for various values of the total angular momentum . The neck snapping, after which the spins remain constant, is indicated by a small vertical bar.

smallness of the form factor prevents the distribution from aligning itself relative to the dinuclear axis. For J = 320 ℏ the window grows wider and the isotropic negative modes are more readily excited; the distribution also follows better the turning dinuclear axis. These features are even more apparent for J = 100 ℏ.

80

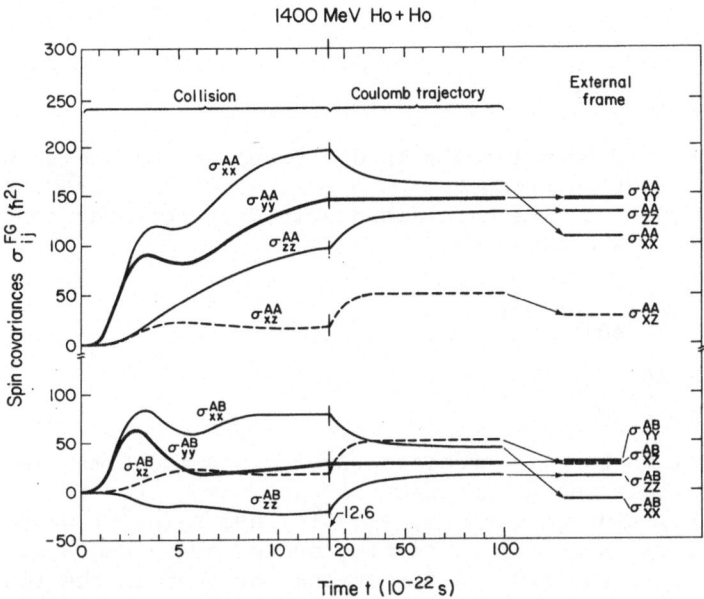

Fig. 1.4. Calculated time evolution of the various spin covariances σ_{ij}^{FG} in the reaction 1400 MeV ^{165}Ho + ^{165}Ho for a total angular momentum of =320 $\hbar$. At the time of neck snapping (t=12.6 · 10^{-22} s) the time scale is changed by a factor of ten. After the asymptotic values have been reached, the effect of transforming to the external reference frame from XYZ is shown.

LECTURE 2: Correlated Fission Angular Distributions

In this lecture we derive simple approximate expressions for the correlated angular distributions of fission fragments from the damped reaction products and discuss calculated results for a case of experimental interest. The presentation follows closely that in ref.6.

2.1. Definitions

The normalized correlated spin distribution for the spins $\vec{S}^A$ and $\vec{S}^B$ of the two reaction products A and B is given by $f_{AB}(\vec{S}^A, \vec{S}^B)$ with $\int d\vec{S}^A \int d\vec{S}^B f_{AB}(\vec{S}^A, \vec{S}^B) = 1$. The normalized spin distributions for the nuclei separately are given by

$$f_A(\vec{S}^A) = \int d\vec{S}^B f_{AB}(\vec{S}^A, \vec{S}^B) \ ,$$

$$f_B(\vec{S}^B) = \int d\vec{S}^A f_{AB}(\vec{S}^A, \vec{S}^B) \ . \qquad (2.1)$$

We now assume that the reaction products A and B may undergo fission after their mutual interaction has ceased. Let the corresponding probabilities be given by $P_A(\vec{\alpha}, \vec{S}^A)$ and $P_B(\vec{\beta}, \vec{S}^B)$ where $\vec{S}^A$ and $\vec{S}^B$ are the spins of the fissioning nuclei and $\vec{\alpha}$ and $\vec{\beta}$ are unit vectors indicating the fission directions, as seen in the respective c.m. frames of the fissioning nuclei. Taking into account the distribution of spins in the fissioning nuclei, we find for the corresponding angular distributions of fission products

$$P_A(\vec{\alpha}) = \int d\vec{S}^A \ f_A(\vec{S}^A) P_A(\vec{\alpha}, \vec{S}^A) \ ,$$

$$P_B(\vec{\beta}) = \int d\vec{S}^B \ f_B(\vec{S}^B) P_B(\vec{\beta}, \vec{S}^B) \ , \qquad (2.2)$$

respectively. The joint probability for A fissioning in the direction $\vec{\alpha}$ and B fissioning in the direction $\vec{\beta}$ is

$$P_{AB}(\vec{\alpha}, \vec{\beta}) = \int d\vec{S}_A \int d\vec{S}^B \ f_{AB}(\vec{S}^A, \vec{S}^B) P_A(\vec{\alpha}, \vec{S}^A) P_B(\vec{\beta}, \vec{S}^B) \ . \qquad (2.3)$$

Since the two nuclear spins are correlated, the detection of a fission product from one of the nuclei introduces a bias on the spin distributions of the other. These biased spin distributions are given by

$$\tilde{f}_A(\vec{S}^A;\vec{\beta}) = \int d\vec{S}^B \, f_{AB}(\vec{S}^A,\vec{S}^B) P_B(\vec{\beta},\vec{S}^B)/P_B(\vec{\beta}) \ ,$$

$$\tilde{f}_B(\vec{S}^B;\vec{\alpha}) = \int d\vec{S}^A \, f_{AB}(\vec{S}^A,\vec{S}^B) P_A(\vec{\alpha},\vec{S}^A)/P_A(\vec{\alpha}) \ , \qquad (2.4)$$

where $\tilde{f}_A(\vec{S}^A,\vec{\beta})$ is the probability that the nucleus A has the spin $\vec{S}^A$ when its reaction partner B is known to fission in the direction $\vec{\beta}$; analogously for $\tilde{f}_B(\vec{S}^B;\vec{\alpha})$; they are thus normalized to unity when integrated over the spin variable.

2.2. The Basic Fission Probabilities

We shall assume that the nuclei A and B always fission, regardless of the magnitudes of their spins. We then have

$$P_A(\vec{\alpha},\vec{S}^A) = \left[(2\pi)^{3/2} \frac{K_A}{S^A} \, \mathrm{erf}\left(\frac{S^A}{\sqrt{2}K_A}\right)\right]^{-1} \exp[-(\vec{S}^A \cdot \overleftarrow{\alpha})^2/2K_A^2] \ ,$$

$$P_B(\vec{\beta},\vec{S}^B) = \left[(2\pi)^{3/2} \frac{K_B}{S^B} \, \mathrm{erf}\left(\frac{S^B}{\sqrt{2}K_B}\right)\right]^{-1} \exp[-(\vec{S}^B \cdot \overleftarrow{\beta})^2/2K_B^2] \ , \quad (2.5)$$

where K_A and K_B are the mean K-values associated with A and B; they depend on the nuclear temperatures but need not be further discussed here. If the spin distribution is fairly narrow the spins in the pre-exponential factor may be replaced by an appropriate mean value:

$$S^A \sim S_A \approx \langle S_Y^A \rangle \left(1 + \frac{\sigma_{XX}^{AA} + \sigma_{ZZ}^{AA}}{2\langle S_Y^A \rangle^2}\right),$$

$$S^B \sim S_B \approx \langle S_Y^B \rangle \left(1 + \frac{\sigma_{XX}^{BB} + \sigma_{ZZ}^{BB}}{2\langle S_Y^B \rangle^2}\right). \qquad (2.6)$$

Furthermore, if these mean values are large in comparison with the respective K-values the error functions are nearly unity and may be ignored. With these assumptions we arrive at the simple approximate expressions

$$P_A(\vec{\alpha},\vec{S}^A) \approx (2\pi)^{-3/2} \frac{S_A}{K_A} \exp[-(\vec{S}^A \cdot \overleftarrow{\alpha})^2/2K_A^2] \ ,$$

$$P_B(\vec{\beta},\vec{S}^B) \approx (2\pi)^{-3/2} \frac{S_B}{K_B} \exp[(-\vec{S}^B \cdot \overleftarrow{\beta})^2/2K_B^2] \ . \qquad (2.7)$$

2.3. Notational Tools

Throughout our discussion of the fragment spin correlations two different spaces are intertwined: (i) the three-dimensional space associated with the nuclear spins, and (ii) the two-dimensional space associated with the fragment labels A and B. It is therefore useful to introduce a six-dimensional superspace in which both of those spaces are embedded. In analogy with our notation in three-dimensional space, we shall use double arrows over quantities associated with the superspace. Thus, we introduce the superspin

$$\overset{\Rightarrow}{S} \equiv (\vec{S}^A, \vec{S}^B) \tag{2.8}$$

and the associated covariance matrix

$$\overset{\leftrightarrow}{\sigma} \equiv \begin{pmatrix} \overset{\leftrightarrow}{\sigma}\,^{AA} & \overset{\leftrightarrow}{\sigma}\,^{AB} \\ \overset{\leftrightarrow}{\sigma}\,^{BA} & \overset{\leftrightarrow}{\sigma}\,^{BB} \end{pmatrix} . \tag{2.9}$$

It is also natural to imbed the individual fragment spins as

$$\overset{\Rightarrow}{S}{}^A \equiv (\vec{S}^A, \vec{0}), \qquad \overset{\Rightarrow}{S}{}^B \equiv (\vec{0}, \vec{S}^B) . \tag{2.10}$$

For pedagogical reasons we adopt the symbol $*$ to denote multiplication of supervariables. Thus, for example, we have $\overset{\Rightarrow}{S}{}^A * \overset{\Rightarrow}{S}{}^B = \vec{S}^A \cdot \vec{0} + \vec{0} \cdot \vec{S}^B = 0$.

With this notation the gaussian approximation to the spin-spin distribution can be written

$$f_{AB}(\overset{\Rightarrow}{S}) = (2\pi)^{-3} |\overset{\leftrightarrow}{\sigma}|^{-1/2} \exp[-\tfrac{1}{2} (\overset{\Rightarrow}{S} - \langle\overset{\Rightarrow}{S}\rangle) * (\overset{\leftrightarrow}{\sigma})^{-1} * (\overset{\Leftarrow}{S} - \langle\overset{\Leftarrow}{S}\rangle)] . \tag{2.11}$$

It is helpful to introduce the reduced directional vectors

$$\vec{a} \equiv \vec{\alpha}/K_A , \qquad \vec{b} \equiv \vec{\beta}/K_B \tag{2.12}$$

and the associated tensors

$$\overset{\leftrightarrow}{A} = \overset{\leftarrow}{a}\,\vec{a} = \overset{\leftarrow}{\alpha}\,\overset{\rightarrow}{\alpha}/K_A^2 , \qquad \overset{\leftrightarrow}{B} = \overset{\leftarrow}{b}\,\vec{b} = \overset{\leftarrow}{\beta}\,\overset{\rightarrow}{\beta}/K_B^2 . \tag{2.13}$$

These quantities are imbedded in superspace as

$$\overset{\Rightarrow}{a} \equiv (\vec{a}, \vec{0}) , \qquad \overset{\Rightarrow}{b} \equiv (\vec{0}, \vec{b}) ,$$

$$\overset{\leftrightarrow}{A} \equiv \overset{\Leftarrow}{a}\,\overset{\Rightarrow}{a} , \qquad \overset{\leftrightarrow}{B} \equiv \overset{\Leftarrow}{b}\,\overset{\Rightarrow}{b} . \tag{2.14}$$

Finally, for calculating the joint fission angular distributions, the following super-tensor is of interest,

$$\overset{\leftrightarrow}{C} = \overset{\leftrightarrow}{A} + \overset{\leftrightarrow}{B} = \begin{pmatrix} \overset{\leftarrow}{\alpha}\,\overset{\rightarrow}{\alpha}/K_A^2 & \overset{\leftrightarrow}{0} \\ \overset{\leftrightarrow}{0} & \overset{\leftarrow}{\beta}\,\overset{\rightarrow}{\beta}/K_B^2 \end{pmatrix}. \tag{2.15}$$

With this notation the basic fission angular distribution (2.7) can be written

$$P_A(\overset{\rightarrow}{\alpha},\vec{S}^A) = (2\pi)^{-3/2}\frac{S_A}{K_A}\exp[-\tfrac{1}{2}\vec{S}^A\cdot\overset{\leftrightarrow}{A}\cdot\overset{\leftarrow}{S}^A] = (2\pi)^{-3/2}\frac{S_A}{K_A}\exp[-\tfrac{1}{2}\overset{\Rightarrow}{S}^A * \overset{\leftrightarrow}{A} * \overset{\Leftarrow}{S}^A],$$

$$P_B(\overset{\rightarrow}{\beta},\vec{S}^B) = (2\pi)^{-3/2}\frac{S_B}{K_B}\exp[-\tfrac{1}{2}\vec{S}^B\cdot\overset{\leftrightarrow}{B}\cdot\overset{\leftarrow}{S}^B] = (2\pi)^{-3/2}\frac{S_A}{K_B}\exp[-\tfrac{1}{2}\overset{\Rightarrow}{S}^B * \overset{\leftrightarrow}{B} * \overset{\Leftarrow}{S}^B],$$

$$\tag{2.16}$$

and their product can be written as

$$P_A(\overset{\rightarrow}{\alpha},\vec{S}^A)\,P_B(\overset{\rightarrow}{\beta},\vec{S}^B) = (2\pi)^{-3}\frac{S_A}{K_A}\frac{S_B}{K_B}\exp[-\tfrac{1}{2}\overset{\Rightarrow}{S} * \overset{\leftrightarrow}{C} * \overset{\Leftarrow}{S}]. \tag{2.17}$$

It is also useful to introduce the reduced spins $s_\alpha \equiv \vec{S}^A\cdot\overset{\leftarrow}{a}$ $= \vec{S}^A\cdot\overset{\leftarrow}{\alpha}/K_A$ and $s_\beta \equiv \vec{S}^B\cdot\overset{\leftarrow}{b} = \vec{S}^B\cdot\overset{\leftarrow}{\beta}/K_B$ which are dimensionless measures of the alignment of the fragment spins with the respective fission directions. For this we introduce the operator

$$\overset{\Rightarrow}{\underset{\leftarrow}{T}} = \begin{pmatrix} \vec{a} \\ \vec{b} \end{pmatrix} = \begin{pmatrix} \overset{\rightarrow}{\alpha}/K_A & \overset{\rightarrow}{0} \\ \overset{\rightarrow}{0} & \overset{\rightarrow}{\beta}/K_B \end{pmatrix} \tag{2.18}$$

which transforms supervectors into vectors in label space. We shall adopt the use of arrows under quantities associated with the label space and the symbol ∘ for the multiplication in label space. (Again, the explicit use of multiplication symbols is redundant since the dimensionality of the matrices involved automatically indicates the type of multiplication involved.) We then have for the reduced spin

$$\underset{\rightarrow}{s} \equiv (s_\alpha, s_\beta) = \overset{\Rightarrow}{S} * \overset{\Leftarrow}{\underset{\rightarrow}{T}}. \tag{2.19}$$

The associated reduced covariance matrix is

$$\underset{\leftrightarrow}{\sigma} \equiv \overset{\Rightarrow}{\underset{\leftarrow}{T}} * \overset{\leftrightarrow}{\sigma} * \overset{\Leftarrow}{\underset{\rightarrow}{T}} = \begin{pmatrix} \sigma_{\alpha\alpha} & \sigma_{\alpha\beta} \\ \sigma_{\beta\alpha} & \sigma_{\beta\beta} \end{pmatrix}, \tag{2.20}$$

where, for example, $\sigma_{\alpha\beta} = (\overset{\rightarrow}{\alpha}/K_B)\cdot\overset{\leftrightarrow}{\sigma}^{AB}\cdot(\overset{\leftarrow}{\beta}/K_B)$. We note that $\underset{\leftrightarrow}{\sigma}$ is symmetric, $\sigma_{\alpha\beta} = \sigma_{\beta\alpha}$. We also note that

$$\overset{\leftrightarrow}{C} = \overset{\Leftarrow}{\underset{\rightarrow}{T}} \circ \overset{\Rightarrow}{\underset{\leftarrow}{T}}. \tag{2.21}$$

2.4. The Fission-Fission Angular Distribution

We first consider the joint fission-fission angular distribution. This is the most complicated case and the other cases can readily be obtained subsequently by appropriate specializations.

Combining eqs.(2.3), (2.11) and (2.16) we have

$$P_{AB}(\vec{\alpha}, \vec{\beta}) = (2\pi)^{-6} \frac{S_A}{K_A} \frac{S_B}{K_B} |\overleftrightarrow{\sigma}|^{-1}$$

$$\times \int d\vec{S} \, \exp[-\tfrac{1}{2}[(\vec{S} - \langle\vec{S}\rangle) * \overleftrightarrow{\sigma}^{-1} * (\overleftarrow{S} - \langle\overleftarrow{S}\rangle) + \vec{S} * \overleftrightarrow{C} * \overleftarrow{S}]]. \quad (2.22)$$

Since the exponent is of second order in the integration variable it is possible to evaluate the integral by bringing the integrand on quadratic form ("completing the square"). The integrand is of the form

$$[\cdot] = (\vec{S} - \langle\vec{S}\rangle - \vec{\Delta}_C) * (\overleftrightarrow{\sigma}^{-1} + \overleftrightarrow{C}) * (\overleftarrow{S} - \langle\overleftarrow{S}\rangle - \overleftarrow{\Delta}_C)$$

$$+ \langle\vec{S}\rangle * \overleftrightarrow{C} * (\overleftrightarrow{I} + \overleftrightarrow{\sigma} * \overleftrightarrow{C})^{-1} * \langle\overleftarrow{S}\rangle \quad (2.23)$$

provided that the induced shift is given by

$$\vec{\Delta}_C = -\langle\vec{S}\rangle * \overleftrightarrow{C} * \overleftrightarrow{\sigma} * (\overleftrightarrow{I} + \overleftrightarrow{C} * \overleftrightarrow{\sigma})^{-1} . \quad (2.24)$$

It expresses the bias introduced in the mean spins as a result of the joint detection of the two fission fragments at $\vec{\alpha}$ and $\vec{\beta}$. After carrying out the $\vec{S}$-integration we are left with

$$P_{AB}(\vec{\alpha}, \vec{\beta}) = (2\pi)^{-3} \frac{S_A}{K_A} \frac{S_B}{K_B} |\overleftrightarrow{I} + \overleftrightarrow{\sigma} * \overleftrightarrow{C}|^{-1/2}$$

$$\exp[-\tfrac{1}{2}\langle\vec{S}\rangle * \overleftrightarrow{C} * (\overleftrightarrow{I} + \overleftrightarrow{\sigma} * \overleftrightarrow{C})^{-1} * \langle\overleftarrow{S}\rangle] . \quad (2.25)$$

The rank of $\overleftrightarrow{C}$ is only two, and therefore it is possible to reduce the six-dimensional determinant and exponent to quantities in the two-dimensional label space. We first note the following identity:

$$\overleftrightarrow{C} * (\overleftrightarrow{I} + \overleftrightarrow{\sigma} * \overleftrightarrow{C})^{-1} = \overleftarrow{T} \circ \overrightarrow{T} * (\overleftrightarrow{I} + \overleftrightarrow{\sigma} * \overleftarrow{T} \circ \overrightarrow{T})^{-1}$$

$$= \overleftarrow{T} \circ \overrightarrow{T} * \sum_{n \geq 0} (-\overleftrightarrow{\sigma} * \overleftarrow{T} \circ \overrightarrow{T})^n$$

$$= \overset{\leftarrow}{\underset{\rightarrow}{T}} \circ \sum_{n \geq 0} (-\overset{\rightarrow}{\underset{\leftarrow}{T}} * \overset{\leftrightarrow}{\sigma} * \overset{\leftarrow}{\underset{\rightarrow}{T}})^n \circ \overset{\rightarrow}{\underset{\leftarrow}{T}}$$

$$= \overset{\leftarrow}{\underset{\rightarrow}{T}} \circ (\underset{\leftrightarrow}{I} + \underset{\leftrightarrow}{\sigma})^{-1} \circ \overset{\rightarrow}{\underset{\leftarrow}{T}} \ , \tag{2.26}$$

where $\underset{\leftrightarrow}{I}$ is the identity in label space. The inverse matrix is readily calculated,

$$(\underset{\leftrightarrow}{I} + \underset{\leftrightarrow}{\sigma})^{-1} = \begin{pmatrix} 1 + \sigma_{\alpha\alpha} & \sigma_{\alpha\beta} \\ \sigma_{\beta\alpha} & 1 + \sigma_{\beta\beta} \end{pmatrix}^{-1} = \frac{1}{d} \begin{pmatrix} 1 + \sigma_{\beta\beta} & -\sigma_{\alpha\beta} \\ -\sigma_{\beta\alpha} & 1 + \sigma_{\alpha\alpha} \end{pmatrix} , \tag{2.27}$$

where $d = |\underset{\leftrightarrow}{I} + \underset{\leftrightarrow}{\sigma}| = (1 + \sigma_{\alpha\alpha})(1 + \sigma_{\beta\beta}) - \sigma_{\alpha\beta}\sigma_{\beta\alpha}$. Thus the exponent in (2.25) is

$$-\tfrac{1}{2} \overset{\rightarrow}{\underset{}{\langle S \rangle}} * \overset{\leftrightarrow}{C} * (\underset{}{I} + \overset{\leftrightarrow}{\sigma} * \overset{\leftrightarrow}{C})^{-1} * \overset{\leftarrow}{\underset{}{\langle S \rangle}} = -\tfrac{1}{2} \underset{\rightarrow}{\langle s \rangle} \circ (\underset{\leftrightarrow}{I} + \underset{\leftrightarrow}{\sigma})^{-1} \circ \underset{\leftarrow}{\langle s \rangle} \ . \tag{2.28}$$

In order to calculate the pre-exponential factor we make use of the identity

$$|\overset{\leftrightarrow}{I} + \overset{\leftrightarrow}{\sigma} * \overset{\leftrightarrow}{C}| = |\overset{\leftrightarrow}{I} + \overset{\leftrightarrow}{C} * \overset{\leftrightarrow}{\sigma}| = |\overset{\leftrightarrow}{I} + \overset{\rightarrow}{\underset{\leftarrow}{T}} * \overset{\leftrightarrow}{\sigma} * \overset{\leftarrow}{\underset{\rightarrow}{T}}| = |\underset{\leftrightarrow}{I} + \underset{\leftrightarrow}{\sigma}| = d \ . \tag{2.29}$$

We thus arrive at the following expression for the joint angular distribution

$$P_{AB}(\vec{\alpha}, \vec{\beta}) = (2\pi)^{-3} \frac{S_A}{K_A} \frac{S_B}{K_B} |\underset{\leftrightarrow}{I} + \underset{\leftrightarrow}{\sigma}|^{-1/2} \exp[-\tfrac{1}{2} \underset{\rightarrow}{\langle s \rangle} \circ (\underset{\leftrightarrow}{I} + \underset{\leftrightarrow}{\sigma})^{-1} \circ \underset{\leftarrow}{\langle s \rangle}] \ . \tag{2.30}$$

2.5. The Individual Fission Angular Distributions

The individual angular distributions of the fission fragments from one of the reaction products are given by eqs. (2.2). Due to the results (2.16), these expressions are of the same form as the expression (2.17) for the joint angular distribution, the only difference being the replacement of the supertensor $\overset{\leftrightarrow}{C}$ of rank two by either of the rank one tensors $\overset{\leftrightarrow}{A}$ or $\overset{\leftrightarrow}{B}$. The results (2.15) can then immediately be taken over for $P_A(\vec{\alpha})$ and $P_B(\vec{\beta})$ with the appropriate replacements of $\overset{\leftrightarrow}{C}$. By proceeding in an analogous manner we then arrive at the results

$$P_A(\vec{\alpha}) = (2\pi)^{-3/2} \frac{S_A}{K_A} (1 + \sigma_{\alpha\alpha})^{-1/2} \exp[-\tfrac{1}{2} \langle s_\alpha \rangle (1 + \sigma_{\alpha\alpha})^{-1} \langle s_\alpha \rangle] \ ,$$

$$P_B(\vec{\beta}) = (2\pi)^{-3/2} \frac{S_B}{K_B} (1 + \sigma_{\beta\beta})^{-1/2} \exp[-\tfrac{1}{2} \langle s_\beta \rangle (1 + \sigma_{\beta\beta})^{-1} \langle s_\beta \rangle]. \tag{2.31}$$

With these results we can derive an expression for the enhancement factor

$$F(\vec{\alpha},\vec{\beta}) \equiv \frac{P_{AB}(\vec{\alpha},\vec{\beta})}{P_A(\vec{\alpha})P_B(\vec{\beta})} = \left(1 - \frac{\sigma_{\alpha\beta}\sigma_{\beta\alpha}}{(1+\sigma_{\alpha\alpha})(1+\sigma_{\beta\beta})}\right)^{-1/2}$$

$$\exp\left[-\frac{1}{2}\left[\sigma_{\alpha\beta}\sigma_{\beta\alpha}\left(\frac{s_\alpha^2}{1+\sigma_{\alpha\alpha}}+\frac{s_\beta^2}{1+\sigma_{\beta\beta}}\right) - (\sigma_{\alpha\beta}+\sigma_{\beta\alpha})s_\alpha s_\beta\right]\frac{1}{d}\right] .$$

$$(2.32)$$

2.6. Biased Spin Distributions

The biased spin distributions can also be derived in a manner analogous to the one used for obtaining the joint angular distribution. Let us consider the biased distribution for the spin of nucleus A; the other one follows analogously. Using (2.4), (2.7) and (2.11) we have

$$P_B(\vec{\beta})\tilde{f}_A(\vec{S}^A;\vec{\beta}) = (2\pi)^{-9/2}\frac{S_B}{K_B}|\overleftrightarrow{\sigma}|^{-1/2}$$

$$\times \int d\vec{S}^B \exp[-\tfrac{1}{2}[\,(\vec{S}-\langle\vec{S}\rangle)*\overleftrightarrow{\sigma}^{-1}*(\vec{S}-\langle\vec{S}\rangle) + \vec{S}*\overleftrightarrow{B}*\vec{S}]] .$$

$$(2.33)$$

In analogy to (2.23) we find for the exponent

$$[\cdot] = (\vec{S}-\langle\vec{S}\rangle-\vec{\Delta}_\beta)*(\overleftrightarrow{\sigma}^{-1}+\overleftrightarrow{B})*(\vec{S}-\langle\vec{S}\rangle-\vec{\Delta}_\beta)$$

$$+ \langle\vec{S}\rangle*\overleftrightarrow{B}*(\overleftrightarrow{I}+\overleftrightarrow{\sigma}*\overleftrightarrow{B})^{-1}*\langle\vec{S}\rangle .$$

$$(2.34)$$

Here the biased shift is given by

$$\vec{\Delta}_\beta = -\langle\vec{S}\rangle*\overleftrightarrow{B}*\overleftrightarrow{\sigma}*(\overleftrightarrow{I}+\overleftrightarrow{B}*\overleftrightarrow{\sigma})^{-1} .$$

$$(2.35)$$

The integrand in the above expression for $\tilde{f}_A$ is thus a six-dimensional gaussian centered at $\vec{S}=\langle\vec{S}\rangle+\vec{\Delta}_\beta$ and with covariance tensor $(\overleftrightarrow{\sigma}^{-1}+\overleftrightarrow{B})^{-1}$. The integration over $\vec{S}^B$ then leaves a distribution in $\vec{S}^A$ which is a three-dimensional gaussian centered at $\vec{S}=\langle\vec{S}^A\rangle+\vec{\Delta}_\beta^A$ where $\vec{\Delta}_\beta^A \equiv (\vec{\Delta}_\beta)^A$ is the A-part of the supervector $\vec{\Delta}_\beta$. Furthermore, the corresponding biased covariance tensor $\overleftrightarrow{\sigma}^{AA}$ is the AA part of $(\overleftrightarrow{\sigma}^{-1}+\overleftrightarrow{B})^{-1}$.

We now make use of the following relations:

$$|\overset{\leftrightarrow}{I} + \overset{\leftrightarrow}{B} * \overset{\leftrightarrow}{\sigma}| = |\overset{\leftrightarrow}{I} + \overset{\leftrightarrow}{\sigma} * \overset{\leftrightarrow}{B}|$$

$$= |\overset{\leftrightarrow}{I} + \overset{\leftrightarrow}{B} \cdot \overset{\leftrightarrow}{\sigma}{}^{BB}| = |\overset{\leftrightarrow}{I} + \overset{\leftrightarrow}{\sigma}{}^{BB} \cdot \overset{\leftrightarrow}{B}|$$

$$= 1 + \frac{\vec{\beta}}{K_B} \cdot \overset{\leftrightarrow}{\sigma}{}^{BB} \cdot \frac{\overset{\leftarrow}{\beta}}{K_B} = 1 + \sigma_{\beta\beta} \equiv d_\beta \ , \tag{2.36}$$

$$(\overset{\leftrightarrow}{I} + \overset{\leftrightarrow}{B} * \overset{\leftrightarrow}{\sigma})^{-1} = \overset{\leftrightarrow}{I} - \frac{1}{d_\beta} \overset{\leftrightarrow}{B} * \overset{\leftrightarrow}{\sigma} \ . \tag{2.37}$$

It is then readily found that

$$\overset{\Rightarrow}{\Delta}_\beta = - \langle \overset{\Rightarrow}{S} \rangle * \overset{\leftrightarrow}{B} * \overset{\leftrightarrow}{\sigma} \frac{1}{d_\beta}$$

$$= \left(-\frac{s_\beta}{d_\beta} \frac{\vec{\beta}}{K_B} \cdot \overset{\leftrightarrow}{\sigma}{}^{BA} , -\frac{s_\beta}{d_\beta} \frac{\vec{\beta}}{K_B} \cdot \overset{\leftrightarrow}{\sigma}{}^{BB} \right) = (\vec{\Delta}_\beta^A , \vec{\Delta}_\beta^B) \ . \tag{2.38}$$

Thus, the detection of a fission fragment from B in the direction $\vec{\beta}$ shifts the mean spin of A to

$$\overset{\sim}{\vec{S}}{}^A_\beta = \langle \vec{S}^A \rangle - \vec{\Delta}_\beta^A$$

$$= \langle \vec{S}^A \rangle - \langle \vec{S}^B \rangle \cdot \frac{\overset{\leftarrow\to}{\beta\beta}}{K_B^2} \cdot \overset{\leftrightarrow}{\sigma}{}^{BA} \left(1 + \frac{\vec{\beta}}{K_B} \cdot \overset{\leftrightarrow}{\sigma}{}^{BB} \cdot \frac{\overset{\leftarrow}{\beta}}{K_B} \right)^{-1} \ . \tag{2.39}$$

Furthermore, the mean spin of those target-like nuclei which fission in the direction of $\vec{\beta}$ is given by

$$\overset{\sim}{\vec{S}}{}^B_\beta = \langle \vec{S}^B \rangle - \vec{\Delta}_\beta^B = \langle \vec{S}^B \rangle \left[\overset{\leftrightarrow}{I} - \frac{\overset{\leftarrow\to}{\beta\beta}}{K_B^2} \cdot \overset{\leftrightarrow}{\sigma}{}^{BB} \left(1 + \frac{\vec{\beta}}{K_B} \cdot \overset{\leftrightarrow}{\sigma}{}^{BB} \cdot \frac{\overset{\leftarrow}{\beta}}{K_B} \right)^{-1} \right] \ . \tag{2.40}$$

It also readily follows that

$$(\overset{\leftrightarrow}{\sigma}{}^{-1} + \overset{\leftrightarrow}{B})^{-1} = \overset{\leftrightarrow}{\sigma} * (\overset{\leftrightarrow}{I} + \overset{\leftrightarrow}{B} * \overset{\leftrightarrow}{\sigma})^{-1}$$

$$= \overset{\leftrightarrow}{\sigma} - \frac{1}{d_\beta} \overset{\leftrightarrow}{\sigma} * \overset{\leftrightarrow}{B} * \overset{\leftrightarrow}{\sigma} \ , \tag{2.41}$$

so that the biased covariance for the $\vec{S}^A$ distribution is given by

$$\overset{\sim}{\overset{\leftrightarrow}{\sigma}}{}^{AA}_{\beta} = \overset{\leftrightarrow}{\sigma}{}^{AA} - \frac{1}{d_\beta} \overset{\leftrightarrow}{\sigma}{}^{AB} \cdot \overset{\leftrightarrow}{B} \cdot \overset{\leftrightarrow}{\sigma}{}^{BA}$$

$$= \overset{\leftrightarrow}{\sigma}{}^{AA} - \overset{\leftrightarrow}{\sigma}{}^{AB} \cdot \frac{\overset{\leftrightarrow}{\beta\beta}}{K_B^2} \cdot \overset{\leftrightarrow}{\sigma}{}^{BA} \left(1 + \frac{\vec{\beta}}{K_B} \cdot \overset{\leftrightarrow}{\sigma}{}^{BB} \cdot \frac{\overset{\leftarrow}{\beta}}{K_B} \right)^{-1} \qquad (2.42)$$

The biased spin distribution can therefore be written

$$\tilde{f}_A(\vec{S}^A; \vec{\beta}) = (2\pi)^{-3/2} |\overset{\sim}{\overset{\leftrightarrow}{\sigma}}{}^{AA}_{\beta}|^{-1/2} \exp[-\tfrac{1}{2}[\vec{S}^A - \overset{\sim}{\vec{S}}{}^A_\beta) \cdot (\overset{\sim}{\overset{\leftrightarrow}{\sigma}}{}^{AA}_\beta)^{-1} \cdot (\overset{\leftarrow}{S}{}^A - \overset{\sim}{\overset{\leftarrow}{S}}{}^A_\beta)]].$$

$$(2.43)$$

2.7. Illustration: 8.5 MeV/n U + Pb

We now wish to illustrate the theory by considering the reaction 8.5 MeV/n ^{238}U + ^{208}Pb which is of present experimental interest.[7] We have carried out dynamical trajectory calculations for specified J-values and subsequently integrated over those to obtain the final spin distribution as a function of the kinetic energy loss TKEL. We consider in some detail the results for a moderate energy loss, TKEL = 140 MeV, and a large one, TKEL = 280 MeV. For a specified energy loss, the joint fission-fission angular distribution $P_{AB}(\vec{\alpha}, \vec{\beta})$ is a function of the four angles $\vec{\alpha} = (\theta_A, \varphi_A)$ and $\vec{\beta} = (\theta_B, \varphi_B)$ and thus not easy to display. We therefore choose to fix the direction $\vec{\beta}$ at a specified value and then study P_{AB} as a function of $\vec{\alpha}$. We wish to re-call that $\vec{\alpha}$ denotes the fission direction for the projectile-like re-action fragment A as seen in its rest frame while $\vec{\beta}$ denotes the fis-sion direction for the target-like reaction fragment B as seen in its rest frame; thus the two directions $\vec{\alpha}$ and $\vec{\beta}$ do not refer to the same inertial frame, and the appropriate transformations need be carried out to relate $P_{AB}(\vec{\alpha}, \vec{\beta})$ to angular distributions obtained in the laboratory system. Our particular choice of frames is made in order to best bring out the physical effects of the correlation between the fragment spins.

In order to give a global impression of the effect of the cor-relation of the two fragment spins $\vec{S}^A$ and $\vec{S}^B$ on the joint fission-fission angular distribution we show in fig.2.1 contour plots of the quantity $P_{AB}(\vec{\alpha}, \vec{\beta}) / P_A(\vec{\alpha}) P_B(\vec{\beta})$. The fission direction of the target-like reaction product has been fixed at either $\vec{\beta} = (\theta_\beta = 45°, \varphi_\beta = 60°)$ or $\vec{\beta} = (\theta_\beta = 45°, \varphi_\beta = -30°)$. The directions $\varphi_\beta = 60, -30°$ correspond to the major and minor in-plane principal directions, respectively, at at the specified energy loss of TKEL = 140 MeV. [At both energy losses considered, the calculated scattering angle is $\theta_{c.m.} \approx 70°$ so that the major in-plane principal direction is expected to form an angle of approximately $35° (\approx \tfrac{1}{2}\theta_{c.m})$ with the beam direction. In actuality this angle is close to $30°$. Consequently, in the adopted coordinate system, which has the Z-axis perpendicular to the beam and the X-axis along

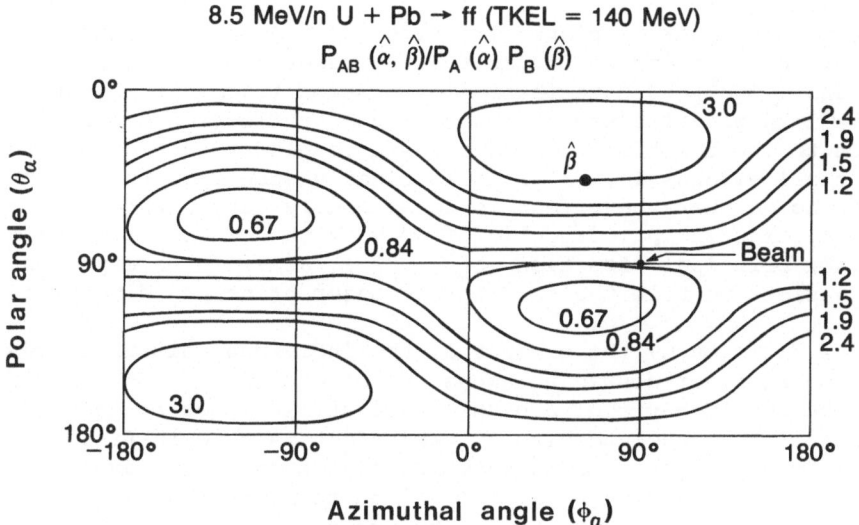

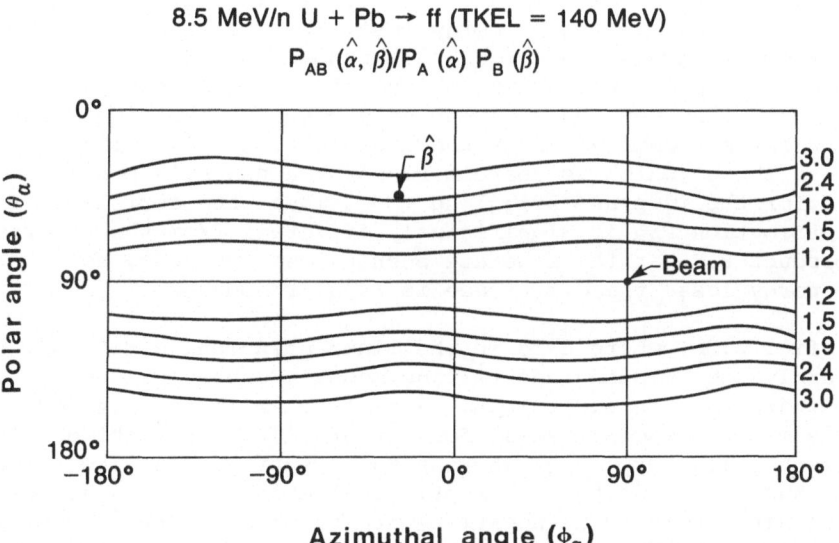

Fig.2.1. Contour plots of the ratio between the joint fission angular distribution $P_A(\hat{\alpha}, \hat{\beta})$ and the product of the two individual fission angular distributions $P_A(\hat{\alpha})$ and $P_B(\hat{\beta})$, for the reaction 8.5 MeV/n ^{238}U + ^{238}Pb at an energy loss of TKEL = 140 MeV. The fission direction $\hat{\beta}$ of the target-like reaction product B has been fixed at either ($\theta_\beta = 45°$, $\varphi_\beta = 60°$ (top) or $\theta_\beta = 45°$, $\varphi_\beta = -30°$) (bottom) and the ratio is displayed as a function of the fission direction $\hat{\alpha} = (\theta_\alpha, \varphi_\alpha)$ for the reaction partner A.

the beam, the major in-plane direction has $\theta_0 \approx 60°$.] When $\vec{\beta}$ is taken to be of (but not perpendicular to) the reaction plane (i.e. when $\theta_\beta \neq 90°$ and $\theta_\beta \neq 0°$) the distribution of fission fragments from the projectile-like reaction product is no longer reflection symmetric with respect to the reaction plane (nor with respect to any other plane, as we shall discuss later.) This is clearly seen from the figure. It is also noted that the effect is considerably larger when $\vec{\beta}$ is chosen in the major principal direction ($\varphi_\beta = 60°$) than when $\vec{\beta}$ is chosen in the minor principal direction ($\varphi_\beta = -30°$), as is to be expected since it is easiest to tilt the spin in the major principal direction.

The division by the product of the individual distributions $P_A(\vec{\alpha})$ and $P_B(\vec{\beta})$ of course enhances the effect in the polar regions where the absolute yields are the smallest. In order to gain an impression of the absolute size of the effect we show in fig.2.2 the absolute yield $P_{AB}(\vec{\alpha}, \vec{\beta})$ as a function of a single angular variable, namely the polar angle θ_α, with φ_α fixed to be the same value as φ_β (since the largest effect is expected when $\vec{\alpha} \approx \vec{\beta}$). Figure 2.2a corresponds to $\vec{\beta} = (45°, 60°)$ while fig. 2.2b has $\vec{\beta} = (45°, -30°)$. The upper portions are for a moderate energy loss of TKEL = 140 MeV while the lower portions are for a large energy loss of TKEL = 280 MeV.

We note that the distribution in fig.2.2 is biased towards $\vec{\beta}$, particularly at the moderate energy loss. This is expected since this particular angular selection will probe the major in-plane components of the spin distribution which arises predominantly from the in-plane wriggling mode and consequently has positive signature. Towards the largest energy losses the distribution will relax more towards equilibrium which is characterized by a slightly positive covariance for the in-plane wriggling mode. Thus, the correlated fission distribution P_{AB} is significantly different from the uncorrelated product $P_A P_B$ at the moderate energy loss, whereas, for the larger energy loss, the difference is less pronounced.

In fig. 2.2b, where $\vec{\beta}$ is in the minor principal direction, we probe mainly the twisting and tilting modes. Generally, the tilting mode dominates at the small TKEL, leading to small positive σ^{AB} along the minor principal axis. For larger TKEL, the tilting relaxation time becomes very long, whereas the twisting relaxation time decreases, and σ_{AB} along the minor principal axis decreases at some point and ultimately turns negative with increasing TKEL. Thus, at the moderate TKEL, where σ^{AB} is still positive along the minor principal axis, the maximum of P_{AB} is shifted slightly towards the direction $\vec{\beta}$, whereas the negative value σ^{AB} at the large TKEL induces a shift of the maximum in the opposite direction.

The fact that the location of the maximum of P_{AB}, as a function of the direction $\vec{\alpha}$ for fixed $\vec{\beta}$, moves toward $\vec{\beta}$ for positive σ^{AB} and away from $\vec{\beta}$ for negative σ^{AB} can qualitatively be understood as a three-step process: (i) Fission into the direction $\vec{\beta}$ imposes a strong bias of the spin $\vec{S}_B$ of nucleus B towards directions perpendicular to $\vec{\beta}$ (cf. eq. (2.40)). (ii) For positive σ^{AB}, this bias of $\vec{S}_B$ induces a bias of $\vec{S}_A$ in the same direction and in the opposite direction for negative σ^{AB}. (iii) The fission fragments emitted from nucleus A will preferably be emitted perpendicular to the spins $\vec{S}_A$ biased in this way, and this favours the direction $\vec{\beta}$ over directions perpendicular to $\vec{\beta}$ for positive σ^{AB} and disfavours it for negative σ^{AB}.

Figure 2.3 shows the average spin and contours for the variance in the two planes containing the Y-axis and, respectively, the minor and major principal axes for the unbiased spin distribution in nucleus A, and for the distribution biased by a detection of fission fragments from nucleus B. The angles of maximum P_{AB} restricted to the planes shown correspond roughly to the directions perpendicular to the biased average spin $\vec{S}_\beta^A$ in the two cases. Since the biased variance tensor does not have the biased average spin direction as a principal axis, P_{AB} is not reflection symmetric with respect to the plane perpendicular to $\vec{S}_\beta^A$, or any other plane.

It is instructive to contrast the above results with the expectations of a statistical model. For this purpose we assume that all spin modes are fully relaxed at the time of neck snapping, as already discussed in Lecture 1. As already pointed out there, the resulting principal system is rotated approximately $-45°$ relative to the dynamically calculated distribution. In order to maximize the effect we therefore choose either $\vec{\beta} = (45°,0°)$ or $\vec{\beta} = (45°,90°)$, corresponding approximately to the major and minor principal directions of the statistical spin distribution.

The results are shown in fig. 2.4. It is clearly seen that the effect is much smaller (nearly an order of magnitude) and probably not experimentally detectable. The results exhibit the qualitative feature of a slight positive shift when biasing along the major principal detection and a slight negative shift when biasing along the minor principal direction, as specified.

In summary, we wish to state the following. Our calculations indicate that the dynamically accumulated correlated structure in the spin-spin distribution of the reaction products gives rise to significant effects in the joint fission-fission angular distribution. Moving one fission detector out of the reaction plane breaks the symmetry with respect to reflections in that plane. The resulting reflection asymmetry of the fission distribution from the other reaction product is fairly sizeable and shows a characteristic variation with direction and energy loss. In contrast, calculations based on the assumption of full statistical relaxation of the spin modes yield very little effect. It would be of great interest to observe such structure. Given sufficient data quality, it should be possible to determine the principal directions and, from the character of the

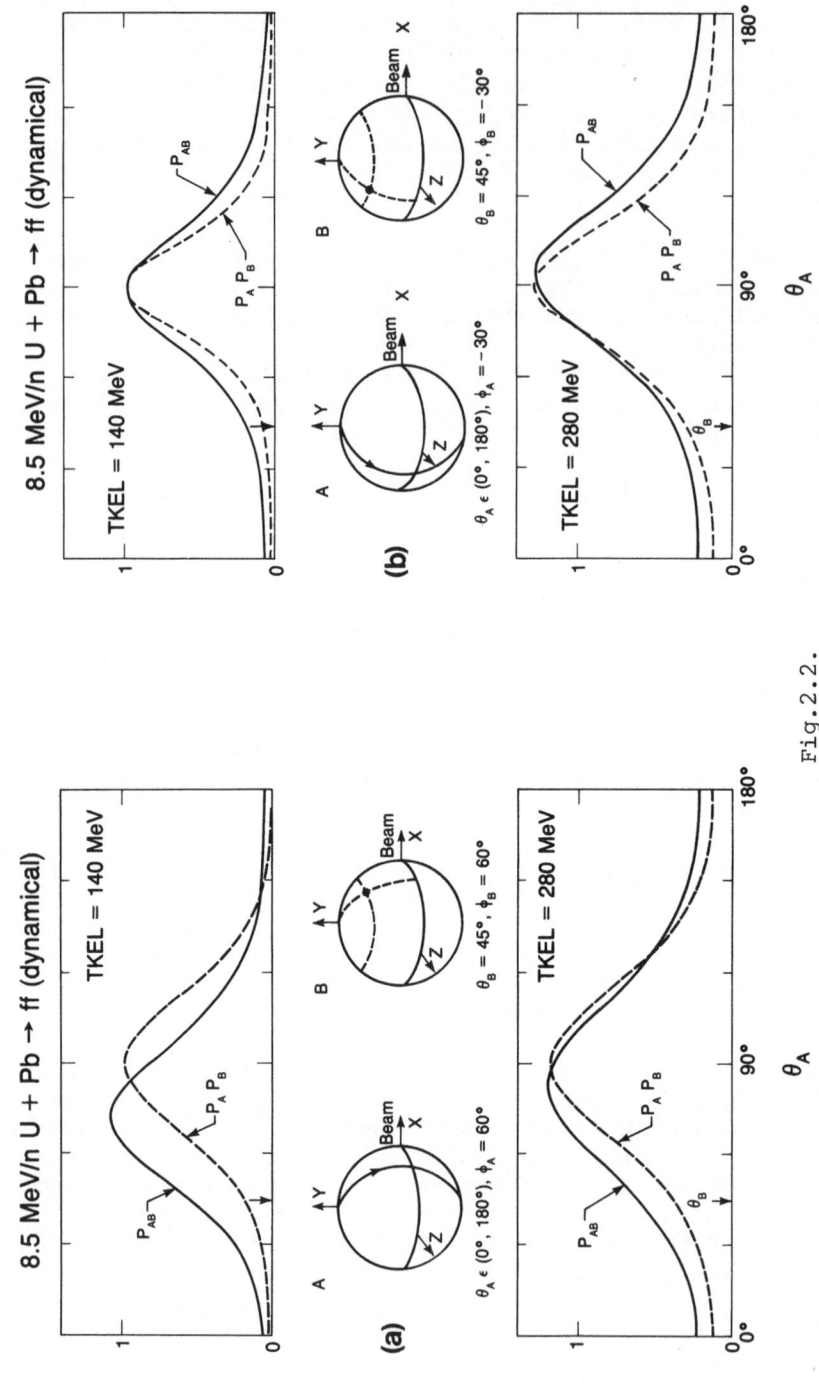

Fig.2.2.

The absolute joint fission probability $P_{AB}(\hat{\alpha}, \hat{\beta})$ for the U + Pb reaction. The direction $\hat{\beta}$ has been fixed at either ($\theta_\beta = 45°, \varphi_\beta = 60°$) (a) or ($\theta_\beta = 45°, \varphi_\beta = -30°$) (b), while the direction $\hat{\alpha}$ is moved from the north pole, through the point $\hat{\alpha} = \hat{\beta}$, and down to the south pole. Two energy losses have been considered. TKEL = 140 MeV and 280 MeV. The dashed curves indicate the product of the individual fission probabilities, $P_A(\hat{\alpha}) P_B(\hat{\beta})$. The curves have been normalized so that isotropy would yield the value one for all $\hat{\alpha}, \hat{\beta}$.

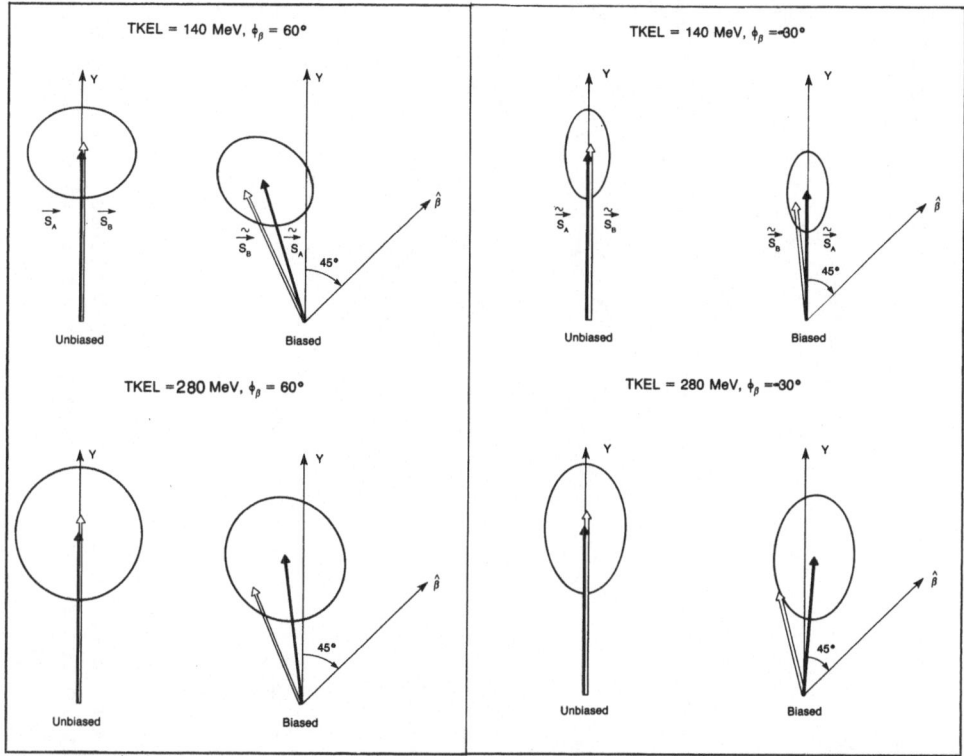

Fig.2.3. Effect on the target-like fragment spin $\vec{S}^A$ of the bias introduced by detection of a projectile fission fragment in the direction $\hat{\beta}$. The figure shows the distribution of $\vec{S}^A$ projected onto the plane containing the reaction normal (the Y-axis) and fission direction $\hat{\beta}$, which in turn has been chosen as either $(\theta_\beta = 45^0, \varphi_\beta = 60^0)$ (left frame) or $(\theta_\beta = 45^0, \varphi_\beta = -30^0)$ (right frame), corresponding to the directions considered in figs.2.1, 2.2. The left-hand sides show the unbiased mean spins and the one-sigma contour for $\vec{S}^A$ while the right-hand sides show the biased mean spins (2.39) and (2.40) and the associated one-sigma contour (as determined from (2.42)). Note that in this figure the labeling of the nuclei unfortunately deviates from our standard notation, in which label A refers to the projectile and B to the target.

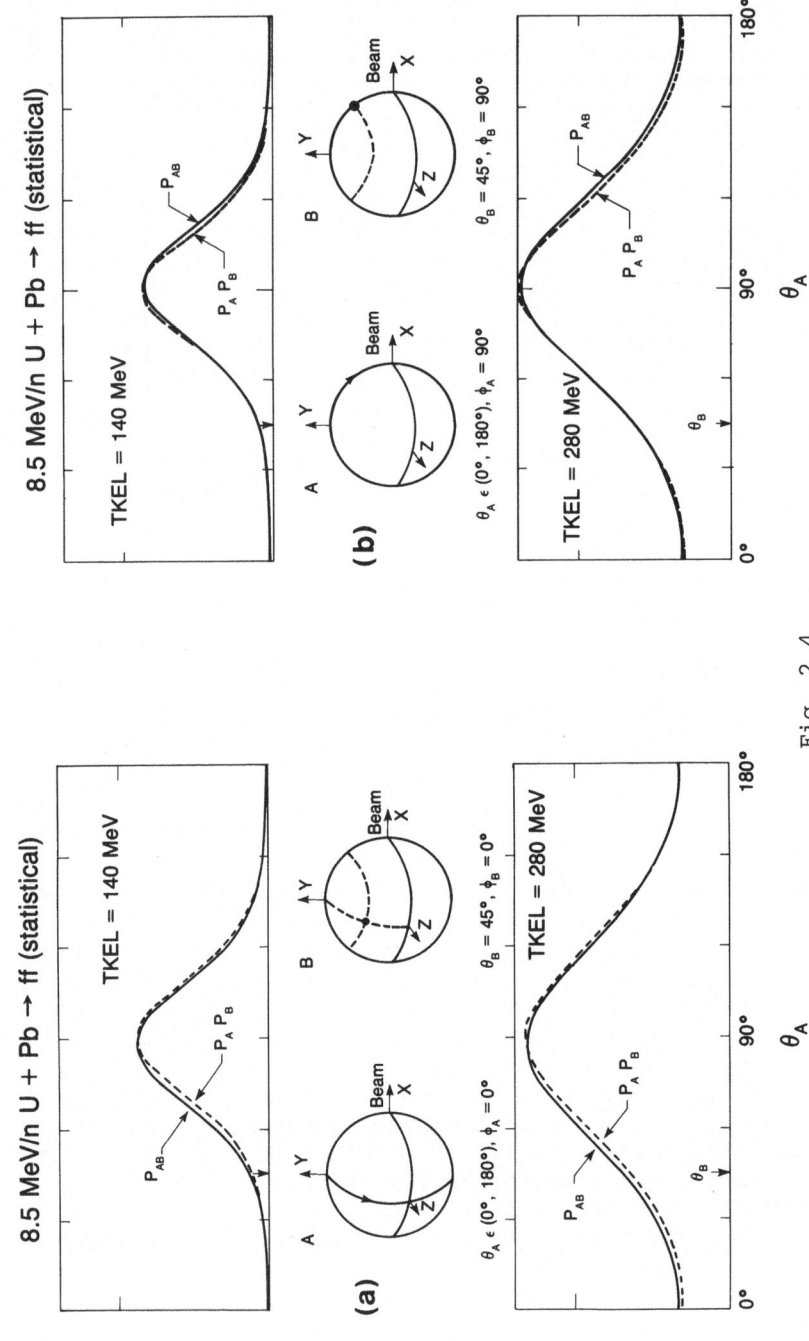

Fig. 2.4.

Similar to fig.2.2 but calculated with a statistical model. The settings of the fission direction β corresponds approximately to the associated major and minor principal directions and are ($\theta_\beta = 45°$, $\varphi_\beta = 0°$) (a) and ($\theta_\beta = 45°$, $\varphi_\beta = 90°$) (b).

correlations, obtain an indication of the relaxation times for the various dinuclear spin modes.

ACKNOWLEDGMENTS

During part of this work, Thomas Døssing was supported by a Niels Bohr Fellowship granted by the Royal Danish Academy of Science.

REFERENCES

1. T. Døssing and J. Randrup, LBL-16825(1983), Nucl. Phys. A433 (1985)215.
2. J. Randrup, Nucl. Phys. A327(1979)490.
3. J. Randrup, Nucl. Phys. A383(1983)468.
4. L.G. Moretto and R.P. Schmitt, Phys. Rev. C21(1980)204.
5. R.P. Schmitt and A.J. Pacheco, Nucl. Phys. A379(1982)313.
6. T. Døssing and J. Randrup, LBL-16826(1983), Nucl. Phys. A433 (1985)280.
7. A. Lazzarini, private communication.

NUCLEAR STRUCTURE AT HIGH SPIN

Bent Herskind

The Niels Bohr Institute
University of Copenhagen
Copenhagen, Denmark

PREFACE

A review of nuclear structure of high spin states is given in these lectures with emphasis on the latest development in γ-ray spectroscopy. It is one of the topics in nuclear structure physics which has attracted most attention recently. Therefore, several complete reviews as well as innumerous papers on selected topics have been published within the last year. I shall therefore in these notes prefer to refer to these reviews and original papers rather than try to make yet another compendium.

A table of contents of the lectures is given below with direct references to the papers which form the basis of the lectures.

Several new or illustrative figures were used to summarize the present data and knowledge. In the cases where these figures were found particularly useful and not easily found in the given references, they are included below with a connecting text and caption for reference.

TABLE OF CONTENTS

	Ref.	Fig.

1. Introduction
   Why is it of particular interest to   1-21   1-3
   study high spin states?

2. Formation and decay
   Angular momentum limits   22-28   4-6
   Temperature effects on the γ-decay
   The path ways and competitions   29-30

3. Experimental techniques
   Detector arrays, 4π systems, hybrid   31-39   7
   systems, example of achievements

4. Moment of inertia
   Definitions, predictions   2,40-42
   Results from discrete line studies   11-16
   Apparent disagreement with theory
   $J^{(2)}_{band}$ deduced from $E_\gamma \cdot E_\gamma$ correlations   12,43

   $J^{(2)}_{eff}$ deduced fron $E_\gamma \cdot$ multiplicity   11
   correlations
   Comparisons experiment-theory

5. Pathways in the quasi-continuum
   $E_\gamma \cdot E_\gamma$ correlation data   43-46   8-9
   Pathway simulations

6. Superdeformed nuclei
   Predictions by Nilsson and Woods-Saxon   47-49
   model data   20

7. Decline and fall of pairing correlations
   Predictions   15
   Experimental signatures   16
   The unpaired regime

8. Electromagnetic properties   50-52

9. The giant resonances at high spin
   Principles and definitions   2
   Experimental procedures
   Statistical descriptions   53
   Direct decay from compound states   54   55-57
   Shape changes with T and I   55
   The observation of GQR   56
   Theoretical calculation for high spin   60-61

1. INTRODUCTION

The nucleus is a unique quantal system consisting of a
finite number of particles forming a shell structure with spherical symmetry at "magic" numbers of protons and neutrons. Only
a few nucleons in the valence shells can change the nuclear
shape. Just above the N=82 shell, f.ex. the largest density of
neutron configuration is found for prolate shapes with $\varepsilon_2 \sim .25$
(see Fig.1a), and a considerable shell energy is gained. Typical
shell energies are ~10 MeV. Similar energies can be gained by
rotating the deformed nucleus at high rotational frequencies,
$\hbar\omega \sim 1$ MeV, due to the centrifugal and Coriolis forces, and given
as ωj_x in the hamiltonian

$$h^\omega = h_{sp} - \omega j_x$$

If the nucleus is in a highly alignable $i_{13/2}$ orbit as
f.ex. $1/2+[660]$ seen in Fig.1b $\hbar\omega j_x$ can be as large as 5-6 MeV,
similar in size to shell energies, and as important for nuclear
properties at high spin.

The sign of the Coriolis force is opposite for nucleons
moving in the direction and counter to the direction of the
nuclear rotational motion. This causes breaking of the time
reversal symmetry of the orbital motion giving rise to large
signature splittings (see Fig.1b). This effect also reduces the
spacial overlap of the nucleons in time reversed orbits and
thereby decreases the pair correlations. The pair correlation
energy, $\varepsilon_{pair} \sim 2.5$ MeV at $\hbar\omega=0$ is reduced significantly at higher
frequencies and becomes close to zero at $\hbar\omega > .4$ MeV. The nucleonic configurations near the fermi surface are modified and energy shifted by these pair correlations as illustrated in Fig.1c
for a N=98 nucleus. In the cranked shell model calculations
shown in Fig.1c, a constant pairing energy $\Delta = .15 \hbar\omega_0$ is assumed
for simplicity. Self consistent calculations with respect to
pairing and deformation are made for many nuclei. For N=96-98
see ref.[21].

The renewed interest in studying the level structure of
high spin states reflects this fact, namely that it is possible
to study the basic elements of the individual configurations in
the nucleus as function of spin I and rotational frequency $\hbar\omega$,
verified as rotational band structures, yrast and side bands
connected through directly measurable γ-ray transition energies,
$E_\gamma = 2 \cdot \hbar\omega$ with enhanced E2 matrix elements. Hence, detailed information about the nuclear dynamics can be extracted from such
spectra. Within the last year it is f.ex. the study of pairing
correlations for different single particle configurations in the
region between superfluid and normal phase, the quenching of
static pairing interactions for neutrons[15], the shape transi-

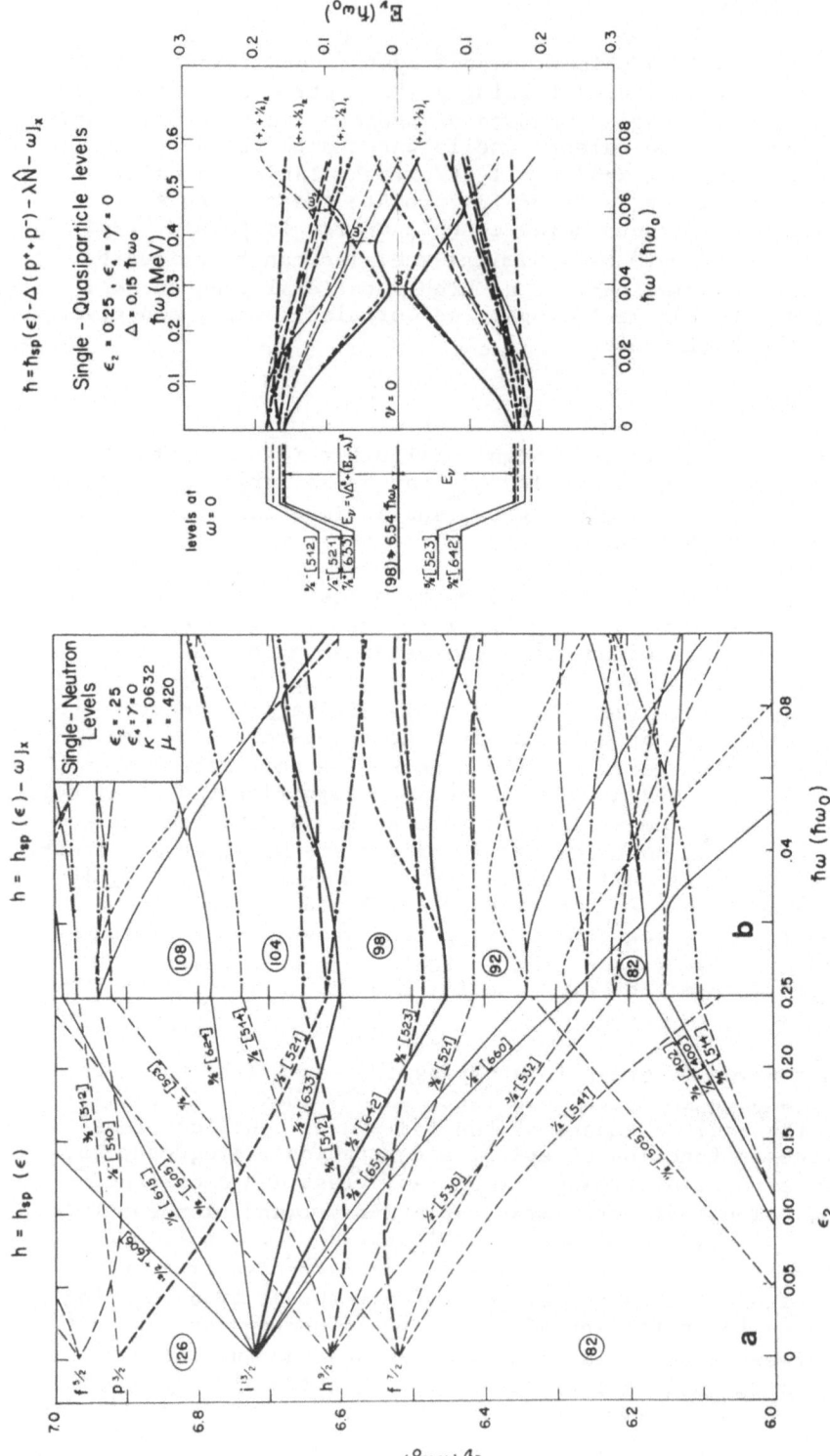

Fig. 1. Neutron configurations in a deformed axial symmetric potential as function of ϵ_2 shown in a, cranked around the x-axis in b, and with pairing correlations included, in reflexed symmetry for quasiparticles in c.

tions of ^{158}Er from prolate to oblate at I~40 $\hbar$[19], the observation of the predicted superdeformed shapes with axis 1:1:2 in ^{152}Dy, ref.[20] and the established stably deformed nuclei ^{168}Hf, ref.[14] and ^{168}Yb, ref.[16] which become the best known almost perfect rotors at the highest frequencies $\hbar\omega$>.35 MeV.

The large changes in structure of the nuclei dependent upon the "nuclear dimensions" or the integer and half integer variables N, Z, I are illustrated in Fig.2 for three selected nuclear systems with increasing number of valence particles ^{147}Gd, ^{158}Er and ^{168}Yb.

For nuclei near closed shells where the deformation driving forces are small the total spin is built from aligned angular momentum of individual particles. A superb example is ^{147}Gd with only 1 neutron outside the semimagic double closed shell of ^{146}Gd as shown in the lower panels of Fig.2. The level scheme is extended up to I=79/2 corresponding to spin alignments of 11 particle-hole configurations of both protons and neutrons[17]. This results in a very irregular yrast line with large fluctuations in the transition energies dE/dI as function of spin I (lower graph), with yrast transitions of both quadrupole and dipole type and close to single particle strength.

It is indeed a fascinating example of extreme symmetry breaking. 11 particles and holes have no time-reversed partners in this highest observed state of I=79/2. The many aligned particles generate an oblate deformation with a significant quadrupole moment[18]. However, no evidence for collective rotation has been found in this nucleus.

For nuclei with more valence particles, a lower energy is favoured by occupation of prolate configurations resulting in a static prolate shape of the nuclei (see Fig.1a). A typical[19] example is shown in the middle panel of Fig.2. The lowest lying positive parity states in ^{158}Er form the yrast decay sequence of enhanced E2 transitions. The spin of the states is built of two components, a collective rotational part R=J·ω and a component consisting of unpaired aligned quasiparticles. The aligned configurations become favoured at specific "backbending" or band-

crossing frequencies $\hbar\omega_c$, the frequencies at which the centrifugal plus Coriolis force counter balance the pair correlation energy for specific pairs of highly-alignable particles. It is seen on the level scheme shown for ^{158}Er that the ground state band is observed up to I=26+. However, at I=12, a band built on an aligned $i_{13/2}$ neutron configuration crosses the ground state band and becomes yrast at higher spin. At $\hbar\omega$=.42 MeV this band is crossed by a 4-quasiparticle configuration built on the lowest lying $\pi h_{11/2}$ pair of protons in addition to $\nu i_{13/2}$ aligned at the lower frequency. Thus, the spin along the rotational axis can be written as a sum of the two components R + Σi = I_x.

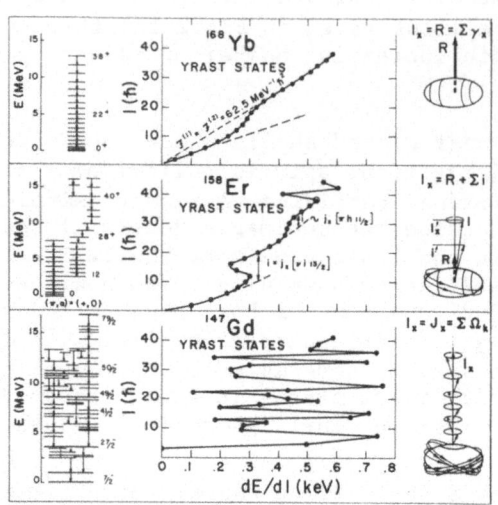

Fig. 2. Level schemes, I vs. dE/dI, and schematics for three striking different nuclei, ^{147}Gd, ^{158}Er and ^{168}Yb.

For rotational nuclei where the major decay routes go via stretched ($\Delta I=2$) E2 transitions the transition energy ΔE_γ gives a direct measure of the rotational frequency

$$E_\gamma = dE/dI \; (\Delta I=2) = 2\hbar\omega$$

The kinematical moment of inertia, $J^{(1)}=I_x/\omega$ related to the motion of the total system, and the dynamical moment of inertia, $J^{(2)}=dI_x/d\omega$, related to the dynamical properties (give information about the local changes) of the fast rotating nucleus, are therefore directly measurable quantities, and a comparison of the two moments of inertia gives information about the total alignment:

$$I_x = R + \sum i = J^{(2)} \cdot \omega + i_o$$
$$J^{(1)} = J^{(2)} + i_o/\omega$$

where i_o is the intercept on the I axis on Fig.2 of the I vs. ω plots.

A smooth band structure is observed for ^{158}Er up to $I=38^+$ which resembles the collective rotational states built on few quasiparticle configurations. Presumably a shape transition occurs at $I\sim40$ to oblate configurations, and the irregular pattern develops like seen in ^{147}Gd at lower spin.

The upper panel of Fig.2 shows data for ^{168}Yb, a nucleus in the middle of the $i_{13/2}$ shell[16]. An even more smooth trend is observed with $J^{(1)}\overset{13/2}{=}J^{(2)}$ at the highest spins indicating that this nucleus performs almost pure collective rotations in the lowest configurations.

2. FORMATION AND DECAY

High-spin states can today be produced in many laboratories by beams of A>30 with energies of 4.7–5.0 MeV/nucleon. The for-

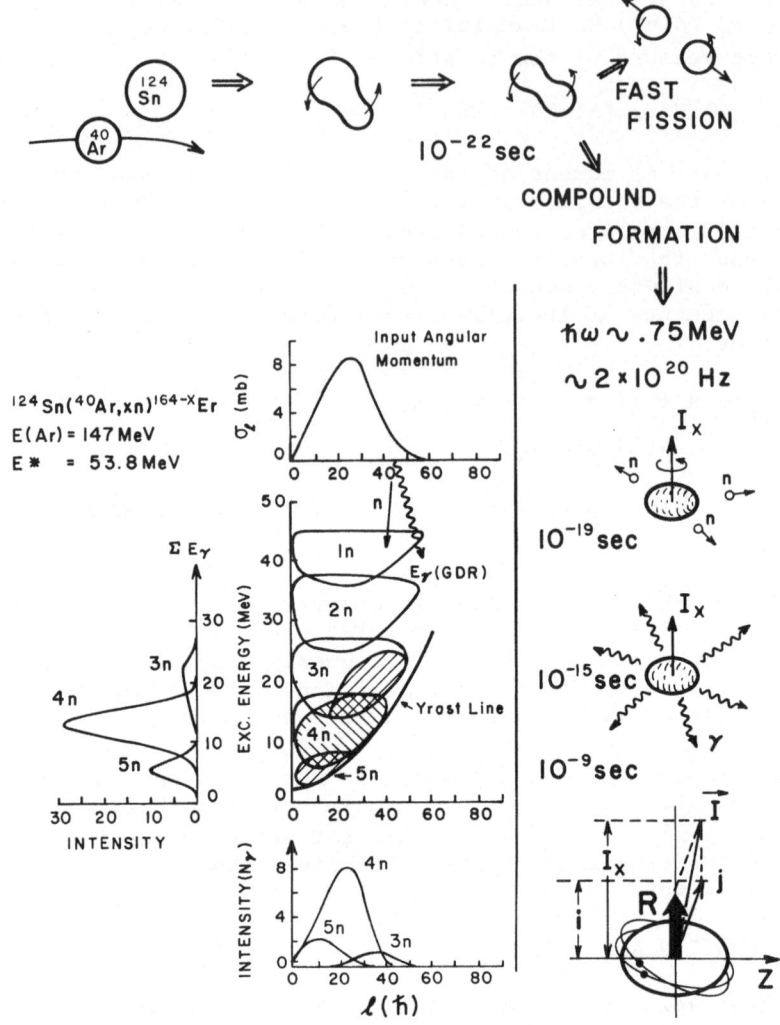

Fig. 3. Schematic illustration of compound formation and decay.
The left side of the figure shows a calculation performed by means of the statistical code GROGY II[23].

mation and decay of the ^{40}Ar + ^{124}Sn reaction are illustrated on
Fig.3 and discussed in detail in Refs.[12],[25]. The angular momentum limit ℓ_{max} is determined from cross-section measurements[22]-[23] and multiplicity measurements[24]-[27]. These fission
limits are found to agree extremely well with the original predictions by Cohen, Plasil and Swiatecki[5], giving compound reaction residues up to $\ell \sim 62$ ℏ for A~150. If lighter ions like ^{12}C
and ^{16}O are used, entrance channel effects put restrictions on

the angular momentum produced in the compound reaction[25],[28]. Semi empirical estimates of these limits can be obtained from the work of Wilczynski and collaborators[28]. Also recently the multiplicity distributions and γ decay from different temperature regions have been surveyed by means of the spin spectrometer in Oak Ridge[29], and selected cases have been studied in greater detail[30].

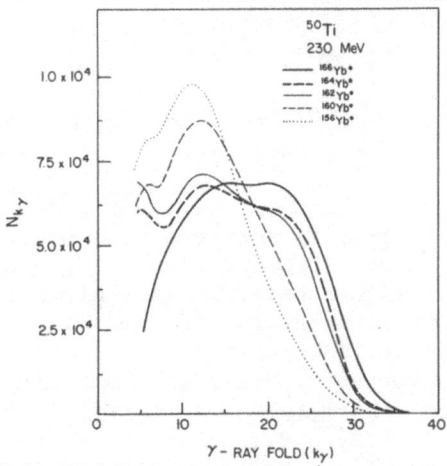

Fig. 4. Total gamma yield as function of fold for different compound nuclei of the Yb isotopes. The curves are normalized to the same number of events.

In the Oak Ridge spin spectrometer, sometimes called Crystal Ball or 4π detector, the sum of the γ-ray energy $\Sigma E_\gamma = H$ and the number of transitions (fold)=K, as well as the transition energy distribution E_γ are measured simultaneously within a given resolution time ~10 ns, by means of 72 NaI detectors forming an almost spherical shell of 18 cm thickness and an inside radius of 18 cm.

The fold distributions measured for a series of Yb evaporation residues after bombarding $^{106-116}$Cd with 230 MeV ^{50}Ti are

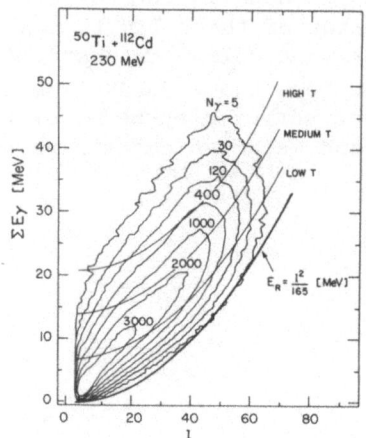

Fig. 5. Total gamma-ray entry state distributions for ^{162}Yb.

shown in Fig.4. It is clearly seen that the neutron rich nuclei show a higher edge on the fold distributions resulting from a significantly higher spin population. A typical contour plot of the H vs. K distribution for the ^{112}Cd + ^{50}Ti reaction is shown in Fig.5. The transition energy spectrum gated from different temperature regions above the yrast line are given in Fig.6. A striking difference is seen in the statistical region $E_\gamma > 2$ MeV where the logarithmic slopes directly reflect the change in effective temperature

$$\frac{d\ell n N_\gamma}{dE_\gamma} = -\frac{1}{T_{eff}} = \frac{5}{E_\gamma} - \frac{1}{T}$$

set by the gating condition $(E - E_{yrast})$ where the average temperature may be estimated from

$$T \sim \sqrt{(E - E_{yrast})/a}$$

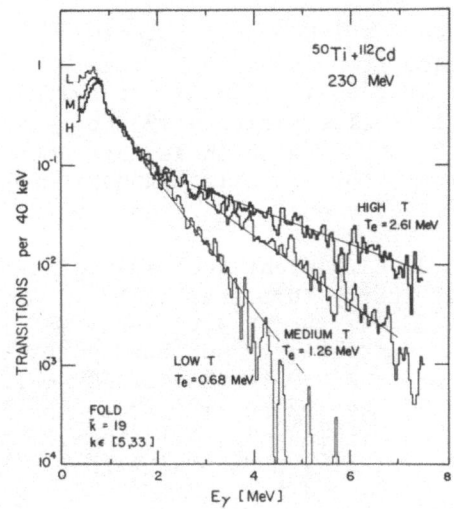

Fig. 6. Unfolded spectra from the decay of ^{162}Yb gated by 3 independent temperature regions shown in Fig. 5. The spectra are integrated over all folds in the range K=5-33. The effective temperature T_e corresponding to the statistical part of each spectrum is indicated.

It is interesting to note that the upper edge of the collective transitions in the E_γ spectrum at $E_\gamma \sim 1-1.5$ MeV seems not to change with temperature.

3. EXPERIMENTAL TECHNIQUES

The rapid experimental advances in gamma spectroscopy is based on the development of several 4π detector systems of various types. The need for measurement of gamma-multiplicity and sum-energy distributions[31-33] in connection with high resolution discrete and continuum spectroscopy, led to the construction of the Spin Spectrometer in Oak Ridge with 72 NaI detectors[34] and the Crystal Ball in Heidelberg with 162 NaI modules[35], both with a total efficiency >95% of 4π. Many important experiments have been made with these instruments already (refs.36-37 and refs. therein), in high spin physics and also in other areas of physics.

For the past 2 years great success has been obtained by means of the Total Energy Suppression Shield Array (TESSA) installed at a beamline at the 25 MeV tandem accelerator, Daresbury Laboratory. The multiplicity sum energy information is in this instrument obtained by an array of 50 small BGO detectors. The high resolution, background reduced, gamma-gamma coincidence spectra are measured simultaneously in two of the six escape suppressed Ge-detectors which "see" the target through six slits in the inner ball.

The beautiful results obtained with this instrument[12,14,15,16,19,20], which however only allows 2-fold photopeak coincidences within a reasonable running time because of a small solid angle of each compton suppressed GeLi 27 cm from the target, have stimulated many groups around the world to design more compact anticompton detectors hoping to obtain even more spectacular results. These detectors can be packed around the target in numbers of 20-30 increasing the probability of obtaining 3- and even 4-fold photopeak coincidences by more than 100 times, which means that much higher selectivity and significantly lower background can be expected. The first compact array with BGO anticomptom shield will get into operation with 20 detectors in January 85 in Berkeley, with an expectation of 5000 triple coincidences/sec. Undoubtedly this will start a new era of gamma spectroscopy. The high cost of these detectors together with the high efficiency making it profitable for many groups to use them, has led to the design of the NORD BALL detector system, where a truncated isosahedron structure forms a standard frame which allows 32 separate detector systems to be mounted in a closed packed geometry around the target. Each separate research group may design special detectors to insert together with the large ensemble of anticompton detectors.

A typical ensemble for high spin spectroscopy is shown in Fig.7 as a cut through the NORD BALL. The cuts show two double-acting anticompton detectors with a thin front (1 cm) and a thick backward Ge detector. The two Ge detectors are run in anticoincidence mode such that they protect each other from compton scattered events[39]. Also shown is a BaF_2-calorimeter consisting of 60 elements of the same shape to form an inner ball structure. It is anticipated that the fast BaF_2 can tolerate 10^7 counts/sec with a resolution time of $.5 \cdot 10^{-9}$ sec. The calorimeter can be used to measure the multiplicity and sum energy simultaneously in a similar way as for the large NaI balls, but with somewhat poorer resolution.

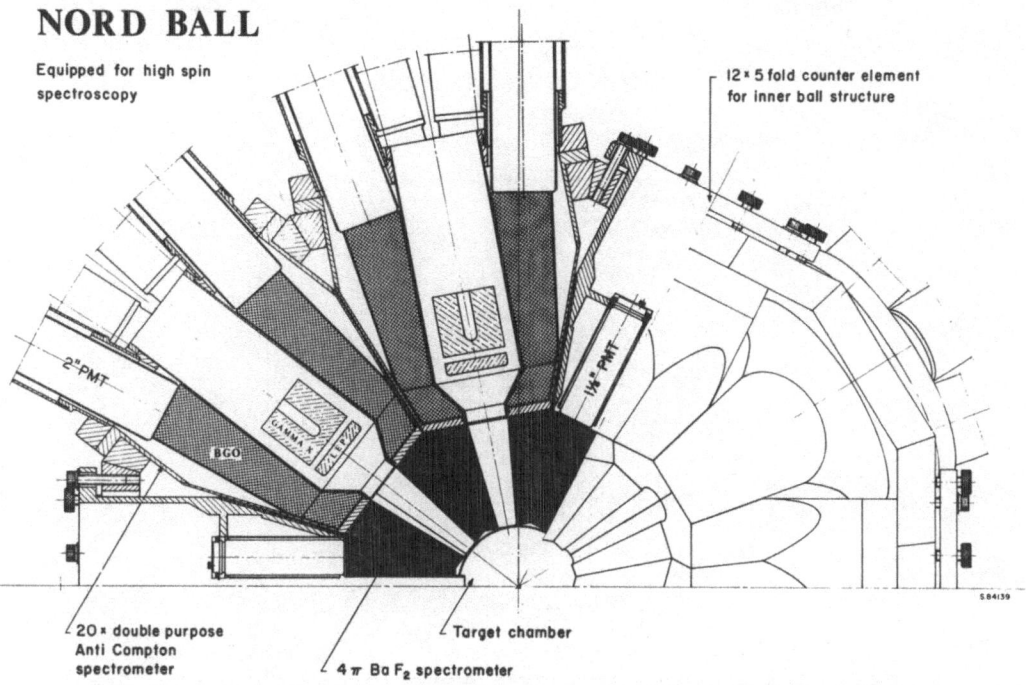

Fig. 7. A cut through the NORDBALL equipped with detectors for a typical high spin spectroscopy experiment; up to 20 double surface anti-compton detectors and a 4π BaF_2 calorimeter.

DAMPING IN THE QUASICONTINUUM

Most of the new results obtained by the new spectroscopic tools discussed in these lectures, confirm to a great extent the theoretical expectations and give a solid ground for further developments for theoretical understanding of the nuclear structure properties at high spin. F.ex. the discovery of the superdeformed ^{152}Dy nucleus[20] predicted with more and more refined theoretical explorations[47,48,49]. Also the overwhelming level scheme systematics obtained for a large number of nuclei are predicted well within few hundred keV by theory. Of course, on the more refined scale many new and interesting features have been found, and much new physics will be learned from this.

However, I should like to emphasize one interesting new result, which at least at first hand, seems rather surprising to current theory. The observation of damping effects in the quasi-continuum.

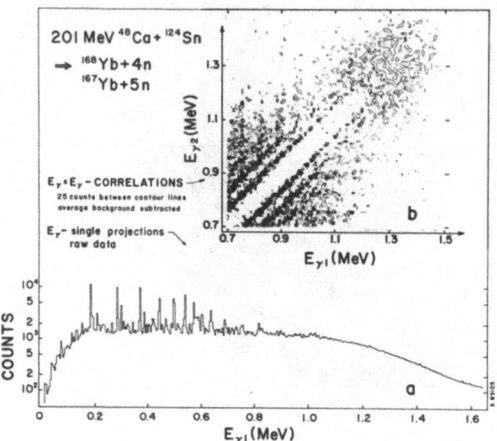

Fig. 8. a. The spectrum of γ-rays observed in a Compton-suppressed Ge detector after the reaction $^{124}Sn(^{48}Ca,xn)^{172x}Yb$ at 201 MeV bombarding energy. b. The $E_{\gamma 1}xE_{\gamma 2}$ correlation spectrum for the same reaction in the region $0.7<E_\gamma<1.5$ MeV.

In an $E_\gamma-E_\gamma$ coincidence experiment one expects a correlation pattern[43,46] with ridge-valley structures along the diagonal $E_1=E_2$ in a 2-dimensional plot (see Fig. 8) with a distance between the inner ridges $W = E_\gamma(I+2) - E_\gamma(I-2) = 8h^2/J_{band}^{(2)}$, due to the rotational band structures. This expectation is based on the fact that the discrete lines in the spectrum show such regular rotational structures.

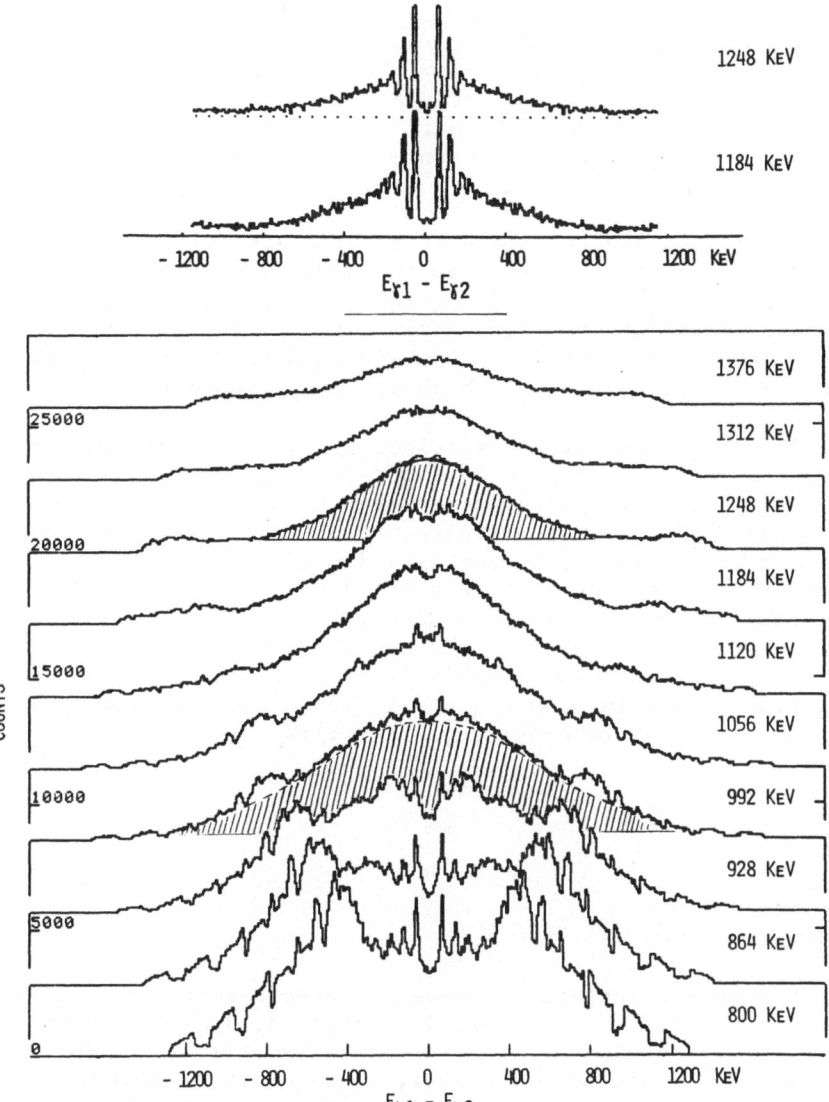

Fig. 9. The lower part shows cuts perpendicular to the $E_{\gamma 1}=E_{\gamma 2}$ line in the true (unfolded and efficiency corrected) $E_{\gamma 1} \times E_{\gamma 2}$ correlation spectrum from the $^{124}Sn(^{48}Ca,4n)$ ^{168}Yb reaction. The width of each cut is 56 keV. Within each spectrum the average of $(E_{\gamma 1}+E_{\gamma 2})/2$ is given. The hatched area at two selected energies corresponds to the "rotational" E2 continuum transitions. The top part shows a computer simulation of the decay process through statistical dipole transitions in competition with rotational E2 transitions of 200 spu.

However, if cuts are made perpendicular to the $E_1=E_2$ diagonal as functions of $(E_1+E_2)/2$, the result in Fig. 9 shows a very large continuum background below the expected ridge-valley structure. The expectation based on a computer simulation[62] assuming in-band transitions with 200 spu in competition with statistical transitions is shown above the experimental data for cuts at $E_1=E_2=1184$ keV and 1248 keV.

It is seen how the ridge structures essentially disappear at $E_\gamma > 1$ MeV in contradiction to the simulated structures and are rather weak even at lower energies. The intensity in the inner ridges compaired to the intensity in the valley, a rough measure of the transitions which proceed within discrete band structures, approaches zero at the highest transition energies.

The effect cannot be explained by variations in the moment of inertia or spin alignments. Therefore the data give evidence for a damping width presumably caused by mixing of the rotational bands at higher temperature. Satisfactory simulation of the data can be obtained by assuming a damping width of 100-200 keV[62]. The effect can also be studied as function of initial temperatures of the gamma cascades by means of the BGO inner ball structure of TESSA II. The preliminary analysis does not show large changes with initial temperature. Theoretical investigations[63] are in progress and indicate that the observed damping width may be expected, after all.

ACKNOWLEDGEMENT

The work discussed in these lectures has been made in collaboration with J.C. Bacelar, C. Ellegaard, J.J. Gaardhøje, J.D. Garrett, G.B. Hagemann, A. Holm and B. Lauritzen, the Niels Bohr Institute, P. Twin and P. Walker, Daresbury Laboratory, R. Chapman, J.C. Lisle, J. Mo and E. Paul, Manchester University, P. Nolan, M.A. Riley, J. Simpson, J.F. Sharpey-Schafer, Liverpool University, P.O. Tjøm, University of Oslo and R. Diamond, M.A. Deleplanque Stephens, A. Macchiavelli and F.S. Stephens, LBL, Berkeley.

Support from the Danish Natural Science Research Council is acknowledged.

REFERENCES

1. Aa. Bohr and B. Mottelson, J. Phys. Soc. Japan, Suppl.
    44:157 (1978).
2. Aa. Bohr and B. Mottelson, "Nuclear Structure", W.A. Benja-
    min Inc., Reading, Mass., Vol.II (1975).
3. R. Bengtsson and S. Frauendorf, Nucl. Phys. A327:139 (1979).
4. I. Hamamoto, NORDITA preprint 81/28 to appear in Heavy Ion
    Science, ed. A. Bromley, Plenum Publ. Co.
5. S. Cohen, F. Plasil, and W. Swiatecki, Ann. Phys. (N.Y.)
    82:557 (1974).
6. R. R. Diamond and F. S. Stephens, Ann. Rev. Nucl. Part. Sci.
    30:85 (1980); and Nature   (1984).
7. M.J.A. de Voigt, J. Dudec and Z. Szymanski, Rev. of Mod.
    Phys. 55:949 (1983).
8. J. D. Garrett, G. B. Hagemann, and B. Herskind, Nucl. Phys.
    A 400:113c (1983).
9. J. D. Garrett, G. B. Hagemann, and B. Herskind, Comm. on
    Nucl. and Part. Phys. XIII:1 (1984).
10. J. D. Garrett, Nucl. Phys. A421:313c (1984).
11. B. Herskind, Nuclear Structure and Heavy Ion Dynamics, Soc.
    Italiana de Fisica, Bologna (1984) p.68.
12. B. Herskind, Proc. International Conference on Nuclear Phy-
    sics, Florence, Vol.II, 117-144. (1984).
13. C. Schuck, N. Bendjaballah, R. M. Diamond, Y. Ellis-Akovali,
    K. H. Lindenberger, J. O. Newton, F. S. Stephens, J. D.
    Garrett, and B. Herskind, Phys. Lett. 142B:253 (1984).
14. R. Chapman, J. C. Lisle, J. N. Mo, E. Paul, A. Simcock, J.
    C. Willmott, J. R. Leslie, H. G. Price, P. M. Walker, J. C.
    Bacelar, J. D. Garrett, G. B. Hagemann, B. Herskind, A.
    Holm, and P. J. Nolan, Phys. Rev. Lett. 51:2265 (1983).
15. B. Herskind, Annals of the Israel Physical Society, Vol.7,
    ed. by G. Goldring and M. Hass (1984) 33-28.
16. J. C. Bacelar, C. Ellegaard, G. B. Hagemann, B. Herskind, A.
    Holm, C.-X. Yang, P. O. Tjøm, and J. C. Lisle, Proc. Fifth
    Nordic Meeting on Nuclear Physics, Jyväskylä, Finland
    (1984), and to be published.
17. G. Sletten, S. Bjørnholm, J. Borggreen, J. Pedersen, P.
    Chowdhury, H. Emling, D. Frekers, R. V. F. Janssens, T. L.
    Khoo, Y. H. Chung, and M. Kortelahti, Phys. Lett. 135B:33
    (1984).
18. O. Häusser, H.-E. Mahnke, J. F. Sharpey-Schafer, M. L.
    Swanson, P. Toras, D. Ward, H. R. Andrews, and T. K. Alex-
    ander, Phys. Rev. Lett. 44:132 (1980).
19. J. Simpson, M. A. Riley, J. R. Cresswell, P. D. Forsyth, D.
    Howe, B. M. Nyako, J. F. Sharpey-Schafer, J. Bacelar, J. D.
    Garrett, G. B. Hagemann, B. Herskind, and A. Holm, Phys.
    Rev. Lett. 53:648 (1984).

20. B.M. Nyako, J.R. Cresswell, P.D. Forsyth, D. Howe, P.J. Nolan, M.A. Riley, J.F. Sharpey-Schafer, J. Simpson, N.J. Ward and P.J.T. Twin, Phys. Rev. Lett. 51:1858 (1984).

21. M. Diebel, Proc. Fifth Nordic Meeting on Nucl. Phys. Jyväskyla, Finland (1984). See also Nucl. Phys. A419:419 (1984).

22. H. C. Britt, B. H. Erkkila, P. D. Goldstone, R. H. Stokes, B. B. Back, F. Folkmann, O. Christensen, B. Fernandez, J. D. Garrett, G. B. Hagemann, B. Herskind, D. L. Hillis, F. Plasil, R. L. Ferguson, M. Blann, H. H. Gutbrod, Phys. Rev. Lett. 39:1458 (1977).

23. Reisdorf, Proc. of Int. Conf. on Fusion Below the Barrier (1984) Cambridge, M.A.

24. J. O. Newton, Phys. Scr. 24:83 (1981).

25. D. L. Hillis, J. D. Garrett, O. Christensen, B. Fernandez, G. B. Hagemann, B. Herskind, B. B. Back, and F. Folkmann, Nucl. Phys. A325:216 (1979).

26. B. Herskind, Proc. of Symposium on Macroscopic Features of Heavy Ion Collisions, Argonne, Ill. (1976) ANL Report/PHY 76-2, Vol.I, p.385.

27. L. Grodzins, S. Gazes, A. Smith, S. Steadman, E. Vulgaris, and J. Wiggins, Ann. of the Israel Phys. Soc. 7:227 (1984).

28. K. Siwek-Wilczynska, E. H. du Marchie van Voorthuysen, J. van Popta, R. H. Siemssen, and J. Wilczynski, Phys. Rev. Lett. 42:1599 (1979); R. H. Siemssen, Nuclear Structure and Heavy Ion Dynamics, Soc. Italiana de Fisica, Bologna (1984), p.

29. J. J. Gaardhøje, J. Beene, F. A. Dilmanian, J. D. Garrett, G. B. Hagemann, M. Halbert, D. C. Hensley, B. Herskind, A. Holm, M. Jäskeläinen, I. Y. Lee, T. Lindblad, P. Nolan, F. Plasil, H. Puchta, D. G. Sarantites, G. Sletten, and R. Woodward, Physica Scripta T5:178 (1983).

30. M. Halbert et al.?, Physica Scripta T5:91 (1983), and references therein.

31. G. B. Hagemann, R. Broda, B. Herskind, M. Ishihara, S. Ogaza, and H. Ryde, Nucl. Phys. A245:166 (1975).

32. P. O. Tjøm, I. Espe, G. B. Hagemann, B. Herskind, and D. L. Hillis, Phys. Lett. 72B:439 (1978).

33. P. Oblozinsky and R. S. Simon, Nucl. Inst. Meth. 223:52 (1984).

34. M. Jäskeläinen, D. G. Sarantites, R. Woodward, F. A. Dilmanian, S. T. Hood, R. Jäskeläinen, D. C. Hensley, M. L. Halbert, and J. H. Burker, Nucl. Inst. Meth. 204::385 (1983).

35. R.S. Simon et al. Proc. of Int. Symp. on Dynamics of NuclearCollective Motion Mt. Fuji (1982).

36. Sarantites et al., Proc. of the Conf. on High angular Momentum Properties of Nuclei, Oak Ridge Nov. 1982 ed. by N.Johnson (Harwood, NY 1983).

37. Metag et al. Proc. 1984 INS-RIKEN Symp. on HI Physics Mt. Fuji, August 27-31, 1984.
38. P. Twin, Proc. Int. Conf. Nucl. Phys. Florence (1983), Vol. II, 527-551.
39. C. Michel, F. Azgui, H. Emling, H. Grein, E. Grosse, H.J. Wollersheim, J.J. Gaardhøje and B. Herskind to be published.
40. G. Leander et al. Nobel Symposium.
41. Sagaria and Døssing, Phys. Lett. 96B:238 (1980).
42. I. Hamamoto, Phys. Lett. 143B:31 (1984).
43. O. Andersen, J.D. Garrett, G.B. Hagemann, B. Herskind, D.L. Hillis, and L.L. Riedinger, Phys. Rew. Lett. 43:687 (1979)
44. M.A. Deleplanque, F.S. Stephens, O. Andersen, J.D. Garrett, B. Herskind, R.M. Diamond, C. Ellegaard, D.B. Fossan, D.L. Hillis, H. Kluge, M. Neiman, C.P. Roulet, S. Shih and R.S. Simon, Phys. Rev. Lett. 45:172 (1980).
45. C. Ellegaard, M.A. Deleplanque, O. Andersen, B. Herskind, F.S. Stephens, R.M. Diamond, H. Kluge, C. Schuck, S. Shih and J.E. Draper, Phys. Rev. Lett. 48:670 (1982).
46. Th. Lindblad, A. Johnson, S.A. Hjorth, C. Linden, O. Andersen, M.A. Deleplanque, J.D. Garrett, B. Herskind and F.S. Stephens, Phys. Scr. 24:184 (1981).
47. K. Neergaard and V.V. Pashkevich, Phys. Lett. 59B:218 (1975)
48. I. Ragnarsson, T. Bengtsson, G. Leander and S. Åberg, Proceedings of Int. Conf. on Band Structure and Nuclear Dynamics, New Orleans (Feb. 1980), North-Holland Publishing Company (1980).
49. Z. Szymanski, J. Dudek, A. Majchofer and W. Nazarevicz, NBI Workshop (Risø) on Nuclear Structure at High Spin (May 1981).
50. G.B. Hagemann, J.D. Garrett, B. Herskind, G. Sletten, P.O. Tjøm, A. Henriques, F. Injebredsen, J. Reckstad, G. Løvhøjden and R.F. Thorsteinsen, Phys. Rev C, 25:3224 (1982).
51. G.B. Hagemann, Annals of the Israel Physical Society, Vol. 7 ed. by G. Goldring and M. Hass (1984) and references there in.
52. I. Hamamoto, Annals of the Israel Physical Society, Vol. 7 ed. by G. Goldring and M. Hass (1984).
53. J.O. Newton, B. Herskind, R.M. Diamond, E.L. Dines, J.E. Draper, K.H. Lindenberger, C. Schuck, S. Shih and F.S. Stephens, Phys, Rev. Lett. 46:1383
54. J.J. Gaardhøje, O. Andersen, R.M. Diamond, C. Ellegaard, L. Grodzins, B. Herskind, Z. Suykowski and P.M. Walker, Phys. Lett. 139B:273 (1984).
55. J.J. Gaardhøje, C. Ellegaard, B. Herskind and S.G. Steadman, Phys. Rev. Lett. 53:148 (1984).

56. J.J. Gaardhøje, C. Ellegaard, B. Herskind, M.A. Deleplanque, R.M. Diamond, E.L. Dines, A. Machiavelli and F.S. Stephens, Phys. Rev. Lett. (1985) in press.
57. J.J. Gaardhøje, XXII Int. Winter Meeting on Nuclear Physics, Bornio, Italy (1984) Vol. 2.
58. J.J. Gaardhøje,Proc. of the ERICE Summer School (1984).
59. K. Snover, Proc. of Masurian Summer School on Nucl. Phys. Mikolajki (1982) Poland.
60. K. Neergaard, Phys. Lett. 110B:7 (1982).
61. J.L. Egido and P. Ring, Nucl. Phys. A338:19 (1982) and Phys. Rev. C25:3339 (1982).
62. J. Bacelar, B. Herskind, G.B. Hagemann, J.D. Garrett, A. Holm, B. Lauritzen, J.C. Lisle and P.O. Tjøm to be printed.
63. R. Broglia, T. Døssing, B. Lauritzen and B. Mottelson, private communication.

CHAOS IN NUCLEI OR

STATISTICAL MECHANICS OF SMALL SYSTEMS: FLUCTUATIONS

Hans A. Weidenmüller

Max-Planck-Institut für Kernphysik

Heidelberg, W. Germany

ABSTRACT

A review of the empirical evidence for stochasticity in nuclear spectra and in nuclear reactions and its description in terms of random-matrix models is followed by a discussion of the connection between such stochasticity and the behaviour of systems which are the quantum analogues of classical chaotic systems. The comparison suggests that the fluctuations observed experimentally, and described in terms of random-matrix models, are universal. This then calls for a systematic approach towards the calculation of fluctuation properties from random-matrix models. It is shown that methods developed over the last few years in the theory of disordered systems combining functional integration with the replica trick and the loop expansion, or with the use of anticommuting (or Grassmann) variables, obey this requirement, and yield novel results. Some of these results are presented, and discussed.

1.   INTRODUCTION

The existence of ordered or regular motion in nuclei is well documented, well understood, and it forms the topic of many textbooks on the subject. The ordered motion is described by a variety of models, notably the shell-model, the collective model, and their various extensions. Grosso modo it seems fair to say that regular nuclear motion can be understood in terms of the mean-field approximation, and the oscillations about the mean field which give rise to collective excitation. The relationship between the mean field and the fundamental force has been the subject of many investigations, and is reasonably well understood both in general terms, and even numerically.

There is another aspect of the nuclear dynamics, perhaps equally important, but usually much less emphasized: The aspect of irregularity, of stochasticity, or of chaos. It is this aspect of the nuclear dynamics which forms the topic of the present lectures. After a brief review of the empirical evidence for stochasticity in nuclear spectra and nuclear reactions, given in section 2, arguments are presented to the effect that stochasticity as seen in nuclei is probably a universal phenomenon in microscopically small systems, and for that reason is of general interest (Section 3). This statement in turn calls for a universal theoretical method of dealing with stochasticity in nuclei (or other small systems). It is shown in section 4 that methods developed recently in field theory and statistical mechanics, in particular in the theory of disordered systems, provide such a method. Novel results obtained in this way are presented, and discussed.

The present lectures do not aim at giving a complete review, and do not have the character of a review article. They are intended to give the reader a survey of the main lines of development as perceived by the author. In order to enable the reader to widen his knowledge of the subject, I have made reference to review articles rather than the original literature wherever possible.

2.   STOCHASTICITY IN NUCLEI

Stochastic behaviour in nuclei is observed both in nuclear spectroscopy, and in nuclear reactions. The situation in nuclear spectroscopy was reviewed a few years ago by Brody et al. |1|, and more recently in ref. |2|. A recent analysis of all available data on nuclear energy levels is contained in ref. |2|, and another review in the talks by M.J. Giannoni at this school. Because of the existence of these reviews, I give in these notes only the minimum information necessary to establish the point of the argument. In nuclear reactions, stochasticity is manifest both in light-ion induced reactions of the compound-nucleus and of the precompound type, and in the dissipative aspects of heavy-ion induced reactions. The latter were reviewed several years back |3| and are therefore not explicitly taken up here; precompound reactions will be suppressed in order to make room for a somewhat extended description of compound-nucleus reactions. This is done because among the first novel results obtained with the method of section 4 is a complete analytical treatment of the compound nucleus.

2.1 Stochasticity in Nuclear Spectra

Stochasticity is observed by investigating the local fluctuations of eigenvalues of fixed spin and parity about the mean behaviour as given by the average level density. Sources of information are mainly sequences of typically 150-200 levels observed as resonances in low energy neutron scattering on heavy nuclei, and sequences of typically 60-80 levels observed as resonances in proton scattering near the Coulomb barrier on nuclei with mass numbers around 70. Needless to say, such experiments require very high energy resolution, a careful determination of the quantum numbers of the resonances seen, and a thorough search for levels with small widths which easily escape detection but when not found spoil the statistics.

Figure 1 shows the two features of central interest. The upper part of the figure contains a histogram of the distribution of spacings of neighbouring levels("nearest neighbour spacing distribution") for all nuclei combined. The characteristic features are: Nearly Gaussian fall-off for large spacings, and a linear rise for small spacings. The lower part of the figure shows, for the same data set, the Δ_3 statistics which is a measure of correlations between level spacings. This quantity is defined as follows. Let $\rho(E)$ be the actual level density, and $n(E) = \int d\varepsilon \rho(\varepsilon)$ be the number of levels with energy below E. Obviously, $n(E)$ is a staircase function. The expression

$$\Delta_3(L,x) = \frac{1}{L} \min_{A,B} \int_x^{L+x} d\varepsilon \; \left| n(\varepsilon) - A\varepsilon - B \right|^2 \qquad (2.1)$$

is optimized with respect to A and B and measures the deviation of $n(\varepsilon)$ from a straight line. This deviation will be large if the spectrum contains lumps of small eigenvalue spacings, followed by lumps of large eigenvalue spacings; the deviation will be small if small spacings are regularly followed by large spacings and vice versa: Small values of $\Delta_3(L,x)$ are an indication of the "stiffness" of the spectrum. Averaging $\Delta_3(L,x)$ over initial points x of integration, one obtains the quantity $\Delta_3(L)$ plotted versus L in the lower half of Figure 1 .

The data shown in Figure 1 must be compared with the solid lines labelled GOE. The latter are the result of a specific model which embodies the assumption of zero information content of eigenvalue fluctuations. It is assumed that the nuclear Hamiltonian, written as a real and symmetric matrix $H_{\mu\nu} = H_{\nu\mu}$ with $1 \lesssim \mu, \nu \lesssim N$ and $N \gg 1$, contains matrix elements $H_{\mu\nu}$ which are uncorrelated random variables with a Gaussian probability distribution and zero mean value. Denoting mean values by a bar, we have

$$\overline{H_{\mu\nu}} = 0 \; ; \quad \overline{H_{\mu\nu} H_{\mu'\nu'}} = \frac{\lambda^2}{N} (\delta_{\mu\mu'} \delta_{\nu\nu'} + \delta_{\mu\nu'} \delta_{\nu\mu'}). \qquad (2.2)$$

These assumptions define an ensemble of matrices first introduced by Wigner and known as the Gaussian Orthogonal Ensemble (GOE). The term orthogonal arises because the distribution is invariant under orthogonal transformations: No basis is preferred over any other. Ensemble-averages of the nearest-neighbour spacing distribution, and of Δ_3 , can be worked out analytically for the GOE

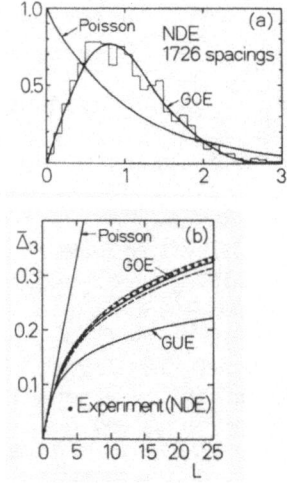

Fig. 1. Upper part: Nearest-neighbour spacing distribution as
determined from a total of 1726 spacings (histogram) and
GOE prediction versus x = s/d with s the actual, d
the mean level spacing. Lower part: The Δ_3 statistic
versus L as explained in the text and the GOE prediction.
For the latter, the dashed lines show the error bars due
to the finite size of the sample. The Poisson distribution
and the GUE distribution shown in the Figure have not been
discussed in the text. This Figure is taken from R.U. Haq,
A. Pandey and O. Bohigas, Phys.Rev.Lett. 48 (1982) 1086.

in the limit $N \to \infty$ and are given in Fig. 1 . The agreement with
the data is striking, particularly in view of the fact that the GOE
is parameter-free. (The quantity λ appearing on the r.h.s. of
Eq. (2.2) defines the average level density, but does not affect
the fluctuations.) Since the GOE embodies the assumption of zero
information content, it appears that (local) fluctuations of nuclear
eigenvalues carry zero information content and, in this sense, are
stochastic.

The GOE also predicts that the projection of an eigenvector
onto an arbitrary but fixed direction in Hilbert space has a Gaussian
probability distribution with mean value zero. Reduced partial neutron
widths measured in conjunction with the energies and spins of resonant
states seen in the scattering of slow neutrons on heavy nuclei, are

essentially squares of matrix elements. Each matrix element involves the wave function of a quasibound state. Given the GOE prediction, we expect the matrix elements to have a Gaussian distribution, or their squares to be Porter-Thomas distributed. Again, this expectation is beautifully borne out by the data.

We know that the GOE is unrealistic. For example, a shell-model calculation in the s-d shell using a residual two-body interaction involves 63 different two-body matrix elements (distinguished by the single-particle orbits and the total spin and isospin of the pair of nucleons). Even if we view these 63 matrix elements as independent random variables, the number 63 is very much smaller than the number $\frac{1}{2}N(N+1)$ of independent random variables in the GOE, with N in the middle of the s-d shell being of the order of several thousand. (We thus understand why the GOE yields for the average level density an energy dependence which is totally unrealistic.) Numerical evidence suggests, however, that both eigenvalue and eigenvector <u>fluctuations</u> are the same for the GOE and for the "two-body random ensemble" (TBRE), are consistent with the data, and in this sense are universal. The GOE has the advantage of being analytically better tractable than the TBRE.

2.2 Stochasticity in Nuclear Reactions

For reasons explained above, we confine ourselves here to light-ion induced reactions and excitation energies less than about 10 MeV above neutron threshold, although the evidence for stochasticity is in no way confined to these reactions. We focus attention on <u>compound nucleus</u> reactions (as opposed to direct reactions). Right above neutron threshold, we encounter isolated resonances: The average level spacing d is large ($\sim$ 10 eV in heavy nuclei) compared to the average level width Γ ($\sim$ 1 eV). Because of the stochastic nature of the resonances, it is impossible to calculate the cross section for neutron scattering exactly as a function of energy. It is therefore meaningful to ask for the mean value and the variance of the cross section, both evaluated by averaging over an interval which contains many resonances.

With increasing excitation energy, both d^{-1} and Γ increase exponentially, so that a few MeV above neutron threshold we have d $\ll \Gamma$. This is the domain of strongly overlapping resonances $|4|$. Fig. 2 shows a typical cross section versus energy. The bumps are not isolated-resonance peaks, but are due to accidental coherent

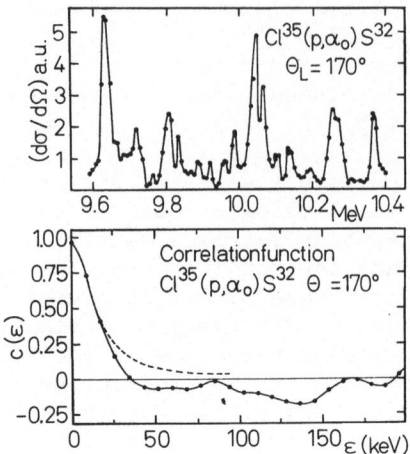

Fig. 2. Upper part: The differential cross section for the reaction
$^{35}Cl(p,\alpha)^{32}S$ leading to the ground state of ^{32}S versus
proton energy in the lab system, for a scattering angle of
170° (lab system). Lower part: The autocorrelation function
(2.3) as obtained from the data (dots connected by a solid
line)versus ε (in keV). The dashed line is a Lorentzian;
this is the theoretical expectation in the absence of
finite-range-of-data errors. This Figure is taken from
ref. |4|. The original paper was written by P. von Brentano,
J. Ernst, O. Häusser, T. Mayer-Kuckuk, A. Richter, and W. von
Witsch, Phys. Lett. 9 (1964) 48.

action of many overlapping resonances, while the valleys signal accidental cancellation. Just as for $d \gg \Gamma$, it is hopeless to calculate
the energy-dependence of the cross section exactly ($\sigma(E)$ is a random
function of energy), and instead one focusses attention on average
and variance. Another observable is the (normalized) autocorrelation
function $C(\varepsilon)$ of the cross section defined by

$$C(\varepsilon) = (<\sigma(E)\sigma(E+\varepsilon)> - <\sigma(E)>^2)/<\sigma(E)>^2 . \qquad (2.3)$$

For $\varepsilon = 0$, it is obviously related to variance and mean value; as
a function of ε it yields information about the correlation width

of the cross section. The lower part of Fig. 2 shows $C(\varepsilon)$ as calculated by averaging the data shown in the upper part in a suitable fashion over energy; this is the meaning of the brackets in Eq.(2.3). The FWHM of $C(\varepsilon)$ defines the correlation width Γ , and $\hbar/\Gamma$ is the mean life-time of the compound nucleus. The stochastic nature of the data shown in Fig. 2 can be ascertained by statistical checks $|4|$. For instance, the cross correlation function of cross sections measured at different scattering angles ascertains that the bumps in the upper part of Fig. 1 are not due to individual resonances with fixed spin and parity.

A theoretical description of the compound nucleus is obtained by generalising the random-matrix description of bound states sketched in section 2.1 . It is assumed that the fluctuations in the cross section for $\Gamma \gg d$ as well as for $\Gamma \ll d$ are due to compound-nucleus resonances, and that these resonances have stochastic features not different from those observed for isolated resonances. Accordingly, we consider a set of $N \gg 1$ bound states (all having the same spin and parity) labelled μ with $\mu = 1, \ldots, N$. These are coupled with each other by a GOE Hamiltonian $H_{\mu\nu}$, $1 \lesssim \mu, \nu \lesssim N$. To obtain a scattering problem, we also consider a set of Λ channels (of the same spin and parity) labelled a,b,c, ... The simplest model (to which we confine ourselves here) is obtained by neglecting all dynamical couplings between channels. Furthermore, we suppress the elastic scattering phase shifts which would only produce trivial modifications of the formulae. The channels are coupled to the levels by real matrix elements $W_{\mu a}$. For simplicity we neglect the energy dependence of the $W_{\mu a}$ due to threshold effects and due to the oscillatory nature of the channel wave functions. It is consistent with this approximation to neglect the shift functions $P \int dE'(\sum_a W_{\mu a} W_{\nu a}) \cdot (E-E')^{-1}$ (with P the principal value) which appear in the evaluation of the scattering matrix. We emphasise that the $W_{\mu a}$ are not random variables but fixed matrix elements. For the model just defined, the S-matrix S_{ab} is given by $|5|$

$$S_{ab}(E) = \delta_{ab} - 2i\pi \sum_{\mu,\nu} W_{\mu a}(D^{-1})_{\mu\nu} W_{\nu b} ; \qquad (2.4)$$

the propagator $D_{\mu\nu}$ has the form

$$D_{\mu\nu} = E\delta_{\mu\nu} - H_{\mu\nu} + i\pi \sum_a W_{\mu a} W_{\nu a} . \qquad (2.5)$$

The energy is labelled E . By definition of the GOE we actually deal with an ensemble of S-matrices rather than with a fixed S-matrix. By construction, $S_{ab}(E)$ is a random function of energy. We aim at calculating the ensemble-averages $\overline{S_{ab}}$ and $\overline{S_{ab}(E_1) S^*_{cd}(E_2)}$. By means of an ergodic theorem not to be discussed here, these are equal to the energy averages $\langle S_{ab} \rangle$ and $\langle S_{ab}(E_1) S^*_{cd}(E_2) \rangle$ for $N \to \infty$. The latter are related to the observables as follows. Aside from kinematical and geometrical factors not introduced here, the cross section is proportional to $|S_{ab}(E)-\delta_{ab}|^2$. It is customary to write S as the sum of a smooth and of a fluctuating part,

$$S_{ab}(E) = <S_{ab}> + S_{ab}^{fl} . \tag{2.6}$$

The smooth part $<S_{ab}>$ of the scattering matrix is related to the fast part (short time duration) of the reaction and describes direct reactions (a ≠ b) and optical-model scattering (a = b). For peda-gogical reasons, we simplify the problem by assuming that direct reactions are absent so that $<S_{ab}>$ is diagonal. [This consistent with the neglect of dynamical couplings between channels in the model of Eq. (2.4).] The average elastic S-matrix $<S_{aa}>$ is given in terms of a suitable optical-model potential. We use this phenomeno-logically given quantity $<S_{aa}>$, $a = 1, ..., \Lambda$ and the equality $<S_{aa}> = \overline{S_{aa}}$ to define the input parameters of the model (2.4). In the limit $N \to \infty$, these are the matrix elements $W_{\mu a}$ and the strength λ of the <u>GOE</u> Hamiltonian (2.2). This may seem a fearfully large number of parameters. It turns out, however, that λ is given in terms of the average level spacing d. . Moreover, orthogonal in-variance of the GOE implies that the $W_{\mu a}$ appear in the observables only in the form $\sum_\mu W_{\mu a} W_{\mu b}$ and therefore need not be specified individually. Diagonality of S implies $\sum_\mu W_{\mu a} W_{\mu b} = \delta_{ab} \sum_\mu W_{\mu a}^2$. The remaining number of parameters, $\sum_\mu W_{\mu a}^2$ with $a = 1, ..., \Lambda$, is exactly equal to the number of average <u>S-matrix</u> elements $<S_{aa}>$. The relevant parameters of the model (2.4) are therefore determined uniquely in terms of $\overline{S_{aa}}$. The average cross section is proportional to $|<S_{aa}> - 1|^2 \delta_{ab} + <|S_{ab}^{fl}|^2>$. It is the aim of the statistical model to calculate the quantities

$$<|S_{ab}^{fl}|^2> = \overline{|S_{ab}^{fl}|^2} = \overline{|S_{ab} - \overline{S_{ab}}|^2} \tag{2.7}$$

in terms of the input given by $<S_{aa}> = \overline{S_{aa}}$. Since $|S_{ab}^{fl}|^2$ is a rapidly varying function of energy, the quantity (2.7) is identified as the compound-nucleus contribution to the average cross section. <u>If we succeed in</u> solving the model (2.4) and in evaluating $\overline{S_{ab}}$ and $\overline{S_{ab}(E_1) S_{cd}^*(E_2)}$, this quantity can be calculated. To evaluate the autocorrelation function (2.3), we obviously would have to calculate the mean value of a product of four S-matrix elements. It will be shown in section 4 that <u>some informat</u>ion on this quantity can already be gleaned from $\overline{S_{ab}(E_1) S_{cd}^*(E_2)}$ for $E_1 \neq E_2$.

<u>It is shown in</u> section 4 that the exact calculation of $\overline{S_{ab}}$ and $\overline{S_{ab}(E_1) S_{cd}^*(E_2)}$ defined by the model (2.4) is possible for any number of channels in the limit $N \to \infty$ if modern techniques of field theory and statistical mechanics are used. Prior to the introduction of these techniques, analytical answers to the problem just defined were available only in the limits, $\Gamma << d$ and $\Gamma >> d$. For details, the reader is referred to ref. |6| .

2.3 Discussion

It was emphasised above that spectral fluctuations and compound-nucleus scattering do not afford the only evidence for stochasticity in nuclear systems. While in some sense (section 4) the compound nucleus can be viewed as an equilibrated system, precompound reactions allow the investigation of the nuclear system on its way towards equilibrium. Deeply inelastic heavy-ion reactions with their characteristic diffusion-like transfer of energy, angular momentum and mass provide other examples of equilibration processes. In addition, these systems show large-amplitude collective dynamics: A few degrees of freedom (c.m. distance, shapes of the two nuclei and the combined system, mass distributions) seem to suffice for a dynamical description. This suggests that the relaxation times associated with these collective variables are longer than those for the remaining degrees of freedom (which seem to act as a heat bath). More recent studies of heavy-ion reactions at several 10 MeV per nucleon reviewed in this school by M. Lefort point to the existence of yet other equilibration phenomena. Pion production in heavy-ion collisions has recently been described in terms of a compound-nucleus-like-process. Fission induced by projectiles of mass 12 or 16 appears to show features of a delayed diffusion process as discussed also by M. Lefort. All this evidence was not presented. Spectral fluctuations and compound-nucleus processes were singled out not only because they are so far the only features to which the calculations of section 4 were applied, but also because the stochasticity displayed there is observed in a system in which all quantum numbers have been determined. It is often argued that stochasticity is only apparent: One measures only inclusive processes, and the ensuing loss of information leads to dissipation-like behaviour in the few degrees of freedom actually investigated. The lectures of R. Balian in this school show clearly the origin of this mechanism. In contrast to precompound and heavy-ion reactions, this argument cannot be applied to the observation of a series of eigenvalues and eigenvectors (section 2.1), or to cross-section fluctuations found with complete resolution of all quantum numbers in incoming and exit channels (section 2.2). [In the latter case and the example given in Fig. 2, one actually averages over the spin projection in the incoming channel. Nobody has ever argued, however, that this is the origin of the phenomenon.] The data discussed in sections 2.1 and 2.2 therefore show that stochasticity is a phenomenon <u>inherent</u> in the nuclear dynamics that has to be understood as such.

How can the stochasticity inherent in the nuclear dynamics be reconciled with the regular nuclear motion as described by the mean-field approach and its extensions? Partial answers in select cases exist: The interplay of direct reactions or of isolated resonances (doorways) [both regular features] with a set of compound nuclear resonances [which are stochastic] can be described by a suitable extension of the model (2.4) and is largely understood. It is found

127

that these regular features induce correlations among fluctuating
observables which would be uncorrelated in their absence |5| .
More generally, regular dynamical features described by the various
nuclear models may, from the point of view of a statistical description, be said to define average values of and correlations
between fluctuating observables. However, no consistent nuclear
theory has yet been formulated that would encompass both the mean-
field approach and fluctuating properties in a coherent fashion and
as an outflow from the underlying forces between nucleons. Random-
matrix models are clearly no more than a phenomenological substitute
for such a theory the development of which in my opinion constitutes
one of the most challenging problems facing nuclear physics.

3.   CLASSICAL CHAOTIC MOTION AND NUCLEAR STOCHASTICITY

    Going beyond the phenomenological description of nuclear
stochasticity in terms of random-matrix models, we ask: Which properties of the nuclear Hamiltonian are responsible for the stochastic
features ? Is the phenomenon of stochasticity confined to nuclei,
or is it perhaps universal ?

    A fully satisfactory answer to these questions is not yet
available. On the basis of experimental evidence from outside nuclear
physics, and of numerical evidence concerning stochastic features of
certain simple dynamical models, a surmise can be formulated, however.
The evidence just referred to is discussed in detail in ref. |2|, see
also the lectures by J.M. Giannoni in this school. I therefore confine myself once again to a very brief review of the theoretical
arguments, leaving aside the data accumulated in atomic and molecular
physics altogether.

    Over the past 15 years or so, the study of chaotic classical
systems has made astonishing progress. These are systems (governed
by Hamilton's equation of motion) which are not integrable and for
which, moreover (and somewhat qualitatively speaking), almost all
trajectories originating in neighbouring points of phase-space
diverge exponentially. Such systems give rise to an interesting
dichotomy: Mathematical existence and uniqueness theorems establish
the unique existence of a trajectory for fixed initial conditions.
Physically speaking, these systems are fully deterministic. On the
other hand, it is impossible to calculate the trajectories: Analytical methods cannot be used because of the non-existence of integrals
of the motion, and numerical methods fail because every rounding
error leads to an exponential divergence of the calculated trajectory
from the correct one. A useful description of such systems starts,
for example, from a set of points in phase-space (rather than a
single point) and describes the time-evolution of this set by a
diffusion-like equation: The system tends to fill phase-space uniformly in an irreversible fashion, it behaves in a stochastic
fashion. It is clear that such systems are of great interest for

the foundations of classical statistical mechanics. (They essentially obey Boltzmann's hypothesis of molecular chaos.) In the present context, however, we are more interested in the quantum analogues of such classical chaotic systems and the possible connection with nuclear stochasticity. Unfortunately, nothing is known analytically and in general about the quantum analogues of classical chaotic systems. The available information comes from the study of simple systems.

Chaoticity can be present already in classical systems with two degrees of freedom. In fact, it appears that classical chaos is not primarily determined by the number of degrees of freedom, but by properties of the Hamiltonian of the system. A celebrated example is Sinai's billiard: Consider a point-particle moving freely in two dimensions within the confines of an elastically reflecting rectangular box. (This system is trivially integrable.) Put a circle (the diameter is smaller than either side of the rectangle) inside the box and assume that the particle is also reflected elastically by the (infinitely heavy) circle. If the particle moves in the area between circle and enclosing box, we have Sinai's billiard, a fully chaotic classical system. Quantisation of this system is straightforward and yields the force-free Schrödinger equation with the boundary condition that the wave function be zero on the surfaces of box and circle. Analytical statements are not available about the eigenvectors and eigenvalues even of such a simple system. Numerical calculations exist, however, the most thorough of these being due to the work of Bohigas, Giannoni, and Schmit. Figure 3 due to these authors is arranged in complete analogy with Figure 1. This time, however, the histogram in the upper part and the points with error bars in the lower part of the figure both result from the properties of 740 consecutive eigenvalues (belonging to wave functions with the same symmetry class) of Sinai's billiard and not from the nuclear data set. The agreement with the GOE (and, by virtue of Fig. 1, with the nuclear data set) is statistically significant, impressive, and convincing. It suggests the following surmise: The stochastic properties of nuclei are due to the fact that the nucleus is the quantum analogue of a classical chaotic system. Just as classical chaotic systems are characterised by the instability of their trajectories under small perturbations, so are quantum analogues of classical chaotic systems characterised by the analogous instability of their eigenvalues and eigenfunctions. Nuclear stochasticity is the manifestation of this instability.

The surmise is based on limited numerical evidence only; much more work is obviously needed to establish it firmly. Taken for granted, the surmise establishes stochasticity as a generic feature of microscopic systems (i.e., systems so small that properties on the scale of an average level spacing can be experimentally studied and/or are significant for the behaviour of observables.) This shows that the nuclear properties discussed in this lecture probably are

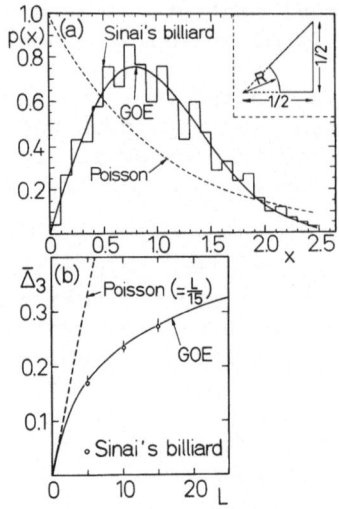

Fig. 3. This figure is arranged as Figure 1, the difference being
that the "data" leading to the histogram in the upper part,
and to the dots with error bars in the lower part, are now
taken from a series of 740 consecutive eigenvalues of
Sinai's billiard. This Figure is taken from O. Bohigas, M.J.
Giannoni and C. Schmit, Phys. Rev. Lett. 52 (1984) 1.

universal and therefore of general physical interest, and explains
the second title of these lectures.

    The insight that nuclear stochastic behaviour is probably
generic makes the need for a unified treatment and understanding of
regular and stochastic nuclear features even more pressing. In the
absence of noticeable progress in this direction, we turn to a more
limited goal. Figures 1 and 3 show that eigenvalue fluctuations for
quantum analogues of classical chaotic systems (and, presumably,
eigenvector fluctuations as well) are satisfactorily described by
the GOE. If these features are generic, we expect that other
"reasonable" random-matrix models will do well likewise. In order
to obtain at least a phenomenological account of the influence of
nuclear stochasticity on observables, we must be in a good position
to work with random-matrix models. Section 4 discusses the new tools
available for this purpose.

4. DISORDERED SYSTEMS AND THE GOE: THE CASE OF COMPOUND-NUCLEUS REACTIONS

The analytical treatment of the GOE has, mainly through the effort of Dyson, Mehta, and Gaudin, been pushed very far, and is nearly complete |1|. The analytical understanding of the TBRE and the related embedded Gaussian Orthogonal Ensembles is less advanced although considerable progress has been made by French and his collaborators |1|. The methods used in either problem find only very limited application in nuclear reaction problems like that of Eq. (2.4) |5|. A method of general applicability in both spectroscopy and reaction theory, useful preferably for a variety of ensembles, is thus called for. The method described below appears to have these features. In describing it, we focus attention on the problem formulated in Eq. (2.4) which receives a complete solution for the first time. We only mention in passing that novel results have also been obtained in statistical spectroscopy. Before explaining the method, we remind the reader of some problems in the theory of disordered systems. (We take the Anderson model as an example.) This is done in order to show where the methods come from, and to suggest that these problems have features in common with the nuclear problem, making the usefulness of the new methods in nuclear physics a little less astonishing.

A full account of the methods used would far exceed the scope of these lectures. I have confined myself to the elucidation of a few points which in my understanding are central, and which can be described without going through the entire algebra of the problem. It is my intention to whet the reader's appetite for more information, not to enable him to fill in the gaps by himself.

4.1 The Anderson Model for Disordered Systems

A completely regular metallic lattice in which the electrons interact with the lattice points but not with one another gives modulated plane waves (Bloch functions) for the electrons. This system has infinite conductivity. A finite conductivity results from interstitial atoms and other irregularities in the lattice. If the density of these irregularities is sufficiently small, perturbative methods can be used to calculate the conductivity. But what if irregularity is present everywhere, and not just a small perturbation: Will conductivity persist or not ? To study this question, Anderson proposed the model sketched in Fig. 4 for the case of two dimensions. Given a regular lattice of cubic symmetry, we allow the electrons to hop from any one site to the neighbouring lattice sites (arrows). The hopping matrix elements

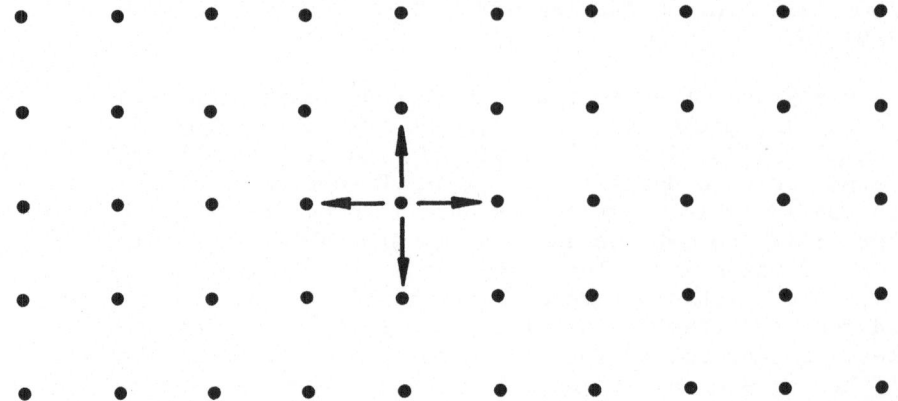

Fig. 4.   Sketch of the Anderson model in two dimensions. The dots
indicate lattice sites, the arrows hopping matrix elements.
The electron's energy at each site is determined by drawing
numbers at random from a given probability distribution.

are the same for the entire lattice. This perfectly regular model
is made irregular by simulating the chemical composition as follows.
The electron's energy at each lattice site is not the same everywhere,
but instead is drawn for each site at random from a given probability
distribution, usually a Gaussian.

In 1979, this model was modified by Wegner |7| who introduced
instead of a single state per lattice site a set of N such states
with stochastic features, and considered the limit N → ∞ . This
brought about a considerable technical advance.

Without going into the literature of the Anderson model (which
is vast) or into its solution (which is still not completely known)
I only mention that it is but one of a large class of models to
describe other disordered systems such as spin glasses etc. Returning
to Fig. 4 we see that Wegner's generalization of the Anderson model,
considered for a lattice with but a single site, is related to the
nuclear problem: N levels coupled by a stochastic Hamiltonian.
In some sense, the nuclear problem is thus an Anderson model of
dimension zero. The methods developed for disordered systems should
therefore be particularly powerful for nuclei, since zero-dimensional
problems are always simpler. This is indeed the case, and this is the
reason why previously intractable nuclear problems can be solved
exactly with these methods.

Interestingly enough, the benefits are not confined to nuclear physics: Testing approximation methods developed in the theory of disordered systems against the exact solution obtained in the nuclear physics context, one obtains a better understanding of the possibilities and limitations of these approximation schemes.

4.2 Compound-Nucleus Scattering

The reader interested in the methods sketched in the present section is referred to the following material: The structure of the saddle-point manifold, and the associated question of the form and convergence of the Hubbard-Stratonovitch transformation, which both are central to the entire approach, is discussed in detail in the work of Schäfer and Wegner |8|. Grassmann variables have found application in many areas of physics, notably in the grand unified theories. A review in the context of disordered systems is given in ref. |9| . The present section is based on work by Verbaarschot, Zirnbauer, and the author |10| .

Why is the evaluation of the model (2.4) (that is the calculation of $\overline{S_{ab}}$ and of $\overline{S_{ab}(E_1) \, S^*_{cd}(E_2)}$ so difficult ? The reason, which applies equally to most observables and most random-matrix models, lies in the appearance of the random variables $H_{\mu\nu}$ in the propagator $D_{\mu\nu}$, that is, in the denominator which makes the calculation of averages for $N \gg 1$ extremely cumbersome.

Progress is made by using functional integrals and by writing the observable of interest as the derivative of a quantity which carries the random variables in the exponent. The S-matrix is written as

$$S_{ab} = \delta_{ab} - i \, \frac{\partial}{\partial J_{ab}} \, \ln Z(E,J) \, \Big|_{J_{cd} = 0} \text{ (all c,d)} \qquad (4.1)$$

where the generating function is given by

$$Z(E, J) = \text{const} \int_{-\infty}^{\infty} \left(\prod_{\mu}^{N} d \, S_\mu \right) \exp \{ \mathcal{L} (\phi) \} . \qquad (4.2)$$

The S_μ with $\mu = 1, \ldots, N$ are real integration variables. The "Lagrangian" $\mathcal{L}$ is given by

$$\mathcal{L} (\phi) = \frac{1}{2} i \sum_{\mu,\nu} S_\mu \, D^J_{\mu\nu} \, S_\nu \qquad (4.3)$$

with

$$D^J_{\mu\nu} = D_{\mu\nu} - 4\pi \sum_{ab} W_{\mu a} \, J_{ab} \, W_{\nu b} . \qquad (4.4)$$

The last term in Eq. (4.4) is the "source term"; logarithmic differentiation with respect to the source of $Z(E,J)$ yields the observable. Different observables for the same ensemble are obtained by keeping the formal structure of Eqs. (4.1) to (4.4) the same and by changing the source term appropriately. Aside from the constant factor in Eq. (4.2) which is immaterial because of the logarithmic derivative in Eq. (4.1), the normalization of $Z(E,0)$ contains the factor $\left[\det(D_{\mu\nu})\right]^{-1/2}$. This is the standard factor which appears in the evaluation of a Gaussian integral of the form (4.2). Although the calculation of the ensemble average of $Z(E,J)$ is quite straight-forward (this is shown below), the calculation of $\overline{S_{ab}}$ is not easy because it involves the calculation of the average of $Z^{-1}(E,0) \left(\frac{\partial}{\partial J_{ab}} Z(E,J)\right)_{J=0}$. Two different ways exist to overcome this problem, and both have been used.

(i) The replica trick uses the identity

$$\ln Z = \lim_{n \to 0} \left[\frac{1}{n} (Z^n - 1) \right] . \tag{4.5}$$

The calculation of $\overline{\ln Z}$ is replaced by the calculation of $\overline{Z^n}$ for integer n . The method has two limitations: Taking the limit $n \to 0$ is possible only if $\overline{Z^n}$ is known analytically . Therefore, the method does not lend itself to a partially numerical approach. More importantly, the limiting procedure $n \to 0$ may be ill-defined if $\overline{Z^n}$ is only known for integer n , and it may lead to incorrect results. This is indeed what happens, as briefly mentioned below. Hence, care is required.

(ii) By introducing additional (anticommuting) integration variables, one normalizes $Z(E,J)$ such that $Z(E,0) = 1$. Then, the Eq. (4.1) involves only the ordinary derivative (as opposed to the logarithmic derivative) of $Z(E,J)$, and the calculation of the average is straightforward. The price one has to pay is the appearance of anticommuting integration variables throughout the entire calculation.

A description of this technique would be far beyond the scope of these lectures. Here, I only wish to show that anticommuting (or Grassmann variables) can indeed be used to normalize $Z(E,J)$ to unity so that their use appears plausible. We define an algebra of anticommuting variables $x,y,z, \ldots$ with $xy = -yx$ so that $x^2 = y^2 = \ldots = 0$. Differentials of these variables are also anticommuting, $dx\,dy = -dy\,dx$, $x\,dx = -dx \cdot x$ and $y\,dx = -dx\,y$. Integration over anticommuting variables is defined by the two conventions,

$$\int dx = 0 \qquad \text{and} \qquad \int x\,dx = (\sqrt{2\pi})^{-1} . \tag{4.6}$$

These conventions imply that

$$\int \exp \left\{ \sum_{i,k}^{N} X_i^* A_{ik} X_k \right\} dX_1^* \, dX_1 \ldots dX_N^* \, dX_N = (-2\pi)^{-N} \cdot \det A \tag{4.7}$$

where $X_1, \ldots, X_N$ and $X_1^*, \ldots, X_N^*$ are independent anticommuting integration variables. The reader is invited to verify Eq. (4.7) in the simplest nontrivial case $N = 2$; the calculation is simple and straightforward.

If now we multiply the square of the function (2.4) [which according to the above is proportional to $(\det D)^{-1}$] with another function formally defined in exactly the same way but with anti-commuting instead of commuting integration variables, the result is a function $Z^G(E,J)$ [where G stands for Grassmann] in which the factor $(\det D)^{-1}$ is cancelled by the factor $(\det D)$ coming from integration over anticommuting variables, see Eq. (4.7), and which can therefore trivially be normalised to unity for $J = 0$. A suitable derivative of this function yields $S_{ab}(E)$ by a relation analogous to Eq. (4.1) . To each level μ , $1 \leq \mu \leq N$, there are now associated four integration variables which we write as $S_\mu, T_\mu, X_\mu, X_\mu^*$. Here, S_μ and T_μ are commuting and X_μ, X_μ^* are anticommuting variables.

We use Eq. (4.1) and calculate $\overline{S_{ab}}$ and $\overline{S_{ab}(E_1)\, S_{cd}^*(E_2)}$. In the case of the replica trick, this amounts to calculating $Z^n(E,J)$ and $\overline{Z^n(E_1,J_1) Z^{*n}(E_2,J_2)}$ for integer n, n_1, n_2 . In the case of Grassmann variables, we calculate $Z^G(E,J)$ and $\overline{Z^G(E_1,J_1) Z^{*G}(E_2,J_2)}$. The averaging is simple in either case and uses the following identity. Let h be a Gaussian-distributed random variable with mean value zero. Then, $\overline{\exp(-h)} = \exp(\frac{1}{2}(\overline{h^2}))$. In the case of the replica trick, Z^n is an integral over the variables S_μ^k with $\mu = 1$, ..., N and $k = 1$, ..., n . The random matrix appears in the exponent in the form $-\frac{1}{2} i \sum_{k=1}^{} (\sum_{\mu,\nu} S_\mu^k H_{\mu\nu} S_\nu^k)$. By virtue of Eq. (2.2), the second moment of this expression is

$$-\frac{1}{2} \frac{\lambda^2}{N} \sum_{k,k'} (\sum_\mu S_\mu^k S_\mu^{k'}) (\sum_\nu S_\nu^k S_\nu^{k'}) . \tag{4.8}$$

While the calculation of the ensemble average is thus trivial (and this applies equally to the case of Z^G), the price we pay is the appearance of a polynomial of 4th order in the exponent which involves the integration variables S_μ^k . The integration over the S_μ^k cannot be done analytically.

This difficulty is circumvented by using the Hubbard-Stratono-vitch transformation. In its simplest form it reads

$$\exp \{- \frac{\lambda^2}{4N} \phi^4\} = \text{const.} \int_\infty^\infty d\sigma \{-\frac{N}{4} \sigma^2 - \frac{i}{4} \lambda\sigma\phi^2\} \tag{4.9}$$

The transformation (4.9) enables us to express the exponential of the term (4.8) as an exponential containing a bilinear form $(\sum_\mu' S_\mu^k S_\mu^{k'})$ of the integration variables only. This is done at the expense of introducing auxiliary integration variables. The number of such variables is determined by the fact that in Eq.(4.8), the variables S_μ^k appear only in the form $(\sum_\mu S_\mu^k S_\mu^{k'})$. (This is a consequence of the orthogonal invariance of $H_{\mu\nu}$.) It follows that the auxiliary integration variables can be arranged in the form of a symmetric matrix $\sigma_{kk'}$, with $1 \leq k,k' \leq n$; each element is real.

For Z^G the integration variables are S_μ, T_μ, X_μ, X_μ^* and evaluation of the 2nd moment leads in analogy to Eq. (4.8) to an expression of fourth order in these variables. The orthogonally invariant bilinear forms involved in this expression can be arranged in the form of a 4×4 matrix,

$$
\begin{matrix}
(\sum_\mu S_\mu^2) & (\sum_\mu S_\mu T_\mu) & (\sum_\mu S_\mu X_\mu) & (\sum_\mu S_\mu X_\mu^*) \\
(\sum_\mu S_\mu T_\mu) & (\sum_\mu T_\mu^2) & (\sum_\mu T_\mu X_\mu) & (\sum_\mu T_\mu X_\mu^*) \\
(\sum_\mu S_\mu X_\mu) & (\sum_\mu T_\mu X_\mu) & \emptyset & (\sum_\mu X_\mu X_\mu^*) \\
(\sum_\mu S_\mu X_\mu^*) & (\sum_\mu T_\mu X_\mu^*) & -(\sum_\mu X_\mu X_\mu^*) & \emptyset
\end{matrix}
\qquad (4.10)
$$

We have used $X_\mu^2 = 0$ and $X_\mu X_\mu^* = - X_\mu^* X_\mu$. Using the Hubbard- Stratonovitch transformation (4.9), we are led to introduce a number of integration variables which can also be arranged in the form of a 4×4 matrix. This matrix has the same symmetries as (4.10): The upper left hand 2×2 block is a symmetric matrix of commuting variables; the lower right hand 2×2 block is a matrix involving only a single non-vanishing commuting variable. Using the terminology of field theory, we refer to these blocks as to the "Boson-Boson" and "Fermion-Fermion" block, respectively. The remaining two 2×2 blocks are simply related by a symmetry transformation and therefore involve both the same four anticommuting integration variables. In all we have thus 4 commuting and 4 anticommuting σ-variables, and their number is completely determined by the orthogonal invariance of the GOE.

Having introduced the Hubbard-Stratonovitch transformation, we see that the integrals over the original variables S_μ etc. are now of Gaussian type and can be done trivially. Using the identity det C = exp { trace ln C } we can cast the result in the form

$$
\overline{Z^n (E,J)} = \text{const.} \int_{-\infty}^{\infty} (\prod_{\substack{kk'=1 \\ k \leqslant k'}}^{N} d\sigma_{kk'}) \exp\{-\frac{N}{4} \text{tr}_k \sigma^2 - \frac{1}{2} \text{tr}_{k\mu} C(\sigma)\}. \qquad (4.11)
$$

(A completely analogous expression is obtained for Z^G.) The matrix C carries indices (μ,ν) and (k,k'); the trace extends over both sets of indices. Integration over the original variables of integration S_μ etc. implies an interchange of the S_μ- and the σ-integrations. It is necessary to ascertain that the relevant integrals are sufficiently convergent to allow for this interchange; this condition is of paramount importance in the case of $S_{ab}S_{cd}^*$

and in fact uniquely determines the path of σ-integrations. This cannot be discussed here $|8|$.

The large dimension $N \gg 1$ of $H_{\mu\nu}$ now appears only in the form of the factor N , and of the trace $C(\sigma)$, in the exponent of Eq. (4.11). In the limit $N \to \infty$, Eq. (4.11) suggests that the integrand has an extremely narrow maximum, so that the integral can be evaluated using the saddle-point method. This is done by finding the extremum of the exponent, expanding the exponent around this point, and expanding the resulting exponential in a Taylor series, keeping only 2nd order terms in the exponent. From the fact that

$$\int_{-\infty}^{\infty} d\sigma \, (\sigma^{2\ell}) \exp\{-\frac{N}{4}\sigma^2\} \, d\sigma \; / \int_{-\infty}^{\infty} d\sigma \, \exp\{-\frac{N}{4}\sigma^2\} \; \alpha \; N^{-\ell} \qquad (4.12)$$

we see that this procedure generates an asymptotic expansion in powers of N^{-1} . It usually suffices $(N \to \infty)$ to keep only the term of lowest order.

This procedure yields $\overline{S_{ab}}$ straightforwardly. A complication of great interest arises in the calculation of $\overline{S_{ab}(E_1)S_{cd}^*(E_2)}$. We demonstrate its origin in the case of the replica trick with $n=1$; for $n > 1$ or the Grassmann case, the argument is analogous. The product $Z(E_1,J_1)Z^*(E_2,J_2)$ involves the random variables $H_{\mu\nu}$ in the exponent in the form $\frac{i}{2} \sum_{\mu\nu}(S_\mu^1 H_{\mu\nu}S_\nu^1 - S_\mu^2 H_{\mu\nu}S_\mu^2)$. The integration variables $S_\mu^1(S_\mu^2)$ refer to $Z(E_1,J_1)$ $(Z^*(E_2,J_2)$, respectively. By analogy with (4.8), the second moment of this expression is found to be

$$-\frac{1}{2} \frac{\lambda^2}{N} \sum_{\mu\nu} (A_{\mu\nu}^{11} - A_{\mu\nu}^{22})^2 \text{ where } A_{\mu\nu}^{ii} = S_\mu^i S_\nu^i , \; i = 1, 2. \qquad (4.13)$$

Since the second moment depends only on $(A^{11} - A^{22})_{\mu\nu}=S_\mu^1 S_\nu^1-S_\mu^2 S_\nu^2$, it is invariant under a transformation of the integration variables which has the form

$$\begin{pmatrix} S_\mu^1 \\ S_\mu^2 \end{pmatrix} \to \begin{pmatrix} \cos h\beta & \sin h\beta \\ \sin h\beta & \cos h\beta \end{pmatrix} \begin{pmatrix} S_\mu^1 \\ S_\mu^2 \end{pmatrix}, \text{ all } \mu , \qquad (4.14)$$

for $-\infty \le \beta \le +\infty$. For $E_1 = E_2$ and $W_{\mu\nu} = 0$, it is easily seen that $Z \cdot Z^*$ depends altogether only on the combination $(S_\mu^1 S_\nu^1 - S_\mu^2 S_\nu^2)$. Introducing new integration variables and choosing the variable β of the transformation (4.14) as one of them, we see that for $E_1 = E_2$ and $W_{\mu a} = 0$, the expression $\overline{Z \cdot Z^*}$ diverges! Convergence is reestablished by the terms which break the "hyperbolic symmetry" of the transformation (4.14), i.e. for $E_1 - E_2 \ne 0$ and $W_{\mu a} \ne 0$. Nevertheless, the singularity of $\overline{ZZ^*}$ at $E_1 = E_2$, $W_{\mu a} = 0$ must persist after the Hubbard-Stratonovitch transformation has been carried out. How can this be reconciled with the use of the saddle-point method discussed above ? Surely, a well defined saddle point cannot give rise to divergent integrals ?

It turns out that at $E_1 = E_2$, $W_{\mu a} = 0$ the saddle-point condition for $\overline{Z\,Z^*}$ yields not a single saddle-point, but rather an _infinity_ of saddle-points: A non-compact manifold. Integration over this manifold yields the singularity. The integrals are made convergent by the symmetry-breaking terms containing (E_1-E_2) or $W_{\mu c}$.

This insight suggests introducing new variables of integration which are adapted to the structure of the saddle-point manifold. One subset of the σ-integration variables denoted σ_M runs, intuitively speaking, in a direction normal to the saddle-point manifold in a direction where the integration is convergent because of a factor $-\frac{N}{4}\sigma_M^2$ in the exponent. These are the "massive modes" with a "mass" proportional to N. (In the calculation of $\overline{S_{ab}}$, _only_ massive modes appear!) As explained above, such integrals can be trivially evaluated using an asymptotic expansion in powers of N^{-1}, and keeping only the term of lowest order. The other subset of integration variables denoted by σ_G defines the points belonging to the saddle-point manifold; for $E_1=E_2$ and $W_{\mu a} = 0$, these correspond to massless modes ("Goldstone modes"). For $E_1 \neq E_2$ and/or $W_{\mu c} = 0$, the σ_G integrals can either be done exactly (this is possible for the Grassmann variables), or the ensuing finite mass of the σ_G-modes can be used to generate an asymptotic expansion in analogy with the N^{-1} expansion described above. In either case, the singularity at $E_1 = E_2$ and $W_{\mu c} = 0$ of $\overline{Z \cdot Z^*}$ is vital: It allows us to probe microscopic fluctuation properties on a scale given by $(E_1-E_2)/d$. This possibility would be obscured if the singularity were absent. Put differently, the fact that $\overline{S_{ab}}$ possesses a singularity-free expansion in powers of N^{-1} shows that this is a "macroscopic" observable: It cannot tell us anything about the fluctuation properties of the system. In contrast, $\overline{S_{ab}(E_1)\,S^*_{cd}(E_2)}$ is sensitive to variations of $(E_1-E_2)/d$ over a scale of order unity (rather than of order N with $N \to \infty$) and therefore _does_ contain information about microscopic fluctuation properties of the system.

In the specific case of $\overline{S_{ab}(E_1)S^*_{cd}(E_2)}$ it turns out that the "mass" of the Goldstone modes is given by

$$M_0 = \sum_c T_c + \frac{2i\pi}{d}(E_2-E_1) \quad . \tag{4.15}$$

Here, the transmission coefficients T_c are defined as usual by the unitarity deficit of the average S-matrix,

$$T_c = 1 - |\overline{S_{cc}}|^2 \quad . \tag{4.16}$$

The coefficients T_c measure the fraction of the flux which is absorbed, and which leads to compound-nucleus formation. The appearance of the transmission coefficients (rather than of the average S-matrix elements themselves) in the expression (4.15), and in the results given below, in itself is not accidental. It is a consequence of the symmetry properties of the saddle-point manifold and thus, eventually, of the orthogonal invariance of the GOE.

Returning to Eq. (4.15) we remark that the energy difference (E_2-E_1) appears in the expression $S_{ab}(E_1) \, S_{cd}^*(E_2)$ only via the quantity M_0. This shows immediately that fluctuations of the scattering matrix occur over a scale Γ given by

$$\Gamma = \frac{d}{2\pi} \sum_c T_c \, . \qquad (4.17)$$

This is the well-known expression for the correlation width of Ericson fluctuations. Here, we see that Γ governs the fluctuation properties of S_{ab} quite universally and not just in the Ericson limit $\Gamma \gg d$. Unfortunately, this insight is not sufficient to establish the correlation width of cross-section fluctuations. These are determined by the behaviour of $|S_{ab}(E_1)|^2 |S_{cd}(E_2)|^2$, and the average of a product of four S-matrix elements has not yet been worked out with our technique.

Before I discuss the results obtained in this way for compound-nucleus scattering, I inject some remarks on the validity of the replica trick. These were obtained recently by J.J.M. Verbaarschot and M.R. Zirnbauer |10| and are based on the fact that the nuclear physics problem can be solved exactly using Grassmann variables. Comparing this exact solution with results obtained via the replica trick, one is led to the following observations. For the one-point functions (the average S-matrix or, more generally, expressions involving a single energy argument) it appears that the replica trick yields always the correct asymptotic expansion in powers of N^{-1}. For the two-point functions (the average product $S_{ab}(E_1)S_{cd}^*(E_2)$ or other expressions involving two energy arguments taken at opposite sides of the real axis) the saddle-point manifold obtained by Grassmann integration is the product of a compact and non-compact manifold. The reason is that the σ_G variables have a matrix structure analogous to (4.10). The Boson-Boson block yields a non-compact saddle-point manifold, while the Fermion-Fermion block yields a compact one. Using the replica trick, one employs only commuting variables of integration. While the analytical structure of the answer is the same as for Grassmann integration, one now finds that the saddle-point manifold is purely non-compact, and does not contain the compact piece coming in the Grassmann case from the Fermion-Fermion block. Therefore, taking the limit $n \to 0$ after working out the answers analytically in the replica trick generally does not give the correct answer. The right answer is obtained, however, if one is interested only in the asymptotic expansion (in powers of M_0^{-1}, for the case of compound-nucleus scattering, for instance). The reason is that the asymptotic expansion of

$$\int_{-L}^{L} d\sigma \exp(- \frac{1}{2} M_0 \, \sigma^2 - \alpha \sigma^3 - \dots)$$

is the same for finite L , and for $L \to \infty$: Both are based on the assumption that $|M_0|$ is so large that the effective range of integration $|M_0|^{-1/2} \to 0$ and is therefore small compared to L . The difference between compact (finite L) and non-compact ($L \to \infty$) saddle-point manifold is therefore lost in the asymptotic expansion. It comes into play only when the physical parameter $|M_0|^{-1/2}$ is of the order of L .

The result obtained for compound-nucleus scattering is

$$S_{ab}(E_1) \, S^*_{cd}(E_2) = S_{ab} \cdot S^*_{cd} \, \delta_{ab} \, \delta_{cd} +$$

$$+ (\delta_{ac} \, \delta_{bd} + \delta_{ad} \, \delta_{bc}) \, f \, (T_a, \, T_b; \, \varepsilon; \, T_1, \, \ldots, \, T_\Lambda) \qquad (4.18)$$

$$+ \, \delta_{ab} \, \delta_{cd} \cdot S_{aa} \cdot S^*_{cc} \cdot g(T_a, \, T_b; \, \varepsilon; \, T_1, \, \ldots, \, T_\Lambda) \, .$$

Here, $\varepsilon = E_2 - E_1$ and the functions f and g are symmetric in the first two arguments and are symmetric in the last Λ arguments. Because of the Kronecker symbols, the function f contributes to both elastic and inelastic scattering, while the function g contributes to elastic scattering only. The replica trick yields asymptotic expansions of f and g in inverse powers of M_0, these have the form

$$f = \frac{T_a \cdot T_b}{M_0} + O(M_0^{-2}) \quad , \qquad\qquad (4.19)$$

$$g = \frac{T_a \cdot T_c}{M_0^2} + O(M_0^{-3}) \quad .$$

Keeping only the term of leading order, we have for $\varepsilon = 0$

$$|S_{ab}|^2 = |S_{aa}|^2 \, \delta_{ab} + (1 + \delta_{ab}) \, \frac{T_a \cdot T_b}{\sum\limits_c T_c} \, . \qquad\qquad (4.20)$$

This is the famous Hauser-Feshbach formula embodying Bohr's picture of an equilibrated compound nucleus: The decay of the compound nucleus is independent of its mode of formation, as is evidenced by the factorization of the last term. This factorization property is lost, when higher-order terms in the expansion (4.19) are taken into account, or when the exact expressions for f and g (see below) are used outside the domain $|M_0| \gg 1$. This is plausible: Equilibration is a thermodynamic concept. It is useful in the thermodynamic limit (Volume $\to \infty$ or level spacing $d \to 0$). In this limit, only the term (4.20) survives. Put differently, higher-order terms in the asymptotic expansion, when important, signal that we are not in the thermodynamic limit $\Gamma \gg d$.

Eq. (4.20) contains an "elastic enhancement factor" (a = b) of value two. While all contributions arising from f have this enhancement factor, contributions from g exist only for elastic scattering and tend to increase the elastic enhancement factor above its asymptotic value 2 attained in the thermodynamic limit.

We finally present the full formula for f (the function g looks similar):

$$f = \frac{1}{4} \int_{-1}^{+1} d\lambda \int_{1}^{\infty} d\lambda_1 \int_{1}^{\infty} d\lambda_2 \; \exp\left\{i\,\frac{\pi\varepsilon}{d}\left(\lambda - \frac{1}{2}\mu_1 - \frac{1}{2}\mu_1\right)\right\}$$

$$\cdot \prod_{c} \left[\frac{1 + \frac{1}{2}T_c(\lambda-1)}{\left[1+\frac{1}{2}T_c(\mu_1-1)\right]^{1/2}\left[1+\frac{1}{2}T_c(\mu_2-1)\right]^{1/2}}\right] \left(\frac{1-\lambda^2}{(\lambda-\mu_1)^2(\lambda-\mu_2)^2}\right)$$

$$\cdot \left\{ T_a \cdot T_b \left[\frac{2(1-\lambda^2)}{\left[1+\frac{1}{2}T_a(\lambda-1)\right]\left[1+\frac{1}{2}T_b(\lambda-1)\right]} + \frac{(\mu_1^2-1)}{\left[1+\frac{1}{2}T_a(\mu_1-1)\right]\left[1+\frac{1}{2}T_b(\mu_1-1)\right]}\right. \right.$$

$$\left. \left. + \frac{(\mu_2^2-1)}{\left[1+\frac{1}{2}T_a(\mu_2-1)\right]\left[1+\frac{1}{2}T_b(\mu_2-1)\right]}\right]\right\} . \tag{4.21}$$

The variables μ_1 and μ_2 are defined by writing $\lambda_1 = \cos h\, \theta_1$ and $\lambda_2 = \cos h\, \theta_2$ and by putting $\mu_1 = \cos h\,(\theta_1+\theta_2)$, $\mu_2 = \cos h\,(\theta_1-\theta_2)$. The structure of Eq. (4.21) is of some interest. The threefold integration reflects the essential structure of the saddle-point manifold, the noncompact manifold $1 \le \lambda_1, \lambda_2 \le \infty$ stemming from the Boson-Boson part, the compact manifold $-1 \le \lambda \le +1$ from the Fermion-Fermion part of the matrix σ_G. The dependence on $\varepsilon = E_2 - E_1$ is contained in the exponential. The product over all channels derives from the matrix elements $W_{\mu a}$ not contained in the source terms while the source terms give rise to the curly bracket. These terms have the required symmetry properties. The round bracket written after the product over all channels comes from integration over the massive modes.

We have checked that Eq. (4.21) together with an analogous expression for g satisfy, when used in Eq. (4.18), the unitary condition $\sum_b S_{ab} S_{bc}^* = \delta_{ac}$; this was done by numerical evaluation of the integrals. We have thereby also convinced ourselves that the expressions for f and g can easily be computed and are therefore practically useful. We have also shown that the asymptotic expansion generated from Eq. (4.21) agrees with the asymptotic expansion (4.19) obtained in the framework of the replica trick. All these indicators suggest that the result is correct. It thus appears that the Grassmann integration method yields a complete analytical solution to the problem of compound-nucleus scattering, a problem that has defied complete analytical treatment for three decades.

5. CONCLUSIONS

In these lectures, I have tried to establish three points.

(i)   Nuclei show inherent stochasticity. This stochasticity can be described successfully in terms of phenomenological random-matrix models.

(ii) The quantum analogues of classical chaotic systems show fluctuation properties of a stochastic nature. On the statistical significance level established so far, these fluctuation properties are indistinguishable from nuclear stochasticity. This suggests that nuclear stochasticity is a universal phenomenon encountered in all small systems which are quantum analogues of classical chaotic systems.

(iii) Methods developed recently in field theory and statistical mechanics, especially in the context of disordered systems, can be successfully used to work out the prediction of random-matrix models in nuclei and, presumably, in other small systems. Some long-standing problems can be solved in this way. Insight of a general nature is also gained in the validity of a frequently applied method, the replica trick.

Without my writing down a list of possibilities the reader will notice that the application of the techniques outlined in section 4 is not restricted to the cases treated so far, but opens fascinating possibilities in several areas of nuclear physics. These exciting prospects, absolutely worthwhile being explored, must not obscure the fact that an understanding of the interplay between regular and stochastic nuclear motion on a fundamental level cannot be reached by an ever so skilful handling of phenomenological random-matrix models, and remains a challenging task for the future.

REFERENCES

|1| T.A. Brody, J. Flores, J.B. French, P.A. Mello, A. Pandey, and S.S.M. Wong, Rev.Mod.Phys. 53 (1981) 385

|2| O. Bohigas and M.J. Giannoni, Lecture Notes in Physics Springer Verlag, Heidelberg, Berlin, New York, Tokyo (1984) Volume 209, 1

|3| H.A. Weidenmüller, in Progress in Particle and Nuclear Physics, Vol. 3 (1980) 49, Pergamon Press, Oxford, New York, Frankfurt, Paris

|4| T. Ericson and T. Mayer-Kuckuk, Ann.Rev.Nucl.Sci. 16(1966) 183

|5| C. Mahaux and H.A. Weidenmüller, Shell-model approach to Nuclear Reactions, North-Holland Publishing Co., Amsterdam 1969.

|6| C. Mahaux and H.A. Weidenmüller, Ann.Rev.Nucl.Part.Sci 29 (1979) 1

|7| F. Wegner, Phys.Rev. B19 (1979) 783

|8| L. Schäfer and F. Wegner, Z.Phys. B38 (1980) 113

|9| K.B. Efetov, Adv. Phys. 32 (1983) 53

|10| J.J.M. Verbaarschot, H.A. Weidenmüller and M.R. Zirnbauer,
Phys.Rev.Lett. 52 (1984) 1597
J.J.M. Verbaarschot and M.R. Zirnbauer, Ann.Phys. (N.Y.),
158 (1986) 78
H.A. Weidenmüller, Ann.Phys. (N.Y.), 158 (1986) 120
J.J.M. Verbaarschot and M.R. Zirnbauer, J.Phys. A
17 (1985) 1093
J.J.M. Verbaarschot, H.A. Weidenmüller and M.R. Zirnbauer,
Phys.Lett. 149B (1986) 263
J.J.M. Verbaarschot, H.A. Weidenmüller and M.R. Zirnbauer,
Physics Reports 129 (1985) 367

SPECTRAL FLUCTUATIONS AND CHAOTIC MOTION

Oriol Bohigas, Marie-Joya Giannoni and Charles Schmit

Division de Physique Théorique[+], Institut de Physique

Nucléaire, 91406 Orsay Cedex, France

1. INTRODUCTION

*For no one successfully investigates the nature
of a thing in the thing itself ; the inquiry must
be enlarged so as to become more general.*

Francis Bacon

It has been known since the early days of nuclear physics the existence of fine structure resonances to which are associated very long lifetimes ($\sim$ six orders of magnitude larger than the time it takes to a nucleon to traverse the nucleus). These compound nucleus resonances have been systematically studied at neutron threshold (~ 6 MeV excitation energy). It has been clear that a "microscopic" (a one by one) study of them was neither possible nor suitable but that an adequate goal was to attain a statistical description. One can distinguish two kinds of properties in a statistical description, global and local ones. An example of the first is the average density $\rho_{av}(E)$ of resonances or quasi-bound levels as a function of excitation energy E. Local spectral properties or spectral fluctuations are connected with the statistical description and characterization of the departures of the microscopic level density $\rho(E)$, a sum of spikes - $\rho(E) = \Sigma \delta(E-E_i)$ -, from its average $\rho_{av}(E)$. The average density $\rho_{av}(E)$ depends on specific nuclear properties like size, shell effects, pairing, etc., and although theories of nuclear level densities are successful at a semiquantitative level, a detailed agreement between theory and experiment is still lacking. In several respects the situation is quite different concerning

---

[+]Laboratoire associé au C.N.R.S.

spectral fluctuations. First, level fluctuations do not depend, for a given nucleus, on the excitation energy (are stationary) and they are presumably the same for all nuclei (except for scale, provided by the mean spacing between resonances). Secondly, there exists a parameter-free theory, the random matrix theory (RMT) which has been tremendously successful in predicting the observed fluctuations. It is most remarkable that the only information which is introduced in RMT are some very general symmetry properties like rotation and/or time reversal invariance of the system under study. However, RMT tells not much about why is it legitimate to substitute the actual nuclear Hamiltonian by an ensemble of stochastic matrices. One usually just invokes that the system should be complicated enough. But then, what is a complicated system ? Is it a heavy nucleus or also a light nucleus ? Or a complex atom ? How far can we go from complexity towards simplicity before the assumptions made in RMT break down ? Is there a clear separation between simple and complicated in our context ? How general are spectral fluctuations ; are they universal ? These are the sort of questions that have been recently discussed in a number of investigations. Attention has been focused on studies of classical dynamical systems, where the concepts of regular and irregular motion and how the onset of chaotic motion takes place play a fundamental role. Whether the notions that are relevant in the study of classical systems are also appropriate when studying a quantum system, like a nucleus, is one of the central questions at issue. And some connections between classical chaotic motion and quantum properties is one of the main outputs of these investigations.

We shall not attempt to give a review of all these topics. Recent reviews on RMT [BFF-81], chaotic motion [Be-83,Ro-85], and their connections [BG-84] are presently available. We rather prefer to give some of the general ideas, to discuss some recent results and to mention open questions.

2. SEPARATION OF AVERAGE BEHAVIOUR AND FLUCTUATIONS

Consider points distributed on the real axis, $E_1, E_2, E_3, \ldots$ The axis may be the energy, the points corresponding to the discrete levels of a quantum system (an atomic nucleus, an atom, a molecule); or the frequency axis, the points corresponding to the normal frequencies of a vibrating membrane ; or the time axis, the points corresponding to successive epochs of occurrence of a given event (times of arrivals on a telephone line in queuing problems, or times of decay of a radioactive source). Let $N(E)$ be the number of points which are below or equal to E. $N(E)$ is a staircase function which increases by one each time that one "crosses" a point. The question now is to separate the staircase function $N(E)$ in a smooth part $N_{av}(E)$ and the remainder will define the fluctuating part $N_{f\ell}(E)$

$$N(E) = N_{av}(E) + N_{f\ell}(E) \ . \tag{1}$$

The average density is given by $\rho_{av}(E) = (d/dE)N_{av}(E)$. Although we know of no general method, let us give several examples for which this separation has been achieved.

a) Prime Numbers

There is no stochasticity in the definition of prime numbers but nothing prevents to study some of their properties in a statistical way. In particular, for the average $N_{av}(E)$ of the staircase function $N(E)$ one can take the Riemann approximation to $N(E)$, which is an entire function of ℓnE with the following expansion

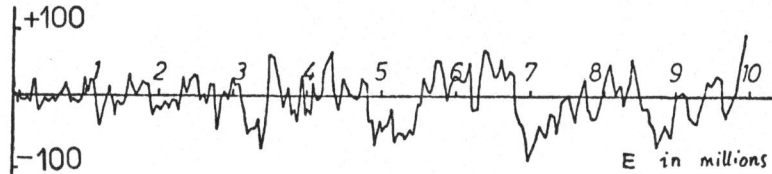

Fig.1 - Plot of $N_{f\ell}(E)$ for prime numbers in the internal $[0,10^7]$.
(Taken from Ref.[Za-77]

$$N_{av}(E) = R(E) = 1 + \sum_{n \geq 1}^{\infty} \frac{1}{n\zeta(n+1)} \frac{(\ell nE)^n}{n!} \quad ; \tag{2}$$

in (2) $\zeta(s)$ is the Riemann zeta function of $s = \sigma + it$, which can be defined by

$$\zeta(s) = \sum_{n \geq 1} \frac{1}{n^s} \tag{3}$$

for $\sigma > 1$ and by analytic continuation for $\sigma \leq 1$, $s \neq 1$. On Fig.1 is plotted $N_{f\ell}(E) = N(E) - R(E)$. As can be seen, no structure is present and only fluctuations are left out.

b) Zeros of the Riemann zeta function

One knows that there is an infinite number of zeros $\rho = \sigma + iE$ of $\zeta(s)$ on the critical line $\sigma = 1/2$. What are the statistical properties of the sequence of successive zeros on this line ? Take only the half line $\sigma = 1/2$, $E > 0$. One has, for the average number of zeros such that their imaginary part is less than or equal to E,

$$N_{av}(E) = \frac{E}{2\pi} \ell n \frac{E}{2\pi} - \frac{E}{2\pi} + S(E) + \frac{7}{8} + 0 \left(\frac{1}{E}\right) \ . \tag{4}$$

The maximum order of S(E) is probably

$$S(E) = 0 \left(\left(\frac{\ell nE}{\ell n \, \ell nE} \right)^{1/2} \right) \qquad (5)$$

c) Eigenmodes of a membrane

Consider the transverse vibrations of a membrane whose boundary B is fixed. The eigenmodes are given by the eigenvalues of the Laplacian

$$(\Delta + E)\psi = 0 \qquad (6)$$

with the condition that ψ vanishes on the boundary B. Eq.(6) possesses an infinite number of eigenvalues $E_1, E_2, .., E_n, ...$ which are real and non-negative and have no accumulation point. The smoothed function $N_{av}(E)$ giving the number of eigenvalues less than or equal to E is given by

$$N_{av}(E) = \frac{SE}{4\pi} - \frac{G\sqrt{E}}{4\pi} + K + 0 \, (E^{-\eta/2} \, \ell n \, \sqrt{E}) \qquad (7)$$

where $0 < \eta \lesssim 1$. In Eq.(7) S is the surface of the area inside the boundary B and G is the perimeter of the boundary. K is a constant term containing complex information on the geometrical and topological properties of the domain.

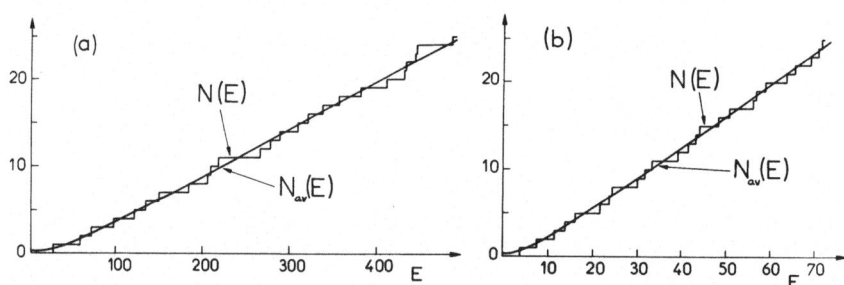

Fig.2 - Plots of N(E) and of $N_{av}(E)$ [see Eq.(7)] for two different boundaries : (a) a quarter of circle, (b) a stadium.

On Fig.2 are compared the exact function N(E) and the smooth function $N_{av}(E)$ given by Eq.(7) for two different shapes, namely a quarter of a circle and a stadium (boundary limited by two parallel straightline segments and two half circles). It can be seen that $N_{av}(E)$ indeed reproduces perfectly the average behaviour of N(E), starting from the bottom of the spectrum, even though Eq.(7) is an asymptotic expression.

d) Eigenvalues of the Schrödinger equation

Consider the Schrödinger equation

$$(\Delta - V(\underset{\sim}{r}) + E)\psi = 0 \qquad (8)$$

and suppose that $V(\underset{\sim}{r}) \to \infty$ at infinity so that the spectrum is discrete. The average number of eigenvalues $N_{av}(E)$ not exceeding E is given asymptotically by

$$N_{av}(E) = \frac{1}{2^d \, \pi^{d/2} \, \Gamma(d/2+1)} \int_{V(\underset{\sim}{r}) \leqslant E} \{E - V(\underset{\sim}{r})\}^{d/2} \, d\underset{\sim}{r} \,, \qquad (9)$$

where d is the dimensionality of the space. Notice that the leading term of (7) is given by (9) for d = 2. To illustrate how

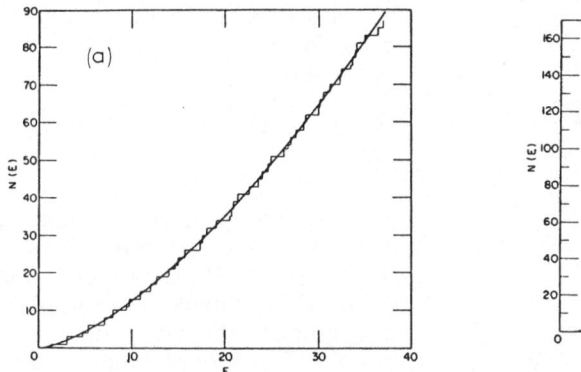 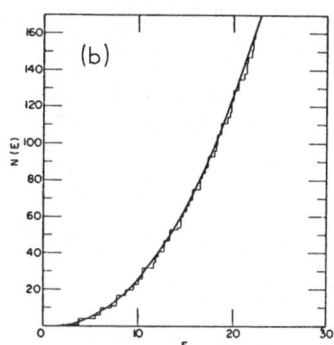

Fig.3 - Plots of N(E) and $N_{av}(E)$ [see Eq.(9)] for a quartic potential :
(a) d = 2 ; (b) d = 3. (Taken from Ref.[HMY-81])

good (9) reproduces the smooth behaviour of N(E), on Fig.3 is plotted N(E) as well as $N_{av}(E)$ for the particular case of the quartic potential

$$V(\underset{\sim}{r}) = \sum_{i=1}^{d} x_i^4 \qquad (10)$$

in two and three dimensions.

3. FLUCTUATION MEASURES

Before studying fluctuations one wants to get rid of $N_{av}(E)$ in order to characterize and compare fluctuation patterns of different systems whose corresponding average behaviours are not the same. For that purpose one can "unfold" the original spectrum $\{E_i\}$ through the mapping $E \mapsto x$

$$x_i = N_{av}(E_i) \qquad i = 1,2,... \qquad\qquad (11)$$

The effect of (11) is that the sequence $\{x_i\}$ has on the average a constant mean spacing (or a constant density) equal to unity, irrespective of the particular form of the function $N_{av}(E)$.

We are now in position to study the statistical laws governing sequences having very different origins, as illustrated on Fig.4, which is inspired from a similar figure of Ref.[BFF-81]. There are displayed six spectra, each containing 50 levels (the spectra have been rescaled to the same spectrum span [0,49], thereby introducing an artificial rigidity, see below). Column (a) corresponds to a Poisson system : Take a random variable s whose probability density $p(x)$ is e^{-x} . Construct a sequence $\{x_i\}$

$$x_1=0 \; , \quad x_{i+1}=x_i+s_i \quad i=1,2,3,.. \qquad (12)$$

where s_i are outcomes of independent trials of the variable s. The resulting spectrum is what is called a Poisson spectrum. Column (b) shows an example of a segment of prime numbers ; column (c) the resonance energies of the compound nucleus observed in the reaction $n+{}^{166}Er$; column (d) the eigenvalues (associated to eigenfunctions with given symmetry) corresponding to the transverse vibrations of a membrane whose boundary is Sinai's (see Section 4) ; column (e) the positive imaginary part of successive zeros of the Riemann zeta function on the critical line ; column (f) an equally spaced sequence of levels (picket fence). Columns (a) and (f) represent two limiting cases, maximum randomness and no randomness at all respectively.

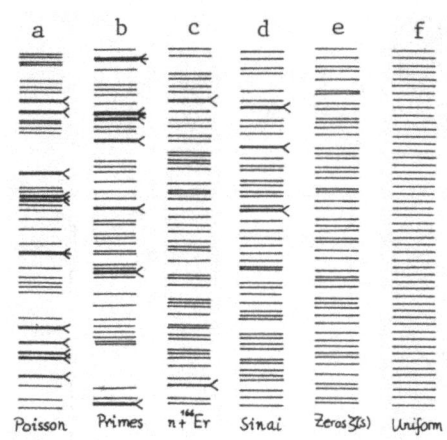

Fig.4 - Segments of "spectra" of different nature. See text for further explanation.

Can one deduce some features just by inspecting Fig.4 ? Arrows indicate spacings $s_i = x_{i+1}-x_i$ which are smaller than 1/4. The Poisson spectrum shows 12 arrows out of 49 spacings, the prime number "spectrum" shows 9 arrows, the Er spectrum only 2 arrows, the frequencies of the membrane 3 arrows, the zeros of $\zeta(s)$ no arrow and, of course, the picket fence no arrow. One therefore sees a statistical similarity between (a) and (b) : large probability of small spacings and occurrence of some large spacings. On the contrary, (c), (d) and (e) show small probability of small and large spacings, the

150

small probability of small spacings being usually referred to in the literature as *level repulsion*. The spectra (c), (d) and (e) deviate from (f) less strongly than (a) and (b). The picket fence (f) is a spectrum that we may qualify as absolutely rigid, in the sense that there is no departure at all from uniformity. Once the position of one level x_i is known, the position of any other level is determined, no matter how far it is from x_i. For this system the correlations between spacings are maximum and it shows perfect short and long range order. At the opposite extreme, the Poisson spectrum contains no correlations between spacings : the knowledge of a stretch of the spectrum puts no restriction on the behaviour of the spectrum beyond the interval considered. In intermediate situations between Poisson and the picket fence the degree of *spectral rigidity* will depend on the nature and strength of the correlations between spacings.

To characterize level fluctuations in a systematic way one deals with the k-level correlation functions $R_k(x_1,...,x_k)$ ($R_k(x_1,...,x_k)$ $dx_1...dx_k$ is the probability of finding one level within each of the intervals $[x_j, x_j+dx_j]$) and measures derived from them. We shall only refer to:

1) The spacing distribution p(x) between adjacent levels. If there is tendency to avoid level clustering (level repulsion) p(x) is small (or zero) for small values of x, in contrast to the Poisson case, for which the spacing distribution $p(x) = e^{-x}$ is maximum at the origin.

2) Quantities related to the number statistic n(L) : given an interval $[\alpha, \alpha+L]$ of length L, it counts the number of levels contained in the interval. The average value of n(L) is L, if the mean spacing is unity. We shall consider higher moments or cumulants of n(L) : variance $\Sigma^2(L)$, skewness $\gamma_1(L)$ and excess $\gamma_2(L)$. Qualitatively we expect that if the spectrum is stiff, the variance of n(L) will be small (in most cases the actual number of levels found in an interval of length L will differ only slightly from L) whereas for a non-rigid or compressible spectrum like Poisson the variance of n(L) will be comparatively large. For a Poisson spectrum one has $\Sigma^2(L) = L$ which tells nothing but the familiar result that in an interval of length L one expects to find $L \pm \sqrt{L}$ levels. At the opposite extreme, for the picket fence, one will have $L \pm 0$. Again we will be interested in what happens in intermediate situations. For a Poisson spectrum one has $\gamma_1(L)=1/\sqrt{L}$ and $\gamma_2(L)=1/L$.

3) The Δ_3 statistic of Dyson and Mehta. It measures, given an interval of length L, the least square deviation of the staircase N(x) from the best straight line fitting it. Its average is related to the variance $\Sigma^2(L)$ of n(L) by

$$\overline{\Delta}_3(L) = (2/L^4) \int_0^L (L^3 - 2L^2 r + r^3)\Sigma^2(r)dr \cdot \tag{13}$$

For a Poisson spectrum $\overline{\Delta}_3(L)$ increases linearly with L : $\overline{\Delta}_3(L)=L/15$. In general $\Sigma^2(L)$ and $\overline{\Delta}_3(L)$ can be expressed in terms of the 2-level correlation function, whereas $\gamma_1(L)$ ($\gamma_2(L)$) depends also on the 3-level (4-level) correlation function.

4. RANDOM MATRIX PREDICTIONS AND UNIVERSALITY OF FLUCTUATIONS

In RMT one considers the Hamiltonian matrix H as an $N \times N$ stochastic matrix (the matrix elements are random variables) ; the random matrix ensemble is specified by the probability density $\mathcal{P}(H)dH$. One is interested in asymptotic results valid for large N.

General underlying space-time symmetries obeyed by the system put important restrictions on the admissible matrix ensembles. We shall refer to results of the Gaussian Orthogonal Ensemble (GOE) and Gaussian Unitary Ensemble (GUE) which correspond to ensembles of real symmetric matrices and of hermitian matrices respectively. GOE applies when time reversal invariance is a good symmetry and GUE when it is not. Besides these general symmetry considerations, no other property of the system is taken into account to define the GOE and GUE.

GOE and GUE eigenvalue fluctuations are presently well known. For GOE, the spacing distribution is very well approximated by

$$p(x) \simeq (\pi/2)x \exp(-(\pi/4)x^2) \qquad (14)$$

which shows level repulsion and is very different from e^{-x} valid for a Poisson spectrum. The level repulsion is stronger for GUE than for GOE : in both cases $p(x)$ vanishes at the origin, but near the origin $p(x) \sim (\pi^2/6)x$ for GOE whereas for GUE one has $p(x) \sim (\pi^2/3)x^2$. For GOE the number variance $\Sigma^2(L)$ for $L \gtrsim 1$ is given by

$$\Sigma^2(L) = (2/\pi^2)\ln L + 0.44 \quad . \qquad (15)$$

In (15) the logarithmic increase is to be compared to the linear increase with L for a Poisson spectrum. One can speak of semicrystalline nature of the spectrum, showing long range order. The GUE spectrum is more rigid than a GOE spectrum : $\Sigma^2(L)$ is half the GOE value plus 1/8. For large L, the ensemble average of $\Delta_3(L)$ is given for GOE by

$$\overline{\Delta}_3(L) \sim (1/\pi^2)\ln L - 0.007 \quad . \qquad (16)$$

GUE shows also a logarithmic increase of $\overline{\Delta}_3(L)$ but with a smaller coefficient than in Eq.(16) (stronger rigidity).

How well do GOE predictions compare with experimental data ? We shall not give a complete account but rather present a few typical examples. The combined set of nuclear resonance - energy data of different nuclei - in short, the nuclear data ensemble (NDE) - has been treated as a sampling of eigenvalues of GOE matrices. Results are reproduced in Fig.5.

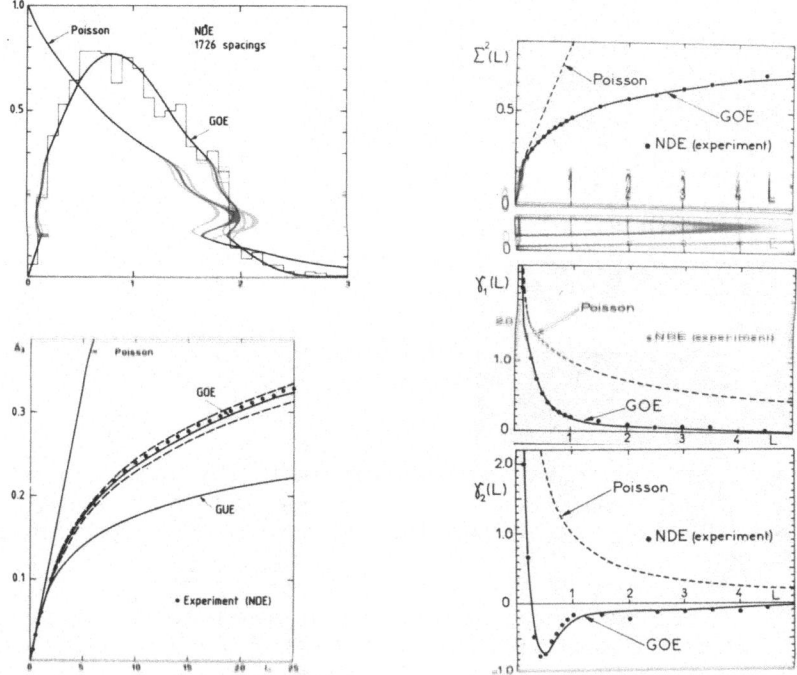

Fig.5 - Results of fluctuation measures of nuclear resonances. Poisson and GOE are given for comparison. (Taken from[HPB-82,BHP-83,BHP-84]).

As can be seen, the fluctuation measures considered, which include a thorough study of 2-point measures and to some extent more than 2-point measures as well, are fully consistent with GOE predictions.

There exists some studies of fluctuations of atomic and molecular levels (see Fig.6). They also seem to be consistent with GOE predictions, although the statistical significance of the agreement is much lower than in the nuclear case.

We therefore see that, on the one hand, the spectra of very different systems (nuclei- light or heavy-, some atoms and molecules), when properly scaled, show identical fluctuation patterns. The scale (average spacing) covers five or six orders of magnitude, when going from a medium nucleus to a complex atom or molecule. Notice that one is considering extremely different systems, some of them governed by short range interactions and others by the Coulomb long range force. On the other hand, these characteristic fluctuation patterns,

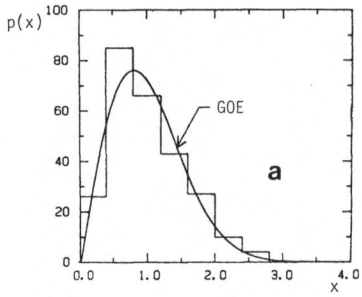
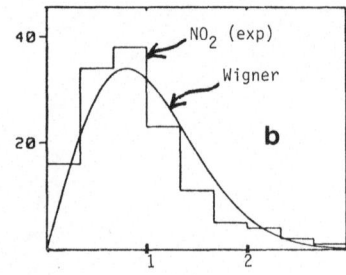

Fig.6 - Spacing distributions :(a) for atoms. (Taken from Ref.[CG-83])
and (b) for NO_2 (Taken from Ref.[HKC-83])

although not specific of, are well reproduced by GOE [BFF-81,BG-84].
Thus, a simple picture emerges : there exists a *universality of level
fluctuations laws,* as well from the experimental than from the theoretical point of view.

Notice that all the systems considered so far have a
large number of degrees of freedom. Is this a necessary condition for
exhibiting GOE-fluctuations ? To answer this question as well as
to obtain some clues on the origin of the universality of fluctuations,
let us briefly describe the results of some numerical experiments
with two-dimensional systems. Consider first the spectrum of the
Laplacian with different shapes of the boundary. Are there characteristic patterns in the spectral fluctuations and, if the answer is in
the affirmative, do they depend on the shape ? We shall consider
three different cases, whose choice will made be clear in the following
section, as illustrated on Fig.7.

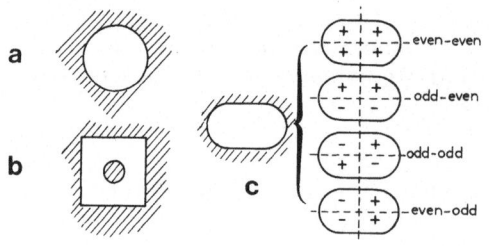

Fig.7 - Shapes of the boundaries
of different membranes
whose spectral fluctuations are discussed in the
text : a) circle ; b) Sinai ;
c) stadium, with the four
symmetry classes of
eigenfunctions.

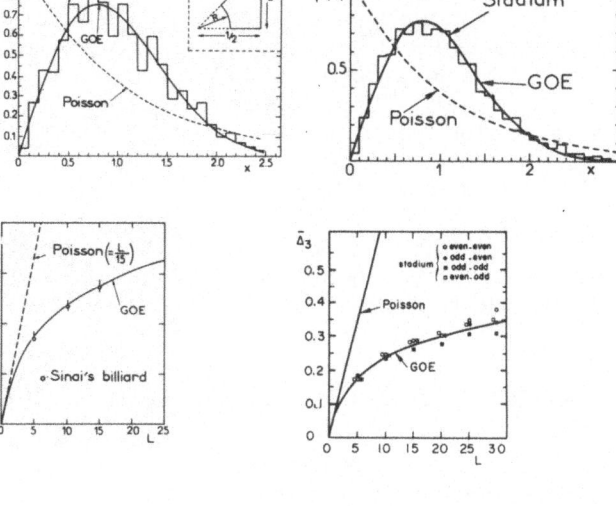

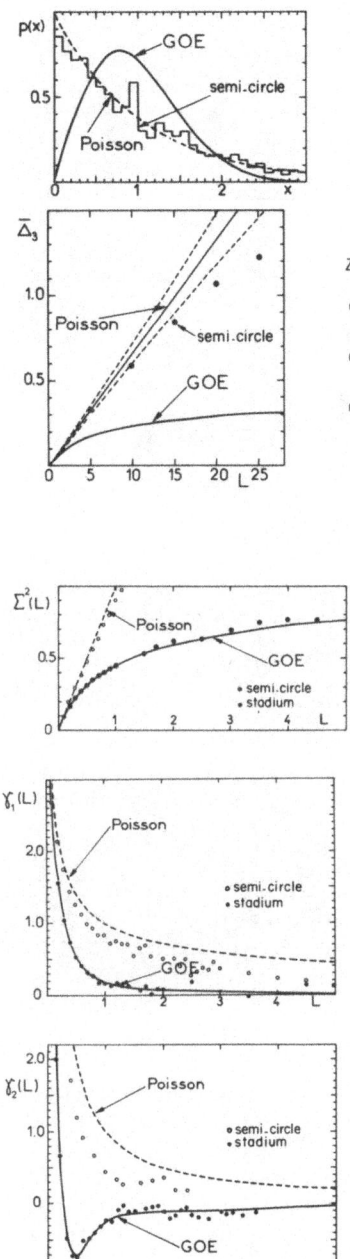

Fig.8 - Spectral fluctuation results for the Helmholtz equation with three different boundaries (see Fig.7). For the stadium, values of $\overline{\Delta}_3$ corresponding to each symmetry class separately (see Fig.7) are given. See text for further explanation. (Taken from Refs.[BGS-84a, BGS-84b]).

Results are presented on Fig.8. Inspection of Fig.8 shows that the spectral fluctuations of the circular (or rather semi-circular, to avoid degeneracies) membrane follow closely Poisson fluctuations. In contrast, the fluctuations corresponding to the stadium and Sinai's membranes follow closely GOE fluctuations. These results came as a surprise, because for years it was thought that GOE fluctuations apply to systems with many degrees of freedom whereas here we are considering two-dimensional systems.

What is then the relevant difference between boundaries like a circle and a stadium or Sinai's shape ? In order to get a hint, let us briefly discuss the corresponding classical systems.

5. REGULAR AND CHAOTIC MOTION IN CLASSICAL SYSTEMS

All conservative Hamiltonian systems with n degrees of freedom have in common three essential properties :
i) for a given set of initial conditions, the dimensionality of the accessible surface in phase space is less or equal to 2n-1 ; since the system is conservative, the energy is constant along this energy surface ;
ii) from Liouville's theorem, we know that the volume element in phase space is conserved ; in other words, the Hamiltonian flow is incompressible.

Apart from these features which are shared by all systems, the motion in phase space can exhibit a great variety of behaviours. For instance, one may ask how does a given volume element evolve with time : does it tend to cover the whole energy surface S_E as time goes to infinity or does it remain in a restricted part of S_E ? Does it conserve approximately its initial shape, or does it display more or less dramatic deformations with time ? According to the answers to such questions, one can define a hierarchy of regularity for dynamical systems, going from

$$\text{integrable} \rightarrow \text{ergodic} \rightarrow \text{mixing} \rightarrow \text{K-system}$$

in the sense of regularity towards chaoticity. The integrable systems, which are the most regular ones and can be used as clocks, possess as many integrals of motion as number n of degrees of freedom. The motion in phase space of a conservative integrable system is restricted to an n-dimensional torus , instead of a (2n-1)-dimensional energy surface for a generic system. For two-dimensional systems (n=2) one therefore has that a generic conservative system will move on a 3-dimensional manifold (or energy surface S_E) embedded in the 4-dimensional phase space, whereas an integrable system will move on a 2-dimensional torus.

In contrast with integrable systems, almost every trajectory of an ergodic system passes through almost every point of the energy surface S_E ("almost every" referring to the Liouville measure on S_E), spending equal times in equal areas : the energy surface is asymptotically uniformly covered by a typical orbit.

Does ergodicity imply chaotic motion ? For the motion to be irregular, erratic, a given volume element has to deform with time in such a way as to allow for instability with respect to a perturbation. One therefore has to ask for a stronger property than simple ergodicity. Mixing systems are such that any volume element tends to "dilute" uniformly in S_E as time goes to infinity, in the same way as a solute dilutes in a solvent if the two liquids are miscible. Consequently, the distance between two points initially close to each other may become arbitrarily large as time is running.

The mixing property, however, tells nothing on the rate of separation of orbits. It only contains the concept of asymptotic equilibration. Ergodic systems which possess the strongest degree of irregularity (K-systems) have a further property besides mixing : their orbits separate exponentially with time. As a consequence of such a dramatic instability, long time predictions on the system are impossible, the memory of the initial state vanishing with time. Notice that such systems are deterministic in the sense that they are governed by causal equations : in principle, they are predictable. However, due to the finite precision available for any practical purpose, one cannot follow their time evolution beyond some critical time. In the language of communication theory, K-systems are sources which continuously produce information (their so-called Kolmogorov entropy is positive), in contrast with integrable systems, whose motion is periodic or quasiperiodic, and for which knowing the story of any orbit during some given time interval is sufficient to determine with probability one its future evolution. Integrable systems considered as sources can be compared with records, which indefinitely repeat the same message, whereas K-systems can be compared with a broadcast station, which is supposed to produce indefinitely new information.

We notice that the classification of possible motions mentioned above is not complete in the following sense : the phase space of a generic system will show in general islands of regularity surrounded by regions of chaoticity, that is, the structure of the phase space will be neither purely regular not purely chaotic (a generic system is neither integrable nor ergodic).

What are, now, the classical (or rather the ray or geometrical optics) analogues of the systems described at the end of the previous section, namely, solutions of the Helmholtz equation with some boundary ? They correspond to billiards, namely to the motion of a free point particle of mass m in the domain S of the plane bounded by a -or a set of- close curve B, with elastic reflections at the boundary B. A billiard is thus a two degrees of freedom system with at least one constant of the motion, the energy $E=(1/2)mv^2$.

A particular property of billiards is that the behaviour of any orbit in phase space depends neither on the mass of the point particle nor on its velocity v and, in contrast with a generic system, all the energy surfaces S_E have the same structure.

The three billiards with boundaries given in Fig.7 are from the point of view of the present discussion quite different. For the circle, apart from energy, one has a further integral of motion, the angular momentum with respect to the center, and the system is integrable. At the opposite extreme of orderliness, the stadium and Sinai's billiards have, besides the energy, no further integral of motion and belong in fact to the class of the most chaotic systems : they are K-systems. Although rigorous tools and results concerning such systems are relatively recent (ergodic theory), some of the main ideas can be traced back to the founders of kinetic theory and statistical mechanics. Already in 1879 Maxwell stated "if we suppose that the material particles...occasionally encounter a fixed obstacle such as the sides of a vessel containing the particles... it is difficult in a case of such extreme complexity to arrive at a thoroughly satisfactory conclusion, but we may with considerable confidence assert that except for particular forms of the surface of the fixed obstacle, the system will sooner or later, after a sufficient number of encounters, pass through every phase consistent with the equation of energy." Thus Maxwell should be credited with conjecturing the existence of ergodic enclosures.

In summary, and remembering the discussion of the previous section, one is led to conjecture that quantum systems whose classical analogue is integrable have spectral fluctuations asymptotically of Poisson type [BT-77,Be-83] whereas quantum systems which are time reversal invariant and whose classical analogue is strongly chaotic show GOE fluctuations. This would then explain the success of GOE in describing, for instance, the fluctuations of the compound nucleus resonances and would establish the universality of the laws of level fluctuations.

6. SOME REMARKS AND OPEN PROBLEMS

To conclude, let us mention a number of problems of current interest in this field. For some of them important progress has been already performed, for others the situation is still at a very primitive stage.

The stadium and Sinai's billiards already discussed are purely chaotic (a single chaotic region fills the complete energy surface) and as mentioned before, generic systems are neither integrable nor ergodic. It is then interesting to study whether for Hamiltonians depending on a parameter governing a transition from pure regularity to pure chaoticity in the classical case, there is also a transition in the spectral fluctuations of the corresponding quantum Hamiltonian (one expects a transition from Poisson fluctuations to GOE fluctuations, for time reversal invariant systems). Let us mention some recent works in this direction, the study of a one-parameter family of billiards [Ro-84] and the study of some other Hamiltonian systems [SVZ-84, HKC-84]. Seligman et al. study the Hamiltonian

$$H = \frac{1}{2}(p_1^2 + p_2^2) + V_1(x_1) + V_2(x) + V_{int}(x_1-x_2) \tag{17}$$

where V_1, V_2 and V_{int} have the same functional form

$$V_j(x) = \lambda_j(x^2 + \mu_j^4 + \nu_j x^6) \qquad\qquad j = 1,2,int \quad . \tag{18}$$

In (18) λ_j, μ_j and ν_j are parameters. With an adequate choice of the numerical values of μ_j, ν_j, λ_1 and λ_2, the properties of the system are studied as a function of a single parameter λ_{int}. For $\lambda_{int}=0$ the system is separable and therefore integrable, whereas for large values of λ_{int} it is presumably purely chaotic. Some results for level fluctuations are reproduced on Fig.9. Fig.9e corresponds to the integrable case and (9a) to the purely chaotic regime. As expected (e) agrees with

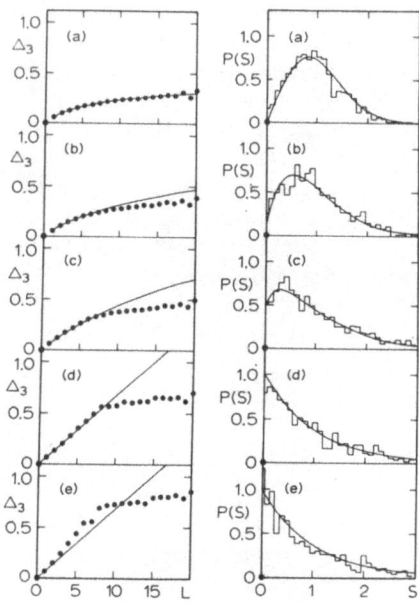

Fig.9 - Results of level fluctuations for the Hamiltonian Eq.(17). The continuous line corresponds in (a) and (e) to GOE and Poisson respectively ; in (b),(c) and (d) to a continuous interpolation between GOE and Poisson. (Taken from Ref. [SVZ-84])

Poisson and (a) with GOE. For intermediate situations, obtained by varying λ_{int}, the fluctuation patterns are intermediate between GOE and Poisson.

The question of identifying the transition parameter which characterizes the spectral fluctuations is discussed in Refs. [SVZ-84,BR-84,IY-84]. A good candidate seems to be the fraction f of phase space filled by chaotic trajectories in the classical case and arguments have been put forward suggesting that the spectral fluctuations result from the superposition of statistically independent GOE and Poisson spectra, with probability f and (1-f) respectively. And the different correlation functions characterizing the fluctuations of spectra resulting from independent superposition of spectra can be expressed in terms of correlation functions of the constituent spectra (Poisson and GOE in our discussion) [Pa-79]. We thus have a definite prediction that should be tested systematically.

So far we have considered only Hamiltonian systems in Euclidean spaces. There is another class of dynamical systems which would be of particular interest to study, because rigorous results are available. Indeed one knows that geodesics in compact spaces of negative curvature constitute a K-system [Or-74] ; it would then be worth, to reinforce the conjecture of universality of fluctuation laws, to study the spectrum of the Laplacian in such spaces, and see whether it shows the GOE-pattern.

In these notes we have limited ourselves to spectral fluctuations, which constitute only one aspect of statistical properties, and no mention has been made to properties related to wave functions (see contribution by Weidenmüller [We-84] to this volume). Fig.10 is reproduced here to give a flavour on the questions that can be asked in our context. Some of the problems raised when discussing spectral properties, for instance characterization of purely chaotic regime, transition from regular to irregular regime, etc.., should and are now being raised for properties related to wave functions (structure of nodal lines, for instance) [MK-79,MD-83,SG-84].

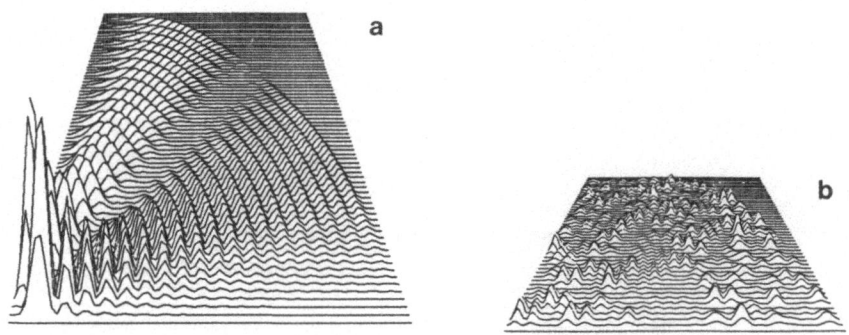

Fig.10 - Examples of intensity distribution $|\psi|^2$ in the positive quadrant for (a) the circular billiard (regular)and (b) the stadium (chaotic). (Taken from Ref.[MD-83])

It is interesting to extend the nature of physical systems in which characteristic level fluctuation patterns can be observed. One may think, in connexion with "ergodic enclosures", of observations made in acoustics (auditoriums) or in electromagnetic oscillations of a cavity. And in fact, there exists a very early attempt to study the statistical properties of normal frequencies of resonant cavities [Sc-54]. One also expects that the electronic specific heat of small metallic particles, at low temperatures, is reduced with respect to the bulk value and that the reduction depends on the fluctuation properties of the one-electron energy spectrum ; however, a detailed comparison between theory and experiment has not yet been done [KKK-84].

A very interesting system, for its fundamental "simplicity", is the hydrogen atom in a magnetic field (see, for instance, the discussion in Ref.[Ga-85])

$$H = \frac{\vec{p}^2}{2} + \gamma^2 (x^2 + y^2) - \frac{1}{r} \qquad ; \qquad (19)$$

(19) is obtained after dropping the uninteresting paramagnetic term and γ governs the intensity of the magnetic field strength. In the limit of small and large values of γ, the problem can be exactly solved (Coulomb problem and Landau problem respectively). The question is now to study the stochastic transition region, from the classical as well as from the quantal point of view. With the improvements of beam experiments on high Rydberg states using laser excitation, one expects that experiments on hydrogen will become possible and that progress is likely to occur in the near future.

We also mention the domain of nucleus-nucleus reactions and nuclear fission. There is likely an ordered regime dominated by symmetries and resulting shell effects, and a chaotic regime which is well described by the liquid drop. The way in which the transition from the ordered to the chaotic regime, with the destruction of nuclear symmetries, takes place, is largely an open question [Sw-84].

It would also be interesting to find systems which exhibit GUE fluctuation patterns (one expects them for time reversal non invariant systems). In this connection, we are presently studying the quantum analogue of a charged point particle moving in a chaotic billiard under the action of a constant magnetic field perpendicular to the plane of the billiard.

It may be also worth mentioning the existence of some formal similarities between apparently disconnected subjects. For instance, the fluctuations of zeros of ζ (s) (see Section 2) may be the same asymptotically as those of the eigenvalues of GUE matrices [Mo-74,Od-84]. And Gutzwiller [Gu-83], when studying the stochastic behaviour of a quantum scattering problem, is naturally led to investigate the complexity of the Riemann zeta function.

From all what preceeds, it is clear that an immediate task is called for : one should go beyond the existing analytical attempts to relate RMT and chaotic motion [Pe-83,Be-84], to put the apparent universality of level fluctuation laws on a firmer basis.

Clearly these notes are to be placed towards the starting stage of a rapidly expanding field, when questions are raised, conjectures are made and tentative answers are given. We are still far from a complete understanding announcing the near death of the subject.

References

[Be-83]   M.V. BERRY, in *Chaotic Behaviour of Deterministic Systems*, Les Houches Summer School Lectures XXXVI, R.H.G. Helleman and G. Joos (Eds.), North-Holland, Amsterdam 1983, p.171.

[Be-84]   M.V. BERRY in *Chaotic Behavior in Quantum Systems*, G. Casati (ed.), Plenum, 1984.

[BFF-81]   T.A. BRODY, J. FLORES, J.B. FRENCH, P.A. MELLO, A. PANDEY and S.S.M. WONG, Rev.Mod.Phys. **53** (1981) 385.

[BG-84]   O. BOHIGAS and M.J. GIANNONI, in *Mathematical and Computational Methods in Nuclear Physics*, J.S. Dehesa et al. (eds.), Lecture Notes in Physics **209**, Springer Verlag,1984,p.1.

[BGS-84a]   O. BOHIGAS, M.J. GIANNONI and C. SCHMIT, Phys.Rev.Lett. **52** (1984) 1.

[BGS-84b]   O. BOHIGAS,M.J. GIANNONI and C. SCHMIT, J.Physique Lett. **45** (1984) L-1015.

[BHP-83]   O. BOHIGAS, R.U. HAQ and A. PANDEY, in *Nuclear Data for Science and Technology*, K.H. Böckhoff (ed.), Reidel, Dordrecht, 1983, p.809.

[BHP-84]   O. BOHIGAS, R.U. HAQ and A. PANDEY, preprint, 1984.

[BR-84]   M.V. BERRY and M. ROBNIK, J.Phys.**A17**(1984) 2413.

[BT-77]   M.V. BERRY and M. TABOR, Proc.Roy.Soc.Lond. **A356** (1977) 375.

[CG-83]   H.S. CAMARDA and P.D. GEORGOPULOS, Phys.Rev.Lett. **50** (1983) 492.

[Ga-85]   J.C. GAY in *Photophysics and Photochemistry in the Vacuum Ultraviolet*, S.P. Mc Glynn et al. (eds.), Reidel, 1985, p.631.

[Gu-83]   M.C. GUTZWILLER, Physica **7D**(1983) 341.

[HKC-83] E. HALLER, H. KÖPPEL and L.S. CEDERBAUM, Chem.Phys. Lett. **101** (1983) 215.

[HKC-84] E. HALLER, H. KÖPPEL and L.S. CEDERBAUM, Phys.Rev. Lett. **52** (1984) 1665.

[HMY-81] F.T. HIOE, E.W. MONTROLL and M. YAMAWAKI, in *Perspectives in Statistical Physics,* H.J. Raveché (ed.), North-Holland, 1981, p.295.

[HPB-82] R.U. HAQ, A. PANDEY and O. BOHIGAS, Phys.Rev.Lett. **48** (1982) 1086.

[IY-84] T. ISHIKAWA and T. YUKAWA, preprint 1984.

[KKK-84] R. KUBO, A. KAWABATA and S. KOBAYASHI, Ann.Rev. Mater.Sci. **14** (1984) 49.

[MD-83] S.W. Mc DONALD, Ph.D. Thesis, University of California, Berkeley, 1983, unpublished.

[MK-79] S.W. Mc DONALD and A.N. KAUFMAN, Phys.Rev.Lett. **42** (1979) 1189.

[Mo-74] H.L. MONTGOMERY, Proc. of Int. Congr. of Mathematicians, Vancouver, 1974.

[Od-84] A. ODLYZKO, private communication, 1984.

[Or-74] D.S. ORNSTEIN, *Ergodic Theory, Randomness, and Dynamical Systems,* New Haven and London, Yale University Press, 1974.

[Pa-79] A. PANDEY, Ann.Phys. **119** (1979) 170.

[Pe-83] P. PECHUKAS, Phys.Rev.Lett. **51** (1983) 943.

[Ro-84] M. ROBNIK, J. Phys. **A17** (1984) 1049.

[Ro-85] M. ROBNIK, in *Photophysics and Photochemistry in the Vacuum Ultraviolet,* S.P. Mc Glynn et al. (eds.), Reidel, 1985, p.579.

[Sc-54] M. SCHRÖDER, Acustica, **4** (1954) 456.

[SG-84] M. SHAPIRO and G. GOELMAN, Phys. Rev. Lett. **53** (1984) 1714.

[SVZ-84] T.H. SELIGMAN, J.J.M. VERBAARSCHOT and M.R. ZIRNBAUER, Phys.Rev.Lett. **53** (1984) 215.

[Sw-84] W.J. SWIATECKI, in *Theoretical Approaches of Heavy Ion Reaction Mechanisms,* M. Martinot , C. Ngô and F. Lepage (eds.), Nucl.Phys. **A428** (1984) p.199c.

[We-84] H.A. WEIDENMÜLLER, these Proceedings.

[Za-77] D. ZAGIER, The Math.Intelligencer **0** (1977) 7.

TIME-DEPENDENT VARIATIONAL PRINCIPLE FOR THE EXPECTATION VALUE OF AN OBSERVABLE: MEAN-FIELD APPLICATIONS [*]

Marcel Vénéroni

Division de Physique Théorique
Institut de Physique Nucléaire

91406 Orsay, France

1. INTRODUCTION

In non equilibrium statistical mechanics one is often interested in making some prediction at a time t_1 from the knowledge of some properties at an earlier time t_0. In quantum statistical mechanics all the knowledge about the state (whether pure or not) of the system of interest is included, at a given time, in the density operator. Let us assume that this system has been prepared in such a way that the density operator $D(t_0)$, characterizing the initial state, is known. At some later time t_1, one intends to perform a measurement of some observable A. The theoretical problem is the prediction of the average value of A at the time t_1. In principle the calculation of this expectation value

$$<A>(t_1;D,t_0) \equiv Tr\ A\ D(t_1,t_0) \tag{1.1}$$

requires only the knowledge of the Hamiltonian H of the system, since the exact evolution of the density operator $D(t,t_0)$ is given by the Liouville-von Neumann equation (2.4).

In practice, the evolution of the system, except for the simplest cases, is so complex that it is out of the question to evaluate exactly the average (1.1). In addition, in most cases, the degrees of freedom are so numerous that the initial value $D(t_0)$ itself is intractable. Recourse to some simplified forms for $D(t,t_0)$ and $D(t_0)$ thus becomes unavoidable. In fact, as has often been

---

[*] These lectures are a condensed version of work done in collaboration with R. Balian, and of related work by P. Bonche and H. Flocard.

emphasized, this simplification is also highly desirable, since exact density operators contain a huge amount of useless information (for instance about high-order correlations). It becomes mandatory to introduce some approximate description, in which the number of degrees of freedom is drastically reduced. Therefore, in almost any physical situation, the question inevitably arises of the optimal choice for the approximate time-dependent density operator. Once this choice is made, it remains to establish (and last but not least to solve) the resulting approximate equations of motion. These are formidable problems for which, at this level of generality, there are no well-defined prescriptions.

We intend to be less ambitious and to address ourselves to the more restricted (but still quite extended) question :

Is it possible, in the framework of some approximation, to obtain a best estimate (in a sense which has yet to be defined) for the expectation value (1.1) of the observable A at the time t_1 from the knowledge of $D(t_0)$?

The following notes will be devoted to the study of an algorithm providing, at least in certain circumstances, a quantitative answer to the preceding question [1]. To attain this goal, we shall write in Section 3 a variational expression which admits $<A>(t_1;D,t_0)$ as its stationary value. The variational objects will be a time-dependent density operator (in the Schrödinger picture) and a time-dependent observable associated with the measurement and defined in a "backward" Heisenberg picture (see Section 2). The particulars of the problem will appear through the boundary conditions : the variational density should be equal to $D(t_0)$ at the time t_0, and the variational observable to A at the time t_1. When the trial spaces for the density operator and for the observable are left unrestricted, exact equations of motion are obtained. Starting with Section 5, various approximations, some well known and some not, will be generated by restricting these trial spaces. A few general features of such approximations are reviewed in Section 5.

In the last part of these notes, we shall concentrate upon a particular class of approximations (mean-field approximations) for which the exact density operator is replaces by an uncorrelated product of single-particle operators. We shall still have the freedom to choose the variational space for the trial observable. The equations of motion provided by the variational principle will depend upon this choice, which will itself be dictated by the structure of the observable A considered. Thus, the approximate dynamics will depend on the question which is posed, namely upon the nature of the measurement to be performed at the final time t_1.

2. PRELIMINARIES : THE BACKWARD HEISENBERG PICTURE

The expectation value (1.1) of the observable A can be formally written

$$<A>(t_1;D,t_0) = \text{Tr } A \, U(t_1,t_0)D(t_0)U(t_0,t_1), \qquad (2.1)$$

in terms of the basic evolution operator

$$U(t_1,t_0) = T \exp[-i\int_{t_0}^{t_1} dt \, H_t] = U^+(t_0,t_1). \qquad (2.2)$$

The Hamiltonian H_t is possibly time-dependent, as recalled by the index t. The expression (1.1) has been implicitly assumed to be given in the Schrödinger picture. In this picture, the density operator

$$D(t_1,t_0) = U(t_1,t_0)D(t_0)U(t_0,t_1) \qquad (2.3)$$

is the solution of the Liouville-von Neumann equation

$$\frac{dD(t,t_0)}{dt} + i \, [H_t, D(t,t_0)] = 0, \qquad (2.4)$$

with the initial condition $D(t_0,t_0) = D(t_0)$.

The Heisenberg representation results from the replacement of (2.1) by the equivalent expression

$$<A> (t_1;D,t_0) = \text{Tr } \hat{A}(t_1,t_0)D(t_0), \qquad (2.5)$$

the Heisenberg transform $\hat{A}(t_1,t_0)$ of $A(t_1)$ has been defined as

$$\hat{A}(t_1,t_0) = U(t_0,t_1) \, A(t_1)U(t_1,t_0), \qquad (2.6)$$

where (2.6) includes the possibility of a time-dependence for the observable A. In order to replace this formal expression by a differential equation, it is traditional in textbooks [2] to take the derivative of (2.6) with respect to t_1. Heisenberg's equation of motion results :

$$\frac{d\hat{A}(t_1,t_0)}{dt_1} + i[\hat{A}(t_1,t_0), \, U(t_0,t_1)H_{t_1}U(t_1,t_0)] = U(t_0,t_1) \frac{\partial A(t_1)}{\partial t_1} U(t_1,t_0). \qquad (2.7)$$

This equation is convenient only if H and A do not depend explicitly on time. Otherwise, equation (2.7) still involves the unknown evolution operator U and it is not a closed equation, in contrast to (2.4).

It should be noticed, however, that the operator (2.6) depends on two times. This point is usually not emphasized, because the Heisenberg transform $\hat{A}(t_1,t_0)$ is implicitly considered in (2.7) as a function of the later time t_1, whereas t_0 is viewed just as some fixed reference time at which the Heisenberg observable $\hat{A}(t_0,t_0)$ reduces to the observable A in the Schrödinger picture . We are going to take advantage of the existence of these two times, and write an alternative equation of motion for $\hat{A}(t_1,t)$ by considering it as a function of the earlier time t, i.e., by interchanging the reference time and the running time. We get then from (2.6) and (2.2)

$$\frac{d\,\hat{A}\,(t_1,t)}{dt} - i[\hat{A}(t_1,t),\,H_t] = 0, \tag{2.8}$$

an equation of motion to be supplemented by the boundary condition

$$\hat{A}(t_1,t_1) = A(t_1). \tag{2.9}$$

Although the equations (2.8) and (2.9) determine after integration the same Heisenberg transform $\hat{A}(t_1,t_0)$ as the conventional Heisenberg equation of motion (2.7), they require the introduction of another picture, in which the time t_1 of observation is considered as fixed, and the time t flows backwards from t_1 to the origin t_0. We shall call (2.8) the "backward Heisenberg equation of motion."

This equation of motion (2.8), to which is associated a "backward Heisenberg picture", is simpler than the equation (2.7) when H, or A, depend on time. In addition, the associated boundary condition is the natural condition in the present context: while the initial state $D(t_0)$ appears as a boundary condition for the Liouville-von Neumann equation governing $D(t,t_0)$, the observable $A(t_1)$ to be measured at the time t_1 appears as a boundary condition for the backward Heisenberg equation governing $\hat{A}(t_1,t)$.

In the following, we shall make use of a representation combining the Schrödinger and the backward Heisenberg pictures. To obtain this representation, we rewrite (2.1) by using the group property satisfied by U, as well as the cyclic invariance of the trace. This yields an expression involving both (2.3) and (2.6) :

$$<A>(t_1;D,t_0) = Tr\ U(t,t_1)\ AU(t_1,t)\ U(t,t_0)D(t_0)U(t_0,t)$$

$$\equiv Tr\ \hat{A}(t_1,t)\ D(t,t_0). \tag{2.10}$$

This expression interpolates between the two pictures (2.1) and (2.5) to which it reduces for $t = t_1$ and for $t = t_0$, respectively.

From t_0 to t, we are in the Schrödinger picture, from t to t_1 in the Heisenberg picture. Although (2.10) does not depend on the intermediate time t, the introduction of this time will prove to be essential in the following. Indeed, we shall obtain $D(t,t_0)$ and $\hat{A}(t_1,t)$ by solving differential equations in which the running time is t with the initial and observation times t_0 and t_1 being fixed. The backward Heisenberg picture arises then quite naturally : $D(t, t_0)$ evolves from $D(t_0)$ onwards according to the Liouville-von Neumann equation, and $\hat{A}(t_1,t)$ evolves from $A(t_1)$ backwards as per equation (2.8).

The above considerations would appear to be somewhat simpler in the Liouville formalism [3], in which states D and observables A are considered as elements of two dual vector spaces with the scalar product

$$\langle A \rangle_D = (A;D) \equiv \text{Tr } A\,D. \tag{2.11}$$

The evolution is generated by the Liouville operator L , acting either on states or on observables according to

$$LD \equiv [H,D] \, , \tag{2.12a}$$

$$AL \equiv [A,H] \, . \tag{2.12b}$$

Although the Liouville formalism would also render the classical limit straightforward, we shall work most of the time in Hilbert space to have things a little more explicit.

3. THE VARIATIONAL PRINCIPLE FOR A STATE AND AN OBSERVABLE

3.1 The action functional

Our aim is to set up a variational formalism which encompasses the dynamical equations of Section 2, and is capable of providing some answers to dynamical questions of the type discussed in the Introduction.

For this purpose, we introduce the action-like functional [1]

$$I \equiv \text{Tr}A(t_1)\mathcal{D}(t_1) - \int_{t_0}^{t_1} dt \left[\text{Tr } A\frac{d\mathcal{D}}{dt} - h(A,\mathcal{D}) \right] , \tag{3.1}$$

where $\mathcal{D}(t)$ and $A(t)$ are two time-dependent operators which are of the same nature as a density operator and an observable, respectively. The independent variational degrees of freedom of (3.1) are the (time-dependent) matrix elements of $\mathcal{D}$ and A.

The scalar h, which we call the pseudo-Hamiltonian, is defined as

$$h(A,\mathcal{D}) = -i \ \text{Tr} \ A[H,\mathcal{D}] = -i(A;L\mathcal{D})$$

$$= -i \ \text{Tr} \ [A,H] \mathcal{D} = -i(AL;\mathcal{D}) \ . \tag{3.2}$$

For notational convenience, we no longer keep the index t of H_t which was introduced in Section 2. When $\mathcal{D}$ and A are Hermitean operators, the pseudo-Hamiltonian h is real.

Let us just rewrite Eqs. (3.1) and (3.2) as

$$I = \text{Tr} \ A(t_1) \ \mathcal{D}(t_1) - \int_{t_0}^{t_1} dt \ \text{Tr} A\left(\frac{d\mathcal{D}}{dt} + i[H,\mathcal{D}]\right) . \tag{3.3}$$

An integration by parts yields equivalently

$$I = \text{Tr} \ A(t_0) \ \mathcal{D}(t_0) + \int_{t_0}^{t_1} dt \ \text{Tr}\left(\frac{dA}{dt} - i[A,H]\right)\mathcal{D}. \tag{3.4}$$

The variations of I resulting from arbitrary variations δA and $\delta \mathcal{D}$ are immediately given by (3.3) and (3.4), respectively. The total variation δI reads

$$\delta I = \text{Tr} \ A(t_0)\delta\mathcal{D}(t_0) + \text{Tr}\delta A(t_1) \ \mathcal{D}(t_1)$$

$$+ \int_{t_0}^{t_1} dt \ \left\{ \text{Tr}\left(\frac{dA}{dt} - i[A,H]\right)\delta\mathcal{D} - \text{Tr}\delta A\left(\frac{d\mathcal{D}}{dt} + i[H,\mathcal{D}]\right)\right\}. \tag{3.5}$$

3.2 The equations of motion and the stationary value of the action

The particulars of the problem stated in the Introduction are now introduced through the boundary conditions imposed on the variables entering the action (3.1). Namely, we require the matrix elements of $\mathcal{D}(t)$ to obey the initial conditions

$$\mathcal{D}(t_0) = D, \tag{3.6}$$

accounting for our information on the state of the system at the initial time t_0. For the matrix elements of $A(t)$, we impose the final conditions

$$A(t_1) = A \ ; \tag{3.7}$$

the observable A which is to be measured at the final time t_1 is thus specified through the boundary condition (3.7).

Our variational principle consists in the requirement that I be stationary with respect to the most general variations $\delta\mathcal{D}(t)$ and $\delta A(t)$ compatible with the boundary conditions (3.6) and (3.7). These boundary conditions imply $\delta\mathcal{D}(t_0) = \delta A(t_1) = 0$, and therefore the two end-point terms of (3.5) vanish. The stationarity of I then provides through (3.5) the equations of motion for $\mathcal{D}(t)$ and $A(t)$, which read

$$\frac{d\mathcal{D}}{dt} + i[H, \mathcal{D}] = 0 ,\qquad (3.8)$$

$$\frac{dA}{dt} - i[A, H] = 0 . \qquad (3.9)$$

These equations are supplemented by the boundary conditions (3.6) and (3.7). Since the time t lies between t_0 and t_1, $\mathcal{D}(t)$ evolves towards the future by starting from the value D at the initial time t_0, whereas $A(t)$ proceeds backwards in time from its final value A at t_1.

The variational equation (3.8) is just the Liouville-von Neumann equation (2.4), with the boundary condition (3.6) suited to the Schrödinger picture. The reader may also have identified (3.9) with the backward Heisenberg equation (2.8), and (3.7) with the appropriate boundary condition (2.9) at the final time t_1. The stationary value I_{st} of I is attained when $\mathcal{D}(t)$ and $A(t)$ coincide with the exact solutions

$$\mathcal{D}(t) = D(t, t_0) = U(t, t_0) D U(t_0, t) , \qquad (3.10)$$

$$A(t) = \hat{A}(t_1, t) = U(t, t_1) A U(t_1, t) , \qquad (3.11)$$

of Eqs.(2.4) and (2.8).

At the stationary point, the integrands of (3.3) or (3.4) vanish, as a consequence of (3.8) or (3.9). This provides the alternative forms

$$I_{st} = \text{Tr } AD(t_1, t_0) \qquad (3.12a)$$

$$= \text{Tr } \hat{A}(t_1, t_0) D . \qquad (3.12b)$$

The stationary value of I, for unrestricted variations of $\mathcal{D}$ and A, is thus precisely our object of interest $<A>(t_1; D, t_0)$, namely, the average value of the observable A at the time t_1 corresponding to an initial (normalized) state D at the time t_0, written either in the Schrödinger or in the backward Heisenberg picture. (As a shorthand notation, we shall write from now on this quantity $<A>$.) As noticed in Eq.(2.10), the use of (3.10) and (3.11) also yields for this quantity

$$I_{st} = <A> = \text{Tr } A(t) \mathcal{D}(t) , \qquad (3.13)$$

where the r.h.s. does not depend on the intermediate time t.

When the Hamiltonian H is time-dependent, the conservation of the pseudo-energy can be established by differentiating (3.2) with respect to time and by using (3.8) and (3.9).

The above formalism holds as well in classical statistical mechanics. The quantities D and A then refer to a density in phase space and to a dynamical variable. The trace is replaced by an integral on phase space, commutators by Poisson brackets (multiplied by i), and the pseudo-Hamiltonian is expressed in terms of the Liouville operator by (3.2).

Our variational principle for a state and an observable is redundant in the sense that exact dynamical laws have been obtained twice, although in two different pictures. This duplication arises from the bilinearity of h in A and D. When we shall build approximate equations of motion for A and D by restricting their trial spaces, the redundancy will disappear in most cases, and we shall need to vary I with respect to both A and D.

3.3 The classical structure

The expression (3.1) is formally reminiscent of the familiar classical action in phase space

$$\int_{t_0}^{t_1} dt \ \sum_i \left\{ p_i \ \dot{q}_i - h(q,p) \right\} , \tag{3.14}$$

where q_i and p_i are the coordinates and momenta, respectively, and h is the Hamiltonian. This formal analogy becomes even stronger if one considers a problem of classical mechanics with mixed boundary conditions, at t_0 for the coordinates q_i and at t_1 for the momenta p_i, instead of boundary conditions on the coordinates at both ends. It is indeed easy to check that the stationarity conditions of

$$I = \sum_i p_i(t_1)q_i(t_1) - \int_{t_0}^{t_1} dt \left\{ \sum_i p_i \dot{q}_i - h(q,p) \right\} , \tag{3.15}$$

with the boundary constraints

$$\delta q_i(t_0) = \delta p_i(t_1) = 0, \tag{3.16}$$

are Hamilton's equations of motion.

The action (3.1) has thus the same structure as the classical action (3.15) suited to the mixed boundary conditions (3.16). The matrix elements of D and A play the roles of coordinates and momenta, respectively. Notice that the pseudo-Hamiltonian (3.2) is quite peculiar since it is bilinear in coordinates and momenta.

3.4 Comparison with some other variational principles

The duality between density operators and observables in statistical mechanics is formally reminiscent of the duality between bras and kets in the quantum mechanics of pure states. As is well-known, one can introduce in the latter case action-like functionals of bras and kets, from which the Schrödinger dynamics is generated by stationarity conditions. For instance, the variation of the expression

$$\int_{t_0}^{t_1} dt <\Phi(t)| \left(\frac{d}{dt} + iH \right) |\Psi(t)> \qquad (3.17)$$

with respect to $\Phi(t)$ provides the Schrödinger equation, while the variation with respect to $\Psi(t)$ provides its Hermitean conjugate (if one ignores the contributions arising from the end points). Another variational principle providing the Schrödinger equation and its conjugate is obtained by considering Φ and Ψ as identical in (3.17). In Ref.[1] some of these variational principles are compared with the variational principle of Section 3 which generates both the Liouville-Von Neumann and the backward Heisenberg equations.

We just want here to stress the analogy between the expression (3.1) and the action used by Lippmann and Schwinger as the starting point of their scattering theory [4]. This analogy is even closer when (3.1) is rewritten in the interaction picture (in fact the interaction picture for the state should be combined with a different picture for the observable). There is however, in spite of the formal similarity, an essential difference : while the Lippmann-Schwinger approach deals with quantum mechanics in Hilbert space, the action (3.1) is associated with statistical mechanics (either quantal or classical) in Liouville space. The stationary value of the action is a transition amplitude in the first case whereas it is the expectation value of an observable in our case. An important new feature is the occurence among the trial objects of a time-dependent observable in addition to the trial state.

A comparison is also made in [1] with variational principles enforcing the normalization of the trial bras and kets, in particular with the action of Ref.[5]. Similarly, modified actions, which involves normalized trial states only, can be build from the action (3.1).

4. THE LINEAR RESPONSE

As a first simple application, let us use the variational principle for a state and an observable to derive the linear response formalism.

We first decompose the Hamiltonian H_t into :

$$H_t = H_0 + H_1(t),\tag{4.1}$$

where $H_1(t)$ is some small time-dependent perturbation. The unperturbed Hamiltonian H_0 may also be time-dependent. We write then the solutions of (3.8) and (3.9) for the full Hamiltonian H as

$$\mathcal{D}(t) = \mathcal{D}_0(t) + \mathcal{D}_1(t) \ , \quad A(t) = A_0(t) + A_1(t) \ ,\tag{4.2}$$

where $\mathcal{D}_0(t)$ and $A_0(t)$ are the solutions $\mathcal{D}$ and A corresponding to H_0. Both perturbed and unperturbed solutions satisfy the boundary conditions

$$\mathcal{D}_0(t_0) = \mathcal{D}(t_0) = D \ , \quad A_0(t_1) = A(t_1) = A.\tag{4.3}$$

We now expand the functional (3.1) up to first order around A_0, $\mathcal{D}_0$ and H_0, by making use of (3.5). We get

$$
\begin{aligned}
\mathcal{I}(A,\mathcal{D},H) - \mathcal{I}(A_0,\mathcal{D}_0,H_0) &\sim \frac{\partial \mathcal{I}}{\partial \mathcal{D}_0}\, \mathcal{D}_1 + \frac{\partial \mathcal{I}}{\partial A_0}\, A_1 + \frac{\partial \mathcal{I}}{\partial H_0}\, H_1 \\
&\equiv \mathrm{Tr}A_0(t_0)\mathcal{D}_1(t_0) + \mathrm{Tr}A_1(t_1)\mathcal{D}_0(t_1) \\
&+ \int_{t_0}^{t_1} dt\, \left\{ \mathrm{Tr}\!\left(\frac{dA_0}{dt} - i[A_0,H_0]\right)\mathcal{D}_1 - \mathrm{Tr}\, A_1\!\left(\frac{d\mathcal{D}_0}{dt} + i[H_0,\mathcal{D}_0]\right) \right\} \\
&- i \int_{t_0}^{t_1} dt\, \mathrm{Tr}A_0[H_1,\mathcal{D}_0] .
\end{aligned}\tag{4.4}
$$

Only the last term, which comes from the derivative of $\mathcal{I}$ with respect to the Hamiltonian, is expected to survive, since $\mathcal{I}(\mathcal{D},A)$ is stationary with respect to variations of $\mathcal{D}$ and A around $\mathcal{D}_0$ and A_0. Actually, it is obvious that all the terms of (4.4), except the last, vanish because of the unperturbed equations of motion (3.8) and (3.9) and the boundary conditions (4.3).

We have thus obtained the linear response for the perturbation H_1 and the observable A, in the form

$$\text{Tr } A\mathcal{D}(t_1) = \text{Tr } A\mathcal{D}_0(t_1) - i \int_{t_0}^{t_1} dt \text{ Tr } A_0(t)[H_1(t),\mathcal{D}_0(t)] . \quad (4.5)$$

The usual linear response formula is recovered by rewriting (4.5) in terms of the observables in the interaction picture :

$$\tilde{A}(t_1,t_0) = U_0(t_0,t_1) A U_0(t_1,t_0) , \quad (4.6a)$$

$$\tilde{H}_1(t,t_0) = U_0(t_0,t) H_1(t) U_0(t,t_0). \quad (4.6b)$$

One gets immediately, by using (3.10) and (3.11), the well known formula

$$\text{Tr } AD(t_1,t_0) = \text{Tr } AD_0(t_1,t_0) - i \int_{t_0}^{t_1} dt \text{ Tr}[\tilde{A}(t_1,t_0),\tilde{H}_1(t,t_0)] D.$$
$$(4.7)$$

Note that the above derivation made no special reference to an unperturbed equilibrium state. The unperturbed state $D_0(t,t_0)$, as well as the unperturbed Hamiltonian H_0, may be time-dependent in the interaction picture. Needless to say, what has been given here is a compact derivation of the linear response (4.7), and not a justification of its validity, which cannot hold for observables A having short characteristic evolution times [6] .

5. VARIATIONAL APPROXIMATIONS : GENERALITIES

5.1. Restricted variational spaces

In this Section, we begin to construct approximation schemes exploiting the stationarity of I. The variational quantities $A(t)$ and $\mathcal{D}(t)$ are now restricted within some trial subsets, and the stationary value of I provides an approximation for $<A>$. Let us review some general features of such reductions.

The variations δA and $\delta\mathcal{D}$ are no longer the most general ones, and the vanishing of δI no longer implies (3.8) and (3.9). Instead, we obtain the equations of motion

$$\text{Tr } \delta A\left(\frac{d\mathcal{D}}{dt} + i [H,\mathcal{D}]\right) = 0 , \quad (5.1)$$

$$\text{Tr } \delta\mathcal{D}\left(\frac{dA}{dt} - i [A,H]\right) = 0 , \quad (5.2)$$

where δA and $\delta\mathcal{D}$ are (at each time t) arbitrary allowed variations.

If the trial set for A happens to be (as in Section 6) a vector space spanned by a given basis of n observables ω^i, the trial observables are linear functions of n independent parameters a_i, and the unknown trial observables have the form

$$A(t) = \sum_i a_i(t) \omega^i .$$

The variational equations (5.1) for $\mathcal{D}(t)$ read

$$\text{Tr } \omega^i\left(\frac{d\mathcal{D}}{dt} + i[H,\mathcal{D}]\right) = 0,$$

and they provide n differential equations for the n parameters of $\mathcal{D}(t)$, in which the motion of $A(t)$ does not enter. However, if the trial set for A is parametrized non-linearly (as in Section 7), the approximate equations of motion for $\mathcal{D}(t)$ are coupled to the trial observable $A(t)$.

Similarly, if the reduced densities $\mathcal{D}$ are parametrized non-linearly, which will be the case in all the examples, the approximate equations of motion (5.2) for $A(t)$ are coupled to the approximate operator $\mathcal{D}(t)$ itself. When both sets of equations (5.1) and (5.2) are coupled (as in Section 7), the mixed boundary conditions (3.6) and (3.7) introduce a further complexity.

The coupling between $\mathcal{D}$ and A is the price paid for the unavoidable contractions of states and observables. The best choice for $\mathcal{D}(t)$ will depend in general on the observable A to be measured as well as on the time t_1 of the measurement. In other words, one should not consider $\mathcal{D}(t)$ as a genuine density operator which could be used for predicting the average value of an arbitrary observable.

5.2 Some general properties of the approximations

We gather here some simple relations, satisfied by the exact solution of Section 3, which remain valid when the restricted trial spaces for $\mathcal{D}$ and A have special properties. In all examples treated below, the restricted space for A will be such as to allow variations δA proportional to A. One of the variational equations (5.1) is then

$$\text{Tr } A\left(\frac{d\mathcal{D}}{dt} + i[H,\mathcal{D}]\right) = 0 , \qquad (5.3)$$

and the stationary value of I is approximately given by (3.3) as

$$<A> = \text{Tr } A \, \mathcal{D}(t_1) , \qquad (\delta A \propto A) . \qquad (5.4)$$

In such cases, it is not necessary to know $A(t)$ for evaluating I_{st} if the equations for $\mathcal{D}(t)$ are decoupled from $A(t)$.

176

Similar considerations hold if the trial subspace $\mathcal{D}$ is such that it allows variations $\delta\mathcal{D}$ proportional to $\mathcal{D}$. We have then

$$<A> = \text{Tr } A(t_0) D \quad , \quad (\delta\mathcal{D} \propto \mathcal{D}) ,\tag{5.5}$$

which may help to evaluate our quantity $<A>$ of interest without solving the approximate equations for $\mathcal{D}(t)$ (at least if the equations for A are not coupled to them).

Allowing variations $\delta\mathcal{D}$ proportional to $\mathcal{D}$ implies that no normalization condition is imposed on the trial space for $\mathcal{D}$. Actually, even in the exact case, we had to perform unrestricted variations of $\mathcal{D}$ to obtain the backward Heisenberg equation (3.9), including variations changing the norm of $\mathcal{D}$. The normalization of $\mathcal{D}$ at all times was only a property of the stationary solution, and appeared as a consequence of the equation of motion (3.8) and of the normalization of the initial condition D. In approximate treatments, the normalization of $\mathcal{D}$ at all times is guaranteed if the allowed variations A include the unit observable I, in which case the equations of motion (5.1) yield

$$\frac{d}{dt} \text{Tr}\mathcal{D} = 0, \quad (\delta A \propto I).\tag{5.6}$$

In all examples given below, both variations $\delta A \propto A$ and $\delta\mathcal{D} \propto \mathcal{D}$ will be allowed. In such cases, besides all the properties (5.3-6), we have

$$<A> = \text{Tr } A(t) \mathcal{D}(t) \quad , \quad (\delta A \propto A, \ \delta\mathcal{D} \propto \mathcal{D}) ,\tag{5.7}$$

for any intermediate time t, as in the exact case (3.13). The equation (5.7) is obtained by adding (5.3) to the corresponding equation for dA/dt ; it interpolates with respect to time between (5.4) and (5.5).

Other general properties concern infinitesimal changes of the boundary conditions. Let us look for the effect $\Delta <A>$ of a small modification ΔD of the initial state, and of a small modification ΔA of the measured observable. (The variations ΔD and ΔA, as well as the induced variations $\Delta\mathcal{D}(t)$ and $\Delta A(t)$, should be compatible with the restrictions imposed on the trial sets of $\mathcal{D}$ and A by the approximation considered). The use of (3.5), together with (5.1) and (5.2), provides

$$\Delta<A> = \text{Tr}A(t_0)\Delta D + \text{Tr}\Delta A\mathcal{D}(t_1).\tag{5.8}$$

The variation $\Delta <A>$ generated by an infinitesimal change ΔD of the initial condition depends only on the (approximate) value of $A(t)$ at the initial time t_0, while the effect of an infinitesimal change ΔA of the observable A depends only on the (approximate) value of $\mathcal{D}(t)$ at the final time t_1.

In the case where $\delta A \propto A$ is allowed, by performing on (5.4) the changes of boundary conditions ΔD and ΔA, we get

$$\Delta <A> = \text{Tr}A\Delta D(t_1) + \text{Tr}\Delta A D(t_1). \tag{5.9}$$

The comparison of (5.9) with the general expression (5.8) suggests the validity in this case of the relation

$$\Delta <A> = \text{Tr}A(t)\Delta D(t) + \text{Tr}\Delta A D(t_1), \quad (\delta A \propto A) \tag{5.10}$$

for an arbitrary intermediate time t. Indeed, by taking the Δ-variation of (5.3) and by using (5.1) and (5.2), one obtains

$$\frac{d}{dt}\text{Tr}A\Delta D = 0, \quad (\delta A \propto A), \tag{5.11}$$

from which (5.10) results.

Similarly, in the case where $\delta D \propto D$ is allowed, one can perform Δ-variations on (5.5), then show that

$$\frac{d}{dt}\text{Tr}\Delta A D = 0, \quad (\delta D \propto D), \tag{5.12}$$

and obtain in this way

$$\Delta <A> = \text{Tr}A(t_0)\Delta D + \text{Tr}\Delta A(t)D(t), \quad (\delta D \propto D), \tag{5.13}$$

for any intermediate time t.

We already have noticed the conservation of the pseudo-Hamiltonian h in the exact evolution. From the definition (3.2) of h, we obtain

$$\frac{dh}{dt} = -i \text{ Tr} \left\{ \frac{dA}{dt}[H,D] + \frac{dD}{dt}[A,H] \right\} . \tag{5.14}$$

In any approximation, the time-derivative of A belongs to the class of allowed variations δA, which enter the approximate equation of motion (5.1). We can, therefore, write the first term of (5.14) as

$$\text{Tr} \frac{dA}{dt} \frac{dD}{dt} .$$

The same remark for dD/dt together with Eq.(5.2) shows that the second term of (5.14) cancels with the first. Therefore, the pseudo-energy h is conserved in any approximation (provided the Hamiltonian H is time-independent) :

$$\frac{dh}{dt} = -i\frac{d}{dt}\text{Tr}A[H,D] = 0 .$$

178

5.3 Symmetries and broken symmetries

In many cases, the contraction of the trial spaces will be responsible for the violation of properties which are fulfilled by the exact solutions. Non-linearity and coupling may creep into the approximate equations of motion (5.1) and (5.2) for $\mathcal{D}(t)$ and $A(t)$, which replace (3.8) and (3.9). Other features of the exact equations may also be violated.

Consider for instance some observable C commuting with H. For the exact solution $\mathcal{D}(t) = D(t,t_0)$, the average of C does not depend on t. However, approximate solutions $\mathcal{D}(t)$ may render Tr C $\mathcal{D}(t)$ time-dependent. The occurence of such a violation does not necessarily indicate an unphysical approximation. A time-dependence for Tr C $\mathcal{D}(t)$, where C is a constant of the motion, will always occur in the context of the evaluation of another average <A>. If we were to use the variational principle to approximately evaluate <C> at t_1, we would have to take a trial subspace for $A(t)$ containing the observable C, and the variational equation (5.2) would admit the constant solution $A(t) = C$. Insertion in (3.4) would provide the exact time-independent expression

$$<C>_{t_1} = Tr\ C\ D = <C>_{t_0} .$$

If, however, the observable A under scrutiny differs from C, this argument does not hold. The approximate solution $\mathcal{D}(t)$ is designed for the prediction of <A>, and it is not legitimate to use it for evaluating <C> through Tr C $\mathcal{D}(t)$.

Other general properties may be violated by approximate solutions. For instance, $Tr\ A(t)\mathcal{D}(t)$ might depend on t, in constrast to the exact equation (3.13). However, in such a case, Tr $A(t)\ \mathcal{D}(t)$ does not coincide with the stationary value of I, which is the only quantity approximating <A>. Also, while the Liouville-von Neumann Eq.(3.8) conserves the eigenvalues of $\mathcal{D}$, this property may be violated by the approximate equation of motion (5.1). One may also think of violating other fundamental symmetries and laws. Nothing prevents an initially pure state ($D^2 = D$) from evolving into a mixture for $\mathcal{D}(t)$; this is acceptable, since the quantity $\mathcal{D}^2(t)$ is never supposed to be evaluated. Neither is it required that $\mathcal{D}(t)$ and $A(t)$ remain Hermitean. Indeed, hermiticity of $\mathcal{D}(t)$ and $A(t)$ means that Tr B $\mathcal{D}(t)$ is real for any Hermitean observable B and that Tr $A(t)$ D is real for any physical state D, while the only requirement existing on $\mathcal{D}(t)$ and $A(t)$ is the reality of I_{st}.

6. SINGLE-PARTICLE OBSERVABLES AND THE TDHF APPROXIMATION

6.1 Single-particle trial sets

We shall focus our interest on many-body fermion systems, having in mind large collective motions, like nuclear fission or collisions between heavy nuclei or atoms. In terms of the creation and annihilation operators associated with an arbitrary basis $\alpha(\alpha = 1, 2, \ldots)$ the Hamiltonian of the system considered reads

$$H \equiv \sum_{\alpha\beta} K_{\alpha\beta} c_\alpha^\dagger c_\beta + \frac{1}{4} \sum_{\alpha\beta\gamma\delta} <\alpha\beta|v|\gamma\delta> c_\alpha^\dagger c_\beta^\dagger c_\gamma c_\delta \ , \tag{6.1}$$

where $K_{\alpha\beta}$ is a matrix element of the kinetic energy, and

$$<\alpha\beta|v|\gamma\delta> = <\alpha(2)\beta(1)|\hat{v}(1,2)|\gamma(1)\delta(2)> - <\alpha(2)\beta(1)|\hat{v}(1,2)|\delta(1)\gamma(2)> \ , \tag{6.2}$$

is an antisymmetrized matrix element of the two-body interaction $\hat{v}$. We shall also denote as V_{12}, the antisymmetrized interaction (6.2), the indices 1 and 2 referring to (β,γ) and (α,δ), respectively.

From now on, we shall always choose as trial class for $\mathcal{D}(t)$ the set of uncorrelated states

$$\mathcal{D}(t) = e^{-\mathbf{M}(t)}, \tag{6.3}$$

$$\mathbf{M}(t) = m(t) + \sum_{\alpha\beta} M_{\alpha\beta}(t) \ c_\alpha^\dagger c_\beta^\dagger \ , \tag{6.4}$$

where the variational parameters are the time-dependent matrix elements $M_{\alpha\beta}$ and the scalar m. Such density operators play a central role in approximations, since they render feasible explicit calculations through the use of the generalized Wick's theorem [7]. Within this choice of $\mathcal{D}(t)$, different mean-field approximations can now be built, depending on the choice of the trial class for A(t).

Suppose first that we are interested in evaluating the average $<\mathbf{Q}>$ at the time t_1 of some single-particle observable

$$\mathbf{Q} \equiv q + \sum_{\alpha\beta} Q_{\alpha\beta} c_\alpha^\dagger c_\beta \ . \tag{6.5}$$

It is then natural to restrict also the time-dependent variational observable within the single-particle class :

$$A(t) = b(t) + \sum_{\alpha\beta} B_{\alpha\beta}(t) \ c_\alpha^\dagger c_\beta \ , \tag{6.6}$$

with the final conditions

$$b(t_1) = q, \quad B(t_1) = Q. \tag{6.7}$$

180

Instead of parameterizing $\mathcal{D}(t)$ in terms of m(t) and M(t), we introduce equivalently the normalization

$$z(t) \equiv \mathrm{Tr}\mathcal{D} = \exp[\, \mathrm{tr}\ln(1+e^{-M}) - m]\, , \tag{6.8}$$

and the contractions

$$\rho_{\alpha\beta}(t) \equiv \overline{c_\beta^\dagger c_\alpha} = \mathrm{Tr}\, c_\beta^\dagger c_\alpha \, \mathcal{D}/\mathrm{Tr}\mathcal{D} = \delta_{\alpha\beta} - \overline{c_\alpha c_\beta^\dagger}\,. \tag{6.9}$$

The matrices ρ and M are connected by

$$\rho = \frac{1}{e^M + 1}\,. \tag{6.10}$$

A particular case of special interest is generated by letting the eigenvalues of M tend to $\pm\infty$; in this limit, $\mathcal{D}$ reduces to the projector on a Slater determinant.

6.2 Variational derivation of the TDHF equations

The next step is to derive the equations of motion for the trial functions b(t), B(t) which parametrize A(t), and z(t), $\rho(t)$ which parametrize $\mathcal{D}(t)$, with the proper final and initial conditions, respectively. It is convenient to start from the expressions (3.3) and (3.4) of the action I in terms of the independent variables b, B, z,ρ.(Traces in the single-particle space will be denoted as tr to distinguish them from traces Tr in the Fock space.) From (6.6), (6.8) and (6.9), we have

$$\mathrm{Tr}A\mathcal{D} = z(b + \mathrm{tr}\, B\rho), \tag{6.11a}$$

and hence

$$\mathrm{Tr}\, \frac{dA}{dt}\mathcal{D} = z\left(\frac{db}{dt} + \mathrm{tr}\, \frac{dB}{dt}\rho\right), \tag{6.11b}$$

$$\mathrm{Tr}A\frac{d\mathcal{D}}{dt} = \frac{dz}{dt}(b + \mathrm{tr}\, B\rho) + z(\mathrm{tr}\, B\,\frac{d\rho}{dt})\,. \tag{6.11c}$$

We also need to evaluate the pseudo-Hamiltonian h defined by (3.2). From the commutator

$$[\, c_\lambda^\dagger c_\mu, H] = \sum_\beta K_{\mu\beta}c_\lambda^\dagger c_\beta - \sum_\alpha K_{\alpha\lambda}c_\alpha^\dagger c_\mu$$

$$+ \frac{1}{2}\sum_{\beta\gamma\delta}\langle\mu\beta|v|\gamma\delta\rangle c_\lambda^\dagger c_\beta^\dagger c_\gamma c_\delta - \frac{1}{2}\sum_{\alpha\beta\gamma}\langle\alpha\beta|v|\gamma\lambda\rangle c_\alpha^\dagger c_\beta^\dagger c_\gamma c_\mu\,,$$

we obtain $[A,H]$, and from the Wick theorem for

$$\mathrm{Tr}\, c_\alpha^\dagger c_\beta^\dagger c_\gamma c_\delta \mathcal{D} = z(\rho_{\gamma\beta}\rho_{\delta\alpha} - \rho_{\gamma\alpha}\rho_{\delta\beta})\,,$$

we deduce

$$h = -i \, TrA[H, \mathcal{D}] = -iz \, tr \, B[W(\rho), \rho] \; , \qquad (6.12)$$

where we recognize the s.p. Hartree–Fock Hamiltonian

$$W_{\alpha\beta}(\rho) = K_{\alpha\beta} + \sum_{\lambda\mu} \langle \lambda\alpha | V | \beta\mu \rangle \, \rho_{\mu\lambda}. \qquad (6.13a)$$

The more compact notations

$$W_1 = K_1 + tr_2 \, V_{12} \, \rho_2 \quad \text{or} \quad W = K + tr_2 \, V\rho \qquad (6.13b)$$

will also be used, where tr_2 denotes the trace on the pair α, δ of indices of $\langle \alpha\beta | V | \gamma\delta \rangle$.

Substituting (6.11) and (6.12) into (3.3), the approximate action takes the s.p. form

$$= z(b + tr \, B\rho) \Big|_{t_1}$$

$$-\int_{t_0}^{t_1} dt \left\{ \frac{dz}{dt}(b + tr \, B\rho) + z \, tr \, B(\frac{d\rho}{dt} + i[W, \rho]) \right\} . \qquad (6.14)$$

The variation of (6.14) with respect to b [with $\delta b(t_1) = 0$ and $z(t_0) = 1$] yields the normalization equation $dz/dt = 0$, and hence

$$z(t) = 1. \qquad (6.15)$$

The variation of (6.14) with respect to the matrix elements of B, together with $\delta B(t_1) = 0$ and with (6.15), gives then the time-dependent Hartree–Fock equations

$$i \frac{d\rho}{dt} = [W(\rho), \rho] , \qquad (6.16)$$

to be solved with the initial condition $\rho(t_0)$.

We have thus recovered within our variational framework the TDHF equations. As a consequence of the linear parametrization of the set $A(t)$ (see Section 5.1), the equations (6.15) and (6.16) for $\mathcal{D}(t)$ are decoupled from the equations of motion for $A(t)$. The average value of any s.p. observable (6.5) is given by

$$\langle A \rangle = Tr \, A \, \mathcal{D}(t_1) = q + tr \, Q\rho(t_1). \qquad (6.17)$$

The equations of motion (6.16) preserve all the eigenvalues of ρ, and hence from (6.3) and (6.10) all the eigenvalues of $\mathcal{D}$, which means that the approximate evolution of $\mathcal{D}(t)$ is unitary.

6.3 The reduced backward Heisenberg equation

It is not necessary, in the TDHF scheme, to look for the approximate evolution of $A(t)$ since (6.17) can be evaluated by solving Eq.(6.16) only. It is nevertheless natural to look into the equations of motion for the dynamic variables $b(t)$ and $B(t)$ which characterize $A(t)$. Moreover, far from being superfluous, the approximate backward Heisenberg equations for $B(t)$ will be an essential ingredient in another problem [8], the evaluation of fluctuations of single-particle observables in a mean-field context.(We shall come back to this topic in Section 8.)

In order to establish directly the desired equations of motion for b and B, we may integrate (6.14) by parts, or equivalently start from (3.4), which by use of (6.11) and (6.12) reads

$$I = z(b + tr\ B\rho)\Big|_{t_0}$$
$$+ \int_{t_0}^{t_1} dt\ z\left\{\frac{db}{dt} + tr\left(\frac{dB}{dt}\rho - i\ B[W(\rho),\rho]\right)\right\}. \tag{6.18}$$

The variation of (6.18) with respect to z yields

$$\frac{db}{dt} + tr\left(\frac{dB}{dt}\rho - i\ B[W,\rho]\right) = 0\ ,$$

an equation which, combined with (6.16), (6.11a) and (6.15), reduces to the property

$$\frac{d}{dt}(b + tr\ B\rho) = \frac{d}{dt}\ Tr\ A(t)\ \mathcal{D}(t) = 0, \tag{6.19}$$

already established on general grounds [Eq.(5.7)].

The variation of (6.18) with respect to ρ yields, taking into account the definition (6.13) of W,

$$0 = tr\left(\frac{dB}{dt}\delta\rho - i\ B[W,\delta\rho]\right) - i\ tr_1\ B[tr_2\ V\ \delta\rho,\rho]\ .$$

By interchanging the particles 1 and 2 in the second term, and by using the cyclic invariance of the trace, we get

$$\frac{dB}{dt} - i[B,W] + i\ tr_2\ V[B,\rho] = 0. \tag{6.20}$$

This linear and homogeneous equation for the matrix elements of B appears as a reduction to the single-particle space of the backward Heisenberg equation, resulting from the choices (6.3) and (6.6) for the trial spaces of $\mathcal{D}$ and A.

The equation (6.20) is the "backward dual" of the linearized version of the TDHF equation (6.16),

$$\frac{d\Delta\rho}{dt} + i[W,\Delta\rho] + i[tr_2 V\Delta\rho,\rho] = 0, \tag{6.21}$$

in which $\rho+\Delta\rho$ is a solution of (6.16) with modified initial conditions. Indeed, one observes that (6.20) and (6.21) can be written (in terms of the same kernel R) as

$$\frac{dB_{\gamma\delta}}{dt} - i\sum_{\alpha\beta} B_{\beta\alpha} R_{\alpha\beta,\gamma\delta} = 0 , \tag{6.20'}$$

$$\frac{d\Delta\rho_{\alpha\beta}}{dt} + i\sum_{\gamma\delta} R_{\alpha\beta,\gamma\delta}\Delta\rho_{\delta\gamma} = 0. \tag{6.21'}$$

The linearized equation (6.21) appears as the time-dependent generalization of the usual RPA, which is recovered by taking in (6.21) a static solution .

The equation of motion for b results from (6.19), (6.20) and (6.16) :

$$\frac{db}{dt} = i\ tr_{12}V_{12}[B_2,\rho_2]\ \rho_1. \tag{6.22}$$

Due to self-consistency, a coupling exists between the scalar part b of the operator $A(t)$ and the proper single-particle part characterized by the matrix B.

6.4 Validity of the TDHF equations

The above derivation exhibits the fact that, within the class of uncorrelated density operators, the time-dependent Hartree-Fock approximation provides the best choice, in the sense of our variational principle, for predicting the expectation value of a single-particle observable. As is well-known, but not always appreciated, the approximate equations of motion depend neither on the particular choice of the s.p. observable A, nor on the observation time t_1. It is worthwhile emphasizing that this feature is not obvious a priori in the present variational framework. It arises from the linear parametrization (6.6) of the trial set for $A(t)$.

According to our variational derivation, the TDHF approximation appears as the best mean field answer to s.p. questions, but only to single-particle questions. If the observable of interest A does not have the s.p. form Q of (6.5), as for instance in the case of a two-body operator, or more simply of the square Q^2 of a s.p. operator, the whole argument of this Section is lost, and there is no reason for the TDHF equation (6.16) to remain valid.

A hint in favour of this statement is also provided by the discussion of conservation laws. As already shown in Section 5.3 for any variational approximation, $<A>$ does not depend on t_1 if the observable A to be measured at the time t_1 is conserved. Here, the equation of motion (6.16) has been adjusted to any s.p. observable A, and therefore the approximate average value

$$<C> = c + tr\ C\ \rho(t_1)\tag{6.23}$$

of any conserved s.p. observable C (taken as the boundary condition A) is independent of t_1. (It is an instructive exercise to check directly this property from the TDHF equation (6.16)).

The proof of Section 5.3 cannot be extended to conserved quantities involving more than one particle, such as the square of the total momentum or angular momentum. It is easy to find examples of conserved quantities C and C^2 such that the TDHF expression

$$<C^2> = (c + tr\ C\rho)^2 + tr\ C\ \rho\ C(1-\rho)\tag{6.24}$$

varies in time (through its last term) when ρ evolves along (6.16). The TDHF approximation may therefore violate conservation laws other than those of s.p. character.

These remarks support our belief that the applicability of TDHF is limited. It seems difficult to trust this approximation beyond the evaluation of averages of s.p. observables. Actually, when the observable A of interest does not belong to the s.p. class, our variational viewpoint leads us to optimize a new action, adapted to the observable A and differing from (6.14). New approximate equations of motion, adapted to the observable of interest, are thus expected.

7. CHARACTERISTIC FUNCTIONS AND TRANSITION PROBABILITIES

7.1 Two problems for a single formalism

The purpose of this Section is to extend the mean-field variational approach of Section 6 to the evaluation of quantities more general than the averages $<Q>$ of s.p. observables. We still wish to take advantage of the Wick theorem. Thus we keep the same uncorrelated trial density operators (6.3-4) as in Section 6 (and hence the same type of initial state). The trial observable $A(t)$, however, will no longer be of the s.p. type. (Since the number of degrees of freedom of $\mathcal{D}(t)$ has not changed, we must limit ourselves to trial observables $A(t)$ which can have the same type of parametrization as a s.p. observable Q.)

Assume for example that we are interested in the prediction of $\langle \mathbf{Q}^2 \rangle$. Just as the form $A = \mathbf{Q}$ for the observable of interest had suggested in Section 6.1 to choose (6.6) as a trial set, it is natural, in this case, to take

$$A(t) = [\mathbf{B}(t)]^2 = [b(t) + \sum_{\alpha\beta} B_{\alpha\beta}(t) \, c_\alpha^+ c_\beta]^2 , \qquad (7.1)$$

with the boundary condition $b(t_1) = q$, $B(t_1) = Q$. The formalism may be developed (through use of Wick theorem) as in Section 6.2, resulting in a set of equations for the independent variables z, ρ which parametrize $\mathcal{D}(t)$ and b, B which parametrize (7.1). Since these parametrizations of $\mathcal{D}(t)$ and $A(t)$ are non-linear, the equations of motion for ρ and B are fully coupled (Section 5.1). The s.p. density ρ suited to the evaluation of $\langle \mathbf{Q}^2 \rangle$ does not obey the TDHF equation.

In most cases, one is actually interested in evaluating a fluctuation $\Delta \mathbf{Q}^2 = \langle \mathbf{Q}^2 \rangle - \langle \mathbf{Q} \rangle^2$ rather than an average $\langle \mathbf{Q}^2 \rangle$. (For instance, in a heavy ion reaction, one would like to predict the mass dispersion of the fragments). The densities ρ appropriate to the evaluation of $\langle \mathbf{Q} \rangle$ and $\langle \mathbf{Q}^2 \rangle$ would then be obtained from two different sets of equations of motion and the evaluation of $\Delta \mathbf{Q}^2$ would require substracting the result of two different approximations. For this reason, we do not consider it worthwhile to persist and to write the equations of motion which would optimize $\langle \mathbf{Q}^2 \rangle$. We shall introduce instead a more natural procedure which allows to evaluate $\Delta \mathbf{Q}^2$ in a single step.

As is well-known, the statistical fluctuation of an observable can be generated by expanding its characteristic function. For a s.p. observable $\mathbf{Q}$, we are thus led to choose for the quantity A at the time t_1 the exponential

$$A = e^{-\lambda \mathbf{Q}} , \quad \mathbf{Q} = q + \sum_{\alpha\beta} Q_{\alpha\beta} c_\alpha^+ c_\beta , \qquad (7.2)$$

and to expand $\ln \langle A \rangle$ in powers of λ. The first order term will provide $\langle \mathbf{Q} \rangle$, for which one expects to find the same TDHF result as in Section 6. The second order term will provide $\Delta \mathbf{Q}^2$, and higher order terms would yield higher moments. Again, the boundary condition (7.2) suggests to restrict the variational operator $A(t)$ to a trial set of the form

$$A(t) = e^{-\mathbf{L}(t)} , \qquad (7.3)$$

where the s.p. operator

$$\mathbf{L}(t) \equiv \ell(t) + \sum_{\alpha\beta} L_{\alpha\beta}(t) \, c_\alpha^+ c_\beta \qquad (7.4)$$

is constrained to take the value $\lambda \mathbf{Q}$ at the time t_1. The choice of an exponential form for $A(t)$, similar to the trial form (6.3) for $\mathcal{D}(t)$, will lead again to a tractable formalism. Indeed, the algebra of exponentials of s.p. operators makes it possible, through an extension of the Wick theorem [9], to express the action I in the s.p. space.

More generally, the same formalism holds for the evaluation of the correlation between two s.p. observables $\mathbf{Q}_1$ and $\mathbf{Q}_2$. It is sufficient to modify the boundary condition imposed on $A(t_1)$ into

$$A = e^{-\lambda_1 \mathbf{Q}_1 - \lambda_2 \mathbf{Q}_2} , \qquad (7.5)$$

and to select in the expansion of the resulting value for $<A>$ the term in $\lambda_1 \lambda_2$. Thus, some two-body quantities such as fluctuations and correlations, may be evaluated approximately but consistently within a s.p. framework.

The ansatz (6.3) for $\mathcal{D}(t)$ and (7.3) for $A(t)$ happens to be of interest in connection with another problem, namely the evaluation of transition probabilities in a mean-field approach. Let us suppose that we are looking for the transition probability from a pure state $|\Psi>$ at the time t_0 to the pure state $|\Phi>$ at the time t_1. Such a transition probability is nothing else but the average at t_1 of the observable

$$A = |\Phi> <\Phi| , \qquad (7.6a)$$

for an initial state

$$D = |\Psi> <\Psi| . \qquad (7.6b)$$

A collision cross-section is for instance obtained by taking for $|\Psi>$ the incoming state of the two projectiles, and for $|\Phi>$ an outgoing state of the final fragments. In the theory of heavy ion collisions, the fragments constituting the initial and final states $|\Psi>$ and $|\Phi>$ have often been described in the mean-field approximation as Slater determinants. The problem of constructing the scattering amplitude in this model has received much attention in recent years. The results obtained by several authors [5,10] can be recovered as a special case of the present formalism, which deals however with transition probabilities rather than amplitudes. In order to fit this problem into the present framework we may notice that the boundary conditions (7.6) for D and A belong to the class of projectors on the Slater determinants, which could be chosen as trials set for both $\mathcal{D}(t)$ and $A(t)$. However, since such projectors are limiting cases of exponentials of s.p. operators, it is not less natural (and it is, in fact, more convenient) to choose as trail sets for $\mathcal{D}(t)$ and $A(t)$ the exponentials (6.3) and (7.3). The resulting equations of motion will be, obviously, identical with the equations suited to the evaluation of characteristic functions. The operators $\mathcal{D}$ and A are parametrized in terms of the trial functions of time m, $M_{\alpha\beta}$, ℓ, $L_{\alpha\beta}$ non-linearly. The equations of motion for these parameters will therefore be coupled.

7.2 Coupled equations for the state and the observable

This Section will be devoted to the evaluation of the stationary value of I for a trial density of the form (6.3) and a trial observable of the form (7.3) . We shall first derive the approximate action I, then write the equations of motion whose solutions give the mean-field expression for <A>.

The two equivalent sets of variables $(m, M_{\alpha\beta})$ and $(z, \rho_{\alpha\beta})$ which parametrize $\mathcal{D}$ are related by Eqs.(6.8) and (6.10). Similarly, we characterize $A(t)$ either by $(\ell, L_{\alpha\beta})$ or by the parameters $(y, \sigma_{\alpha\beta})$ defined as

$$y \equiv \text{Tr } A = \exp\left[-\ell + \text{tr } \ln(1+e^{-L})\right] , \tag{7.7}$$

$$\sigma \equiv \text{Tr } c^{\dagger}cA/\text{Tr}A = \frac{1}{e^{L}+1} . \tag{7.8}$$

The action I, defined in (3.1) and (3.2), involves both products $A\mathcal{D}$ and $\mathcal{D}A$. The product of two exponentials of s.p. operators is still the exponential of a s.p. operator [9] and we can define $\mathbf{L}'$ and $\mathbf{M}'$ as

$$A\mathcal{D} = e^{-\mathbf{L}} e^{-\mathbf{M}} \equiv e^{-\mathbf{L}'}, \quad \mathcal{D}A = e^{-\mathbf{M}}e^{-\mathbf{L}} \equiv e^{-\mathbf{M}'}. \tag{7.9}$$

The variables ℓ' and $L_{\alpha\beta}'$ characterizing $\mathbf{L}'$, and the variables m' and $M_{\alpha\beta}'$ characterizing $\mathbf{M}'$, are expressed in terms of ℓ, L, m, M as

$$\ell' = \ell + m = m' , \tag{7.10}$$

$$e^{-L'} = e^{-L} e^{-M}, \quad e^{-M'} = e^{-M} e^{-L}. \tag{7.11}$$

Equations (7.11) are the transposition in the s.p. space of Eqs.(7.9),which hold in the Fock space. We shall need the trace of $A\mathcal{D}$, or $\mathcal{D}A$, which is evaluated in the same manner as (6.8) :

$$w \equiv \text{Tr } AD = \text{Tr } DA = \exp[-\ell' + \text{tr } \ln(1+e^{-L'})]$$

$$= \exp[-\ell - m + \text{tr } \ln(1+e^{-L} e^{-M})] . \qquad (7.12)$$

We shall also need the two different sets of contractions over DA and AD defined by

$$\rho'_{\alpha\beta} \equiv \text{Tr } DA \ c^{\dagger}_{\beta} c_{\alpha} / \text{Tr } DA, \qquad (7.13a)$$

$$\sigma'_{\alpha\beta} \equiv \text{Tr } AD \ c^{\dagger}_{\beta} c_{\alpha} / \text{Tr } AD. \qquad (7.13b)$$

These contractions are expressed in terms of M' and L' just as ρ and σ were expressed by (6.10) and (7.8) in terms of L and M. By using (7.11) we obtain

$$\rho' = \frac{1}{e^{M'}+1} = \frac{1}{e^{L}e^{M}+1} \ , \quad \sigma' = \frac{1}{e^{L'}+1} = \frac{1}{e^{M}e^{L}+1} \ , \qquad (7.14)$$

or equivalently, in terms of the parameters ρ and σ,

$$\rho' = \left(\frac{1-\sigma}{\sigma} \frac{1-\rho}{\rho} + 1 \right)^{-1} = \rho \ \frac{1}{1-\rho-\sigma+2\sigma\rho} \ \sigma \ , \qquad (7.14a)$$

$$\sigma' = \left(\frac{1-\rho}{\rho} \frac{1-\sigma}{\sigma} + 1 \right)^{-1} = \sigma \ \frac{1}{1-\rho-\sigma+2\rho\sigma} \ \rho \ . \qquad (7.14b)$$

It is easy to check that the complementary contraction $1-\rho'$ is given by

$$1-\rho' = \frac{1}{1+e^{-M}e^{-L}} = (1-\sigma)\frac{1}{1-\rho-\sigma+2\rho\sigma} (1-\rho) \ , \qquad (7.14c)$$

and similarly for $1-\sigma'$. Finally, we shall need to express the normalization w in terms of our variational parameters, z, ρ, y, σ :

$$w = \exp [\ln y + \text{tr } \ln(1-\sigma) + \ln z + \text{tr } \ln(1-\rho) - \text{tr } \ln(1-\rho')]$$

$$= y z \exp \text{tr } \ln(1-\rho-\sigma+2\sigma\rho) \qquad (7.15a)$$

$$= y z \det(1-\rho-\sigma+2\sigma\rho). \qquad (7.15b)$$

Notice that the inversion of the matrix $1-\rho-\sigma+2\sigma\rho$, which was needed to express (7.14), is possible whenever $w = \text{Tr}AD$ does not vanish, a condition always assumed to be satisfied.

We are now in a position to express the action (3.3) explicitly in terms of our independent variables z, ρ, y, σ. The time-derivative term $\mathrm{Tr}\, A\, d\mathcal{D}/dt$ may be obtained from (7.15) by differentiating w with respect to the parameters z and ρ, holding y and σ fixed[1]. The pseudo-hamiltonian term $-i\, \mathrm{Tr} A[H,\mathcal{D}] = -i\, \mathrm{Tr}\, \mathcal{D} A\, H + i\, \mathrm{Tr}\, A\mathcal{D}\, H$ involves two terms, which are readily evaluated by means of the generalized Wick theorem [9] with the weights $\mathcal{D}A$ and $A\mathcal{D}$. Indeed, (7.9) shows that these weights behave as non-normalized and non-Hermitean independent-particle density operators. We get for the two-body Hamiltonian (6.1)

$$\mathrm{Tr}\, \mathcal{D}A H = w\, E(\rho') \, , \tag{7.16}$$

where the energy $E(\rho)$ has the usual mean-field form

$$E(\rho) = \sum_{\alpha\beta} K_{\alpha\beta} \rho_{\beta\alpha} + \frac{1}{2} \sum_{\alpha\beta\gamma\delta} \langle \alpha\beta | V | \gamma\delta \rangle \rho_{\gamma\beta} \rho_{\delta\gamma}$$

$$= \mathrm{tr}\, K\, \rho + \frac{1}{2}\, \mathrm{tr}_1\, \mathrm{tr}_2\, V_{12} \rho_1 \rho_2 . \tag{7.17}$$

A similar expression $wE(\sigma')$ holds for $\mathrm{Tr}\, A\mathcal{D}H$. Altogether, the approximate action takes the form

$$I = w(t_1) - \int_{t_0}^{t_1} dt\, w \left\{ \frac{1}{z} \frac{dz}{dt} + \mathrm{tr} \left[\left(\frac{1-\sigma}{2\sigma-1} + \rho \right)^{-1} \frac{d\rho}{dt} \right] + iE(\rho') - iE(\sigma') \right\} , \tag{7.18}$$

where w, ρ' and σ' are expressed in terms of the independent variables z, ρ, y, σ through (7.15) and (7.14).

The equations of motion for the parameters z,ρ of $\mathcal{D}(t)$ are obtained by expressing the stationarity of I with respect to $A(t)$, i.e., with respect to $y(t)$ and $\sigma(t)$. (A detailed derivation is given in ref.[1].) The equation of motion for ρ reads

$$i\, \frac{d\rho}{dt} = (1-\rho)\, W(\rho')\rho - \rho W(\sigma')(1-\rho) \, , \tag{7.19}$$

whereas the equation of motion for z is

$$i\, \frac{1}{z} \frac{dz}{dt} = E(\rho') - E(\sigma') - \mathrm{tr}(\rho'-\rho)\, W(\rho') + \mathrm{tr}(\sigma'-\rho)\, W(\sigma') . \tag{7.20}$$

The integration of (7.20) over the time t is obvious, starting from the initial value $z(t_0) = 1$. More explicitly, in the case of two-body forces, we can replace $E(\rho')$, $E(\sigma')$, $W(\rho')$ and $W(\sigma')$ by their expressions (7.17) and (6.13).

The equations of motion for the parameters y and σ, obtained by rendering (7.18) stationary with respect to z and ρ, follow directly from (7.19) and (7.20) through interchange of $\mathcal{V}$, M, z, ρ and ρ' with A, $\mathbb{L}$, y, σ and σ'. This yields

$$i \frac{d\sigma}{dt} = (1-\sigma) \, W(\sigma') \, \sigma - \sigma \, W(\rho')(1-\sigma), \tag{7.21}$$

$$\frac{1}{y} \frac{dy}{dt} = \frac{1}{2} \, i \, \mathrm{tr}_{12} V_{12} \left[(\sigma'-\sigma)_1 (\sigma'-\sigma)_2 - (\rho'-\sigma)_1 (\rho'-\sigma)_2 \right]. \tag{7.22}$$

As expected, we have obtained equations of motion (7.19) and (7.21) for ρ and σ, which are coupled to each other through the definitions (6.13) of W, and (7.14) of ρ' and σ'. Once these equations are solved, z and y result by integration of (7.20) and (7.22). Recall that ρ and z satisfy initial conditions whereas σ and y satisfy final conditions.

The equations of motion may also be expressed in terms of the variables m, M, ℓ, L which equivalently parametrize $\mathcal{V}$ and A. By using (6.8), (7.19) and (7.20), we get

$$i \frac{dm}{dt} = -E(\rho') + E(\sigma') + \mathrm{tr}\, W(\rho')\rho' - \mathrm{tr}\, W(\sigma')\sigma'. \tag{7.23}$$

The antisymmetry of this expression in the interchange of A and $\mathcal{V}$ leads to

$$\frac{d\ell}{dt} = -\frac{dm}{dt}. \tag{7.24}$$

By using (6.10) and (7.19), we get

$$i \frac{d}{dt} e^{-M} = W(\rho') \, e^{-M} - e^{-M} \, W(\sigma'), \tag{7.25}$$

and symmetrically

$$i \frac{d}{dt} e^{-L} = W(\sigma') \, e^{-L} - e^{-L} \, W(\rho'). \tag{7.26}$$

We have thus obtained two equivalent sets of equations of motion (7.19-22) or (7.23-26), to be solved with the mixed boundary conditions accounting for the particulars of the problem. The desired quantity I_{st}, which is an approximation for the average (at the time t_1) of an observable A of the form (7.2), (7.5) or (7.6a) (the given initial state being uncorrelated) can be obtained by inserting the solution of these coupled equations of motion into (7.18). Actually, I_{st} is given more simply by (5.7), i.e., by

$\text{Tr}A(t)\mathcal{D}(t)$ at any time t. Our final result for <A> is therefore w, given either by (7.12) and (7.23-26) or by (7.15) and (7.19-22).

The equations of motion for ρ' and σ' resulting from the set (7.25), (7.26) and from the definitions (7.14) are particularly simple. By combining (7.25) and (7.26), we get

$$i \frac{d}{dt} e^{-M} e^{-L} = [W(\rho'), e^{-M} e^{-L}] \ ,$$

which together with (7.14c) provides

$$i \frac{d\rho'}{dt} = [W(\rho'), \rho'] \quad . \tag{7.27a}$$

Similarly, we have

$$i \frac{d\sigma'}{dt} = [W(\sigma'), \sigma'] \quad . \tag{7.27b}$$

The equations (7.27) have exactly the same form as the usual TDHF equations, though with non-Hermitean effective Hamiltonians $W(\rho')$ and $W(\sigma')$. They look more attractive than the set (7.19), (7.21), since they appear decoupled. However, this decoupling is fictitious. The boundary conditions on $\rho(t_0)$ and $\sigma(t_1)$ cannot be expressed conveniently in terms of ρ' and σ'. Moreover, the set of equations (7.27) is not complete, because the expressions (7.14a,b) of ρ' and σ' in terms of ρ and σ cannot in general be inverted in a unique fashion.

If at some time, $\mathcal{D}$ and A are Hermitean operators, and hence ρ and σ are Hermitean matrices, the matrices ρ' and σ' expressed by (7.14) are Hermitean conjugate. The s.p. Hamiltonians $W(\rho')$ and $W(\sigma')$ are also conjugate in (7.19-22). Therefore $d\rho/dt$ and $d\sigma/dt$ are Hermitean, and dz/dt and dy/dt are real : the Hermicity of $\mathcal{D}$ and A is preserved by the evolution.

Among the possible violations of conservation laws in the evolution (7.19) (7.20) of $\mathcal{D}(t)$, the time dependence of the norm $z = \text{Tr}\mathcal{D}(t)$ is noticeable. Even though D is normalized at the initial time, the optimal trial density operator $\mathcal{D}(t)$ may have a varying norm, and it cannot therefore be used for evaluating the expectation value $z = \text{Tr}\mathcal{D}$ of the unit observable I if A differs from I. Indeed, $\mathcal{D}(t)$ has been optimized for the evaluation at t_1 of $<e^{-\lambda Q}>$, so that the property $z(t_1) = 1$ is required only in the limit $\lambda = 0$, which corresponds to the trivial prediction of the expectation value 1 of <I>. We shall check below that z remains actually constant to zeroth order in λ.(If we insist on requiring that $\mathcal{D}(t)$ remains normalized at all times, we can still manage to obtain the same final results by relying on a variant of our variational principle enforcing the normalization ; for a discussion of this point see Section 8.3 of [1].)

7.3 The limiting case of transition probabilities

One of the problem which we had in mind in Section 7.1 was the evaluation of a transition probability from an initial state $|\Psi>$ to a final state $|\Phi>$, both approximated by Slater determinants. In the parametrizations (6.3) for $\mathcal{D}$ and (7.3) for A, we have to let all the eigenvalues of the matrices $M(t_0)$ and $L(t_1)$ tend to $\pm\infty$, with the normalization factors $m(t_0)$ and $\ell(t_1)$ also tending to $+\infty$ in such a way as to keep the traces $z(t_0)$ and $y(t_1)$ equal to 1. The operator $\mathcal{D}(t_0)$ [or $A(t_1)$] then reduces to the projector on the Slater determinant $|\Psi>$ [or $|\Phi>$] built from the eigenvectors of $M(t_0)$ [or of $L(t_1)$] associated with the eigenvalues $-\infty$. Such a limiting process is performed conveniently by using the parametrization of $\mathcal{D}(t)$ and $A(t)$ in terms of ρ, z, σ and y. It amounts to letting all the eigenvalues of ρ and σ tend to 0 or 1, a property characterized by

$$\rho(1-\rho) = 0, \quad \sigma(1-\sigma) = 0 .\tag{7.28}$$

In contrast to the general case of Section 7.2, the knowledge of $\rho' = \sigma'^\dagger$ happens to be sufficient to characterize ρ and σ when $\rho^2 = \rho$ and $\sigma^2 = \sigma$. The equations of motion (7.27) for ρ' and σ' are therefore now equivalent to the equations (7.19) and (7.21). However, the fact that the initial and final conditions fix $\rho(t_0)$ and $\sigma(t_1)$ makes it desirable to work still with ρ and σ.

The equation of motion (7.19) for ρ simplifies when $\rho^2 = \rho$ at the initial time (and hence at all times). This equation,

$$i \frac{d\rho}{dt} = [\tilde{W}(\rho;\sigma), \rho] ,\tag{7.29}$$

involves a self-consistent s.p. hamiltonian defined by

$$\tilde{W}(\rho;\sigma) = (1-\rho) W(\rho')\rho + \rho W(\sigma') (1-\rho),\tag{7.30}$$

where ρ' and σ' are expressed in terms of ρ and σ through equations (7.14) (which can be slightly simplified when $\rho^2 = \rho$ or $\sigma^2 = \sigma$).The hamiltonian $\tilde{W}$ is Hermitean, consistently with the Hermiticity at each time of ρ and σ. It involves only hole-particle and particle-hole matrix elements. The constancy in time of the eigenvalues (0 or 1) of ρ is obvious from (7.29).

Symmetrically, σ follows the equation

$$i \frac{d\sigma}{dt} = [\tilde{W}(\sigma;\rho),\sigma] ,\tag{7.31}$$

provided $\sigma(1-\sigma)$ vanishes at the initial time (and hence at all times).

Our final result, the <u>transition probability</u>, is given by (7.15) which reads

$$w = \text{Tr}A(t)\mathcal{D}(t) = \left| <\Phi(t)|\Psi(t)> \right|^2$$

$$= y\, z\, \det(1-\rho-\sigma+2\sigma\rho)\Big|_t \quad . \tag{7.32}$$

This expression does not depend on the intermediate time t. It should be evaluated after having solved the coupled equations (7.29), (7.31) for ρ and σ, and the equations (7.20) and (7.22) for z and y. The determinant in (7.32) can be simplified by using the identity

$$(1-\rho-\sigma+2\sigma\rho)(1-\rho-\sigma+2\rho\sigma) = 1 - (\rho-\sigma)^2$$

$$= (1-\sigma+\sigma\rho)(1-\rho+\rho\sigma),$$

a consequence of (7.28). Thus we get

$$w = yz\, \det(1-\sigma+\sigma\rho). \tag{7.33}$$

Although the set of equations (7.29-33) suffices in principle to provide the mean-field transition probability, it may be convenient to rewrite it in terms of <u>orbitals</u> rather than in terms of the matrices ρ, σ, ρ', σ'. However the orbitals characterizing these matrices are not uniquely defined. In the present context, the most natural parametrization is given by two orthornormal sets of orbitals, which are eigenfunctions of ρ (or M) and of σ (or L), respectively. The existing mean-field approximations for transition amplitudes [5,10] make use of an alternative parametrization of the Slater determinants $|\Psi>$ and $|\Phi>$. The equations resulting from these two choices of orbitals are given in ref.[1].

8. FLUCTUATIONS OF SINGLE-PARTICLE OBSERVABLES

In recent years, the TDHF approximation has been widely used to describe large amplitude nuclear motions, in particular heavy ion reactions. Single particle quantities, such as the large kinetic energy losses observed in these reactions, have been successfully reproduced without adjustable parameters [11]. However, the TDHF method grossly underestimates the widths of the mass distribution of the final fragments [11]. This discrepancy is often used as an argument against the validity of mean-field approaches, which in this view would only be able to predict the average values of s.p. operators. In Section 7.2 a <u>consistent</u> method has been presented for evaluating, within the mean field framework, the characteristic function $<e^{-\lambda Q}>_{t_1}$ of the single particle observable Q at the final time t_1. The evaluation of characteristic functions for several single-particle observables would require a mere change of this final condition into (7.5).

However, the equations of Section 7.2 are intricate and their solution appears to be difficult. In the present Section, we focus our interest on the first few moments, namely the average $\langle \mathbf{Q} \rangle$ and especially the fluctuation $\langle \mathbf{Q}^2 \rangle - \langle \mathbf{Q} \rangle^2$. Then, we need only the first terms of the expansion of $\ell n \langle e^{-\lambda \mathbf{Q}} \rangle$ in powers of λ. (Or of $\ell n \langle e^{-\lambda_1 \mathbf{Q}_1 - \lambda_2 \mathbf{Q}_2} \rangle$ in powers of λ_1 and λ_2 if we are interested in the correlation $\langle \mathbf{Q}_1 \mathbf{Q}_2 \rangle - \langle \mathbf{Q}_1 \rangle \langle \mathbf{Q}_2 \rangle$.) As we shall see, such expansions bring in considerable simplifications.

8.1 A simple formula for fluctuations

First let us recall [see eq.(5.7)] the relation

$$\langle A \rangle = \langle e^{-\lambda \mathbf{Q}} \rangle_{t_1} = \mathrm{Tr} A(t,\lambda) \mathcal{D}(t,\lambda), \qquad (8.1)$$

where [eqs.(7.2-4)]

$$A(t,\lambda) = e^{-\mathbb{L}}, \quad \text{with} \quad L(t_1) = \lambda Q, \quad \ell(t_1) = 0. \qquad (8.2)$$

Then let us expand the single-particle operator $\mathbf{L}$ in powers of λ up to second order :

$$\mathbf{L}(t,\lambda) = \mathbf{L}^{(0)} + \lambda \mathbf{L}^{(1)} + \frac{\lambda^2}{2} \mathbf{L}^{(2)}. \qquad (8.3)$$

We expand also $\mathcal{D}(t,\lambda)$ and $A(t,\lambda)$:

$$\mathcal{D}(t,\lambda) = \mathcal{D}^{(0)} + \mathcal{D}^{(1)} + \mathcal{D}^{(2)}, \qquad (8.4)$$

$$A(t,\lambda) = A^{(0)} + A^{(1)} + A^{(2)}, \qquad (8.5)$$

with

$$A^{(0)} = \mathbf{I} - \mathbb{L}^{(0)} + \frac{1}{2} \mathbb{L}^{(0)} \mathbb{L}^{(0)}, \qquad (8.6a)$$

$$A^{(1)} = -\lambda \mathbf{L}^{(1)} + \frac{\lambda}{2} (\mathbb{L}^{(0)} \mathbb{L}^{(1)} + \mathbb{L}^{(1)} \mathbb{L}^{(0)}). \qquad (8.6b)$$

Since we are particularly interested in the fluctuation $\Delta \mathbf{Q}^2 = \langle \mathbf{Q}^2 \rangle - \langle \mathbf{Q} \rangle^2$, we will also need the cumulant generating function $\varphi(\lambda)$ (or second characteristic function) :

$$\varphi(\lambda) = \ell n \langle A \rangle = \ell n \langle e^{-\lambda \mathbf{Q}} \rangle_{t_1} = -\lambda \langle \mathbf{Q} \rangle_{t_1} + \frac{1}{2} \lambda^2 \Delta \mathbf{Q}^2 \Big|_{t_1} + \dots \qquad (8.7),$$

which is such that

$$\langle \mathbf{Q} \rangle = - \frac{\partial \varphi}{\partial \lambda} \Big|_{\lambda=0} \quad \text{and} \quad \Delta \mathbf{Q}^2 = \frac{\partial^2 \varphi}{\partial \lambda^2} \Big|_{\lambda=0} . \qquad (8.8)$$

It is convenient to adopt a "mixed" parametrization such that $\mathcal{D}$ is characterized by z and ρ [eqs.(6.8-10)], while A is parametrized by ℓ and L [eqs.(7.3-4)]. The corresponding set of equations of motion, to be expanded in powers of λ, includes (7.19), (7.20), (7.26) and (7.23-24). The equations (7.14) are written accordingly as

$$\rho' = \rho \frac{1}{e^{L}(1-\rho)+\rho} \simeq \rho - \rho L(1-\rho) - \frac{1}{2}\rho L^2(1-\rho) + \rho L(1-\rho)L(1-\rho),$$
$$\tag{8.9a}$$

$$\sigma' = \frac{1}{(1-\rho)e^{L}+\rho}\rho \simeq \rho - (1-\rho)L\rho - \frac{1}{2}(1-\rho)L^2\rho + (1-\rho)L(1-\rho)L\rho .$$
$$\tag{8.9b}$$

Let us examine what happens in zeroth order. For $\lambda = 0$, we have to solve the equations of motion with the final condition $A(t_1) = I$. This means that we wish to predict the (trivial) characteristic function $\langle A \rangle = 1$ for the observable $\mathbf{Q} = 0$. The general argument of Section 5.3 holds since $Q = 0$ is a conserved observable, and the solution of (7.23-24), (7.26) is obviously

$$\ell^{(0)} = L^{(0)} = 0, \text{ or } A^{(0)}(t) = I.$$
$$\tag{8.10}$$

Then both quantites ρ' and σ' reduce in zeroth order to

$$\rho' = \sigma' = \rho ,$$
$$\tag{8.11}$$

and the equation (7.19) for ρ becomes

$$i\frac{d\rho}{dt} = (1-\rho)W(\rho)\rho - \rho W(\rho)(1-\rho) = [W(\rho),\rho] ,$$
$$\tag{8.12a}$$

which is the TDHF equation. From (7.20) we recover the normalization condition

$$\frac{dz}{dt} = 0.$$
$$\tag{8.12b}$$

The first stage of the expansion of the characteristic function led us to a <u>decoupling</u> of the equations of motion into the trivial solution $A(t) = I$ for the observable, and into the TDHF equations (8.12) for $\mathcal{D}(t)$. It may look odd that the TDHF equations arise here for the prediction of $\langle I \rangle = 1$, rather than for the prediction of single particle averages as in Section 6. It should be remembered however that the rules of the game have been changed. We are now asking the best mean field prediction for $\langle e^{-\lambda \mathbf{Q}} \rangle$, not just for $\langle \mathbf{Q} \rangle$ anymore.

This average value $\langle \mathbf{Q} \rangle$ is obtained as the coefficient in (8.1), or (8.7), of the first-order term in λ. However, owing to the stationarity of I, $<\mathbf{Q}>$ can be reached indirectly by solving the equations of motion in zeroth-order only. Indeed, since $\delta A \propto A$ is allowed, one can use (5.10) to write

$$\frac{\partial <A>}{\partial \lambda} = \text{Tr} \frac{\partial A}{\partial \lambda} \mathcal{D}(t_1) = -\text{Tr} \, \mathbf{Q} \, e^{-\lambda \mathbf{Q}} \mathcal{D}(t_1), \tag{8.13}$$

and therefore

$$<\mathbf{Q}> = -\frac{\partial <A>}{\partial \lambda}\bigg|_{\lambda=0} = -\frac{\partial \varphi}{\partial \lambda}\bigg|_{\lambda=0} = \text{Tr} \, \mathbf{Q} \mathcal{D}^{(0)}(t_1) = \text{tr} \, Q \rho(t_1) \ , \tag{8.14}$$

where $\rho(t_1)$ is the TDHF solution at the final time t_1. One has recovered the TDHF result (6.16-17).

From (8.9) and (8.10), we get at first order in λ

$$\rho' = \rho - \lambda \rho L^{(1)}(1-\rho), \qquad \sigma' = \rho - \lambda(1-\rho)L^{(1)}\rho. \tag{8.15}$$

Although this is not needed to compute $<\mathbf{Q}>$, let us expand the equation (7.26) up to first order. One finds

$$\frac{dL^{(1)}}{dt} - i[L^{(1)}, W(\rho)] + i \, \text{tr}_2 \, V[L^{(1)}, \rho] = 0, \tag{8.16}$$

which is the same equation as (6.20), $L^{(1)}(t)$ replacing $B(t)$. We recall (Section 6.3) that (8.16) is the reduced backward Heisenberg equation, which is dual to the time-dependent RPA equation (6.21).

We come now to our main problem, the evaluation of the fluctuations $\Delta \mathbf{Q}^2$ at the time t_1. We need [eq.(8.8.)] to compute

$$\frac{d^2\varphi}{\partial \lambda^2} = \frac{\partial^2 <A>}{\partial \lambda^2} \frac{1}{<A>} - \left(\frac{\partial <A>}{\partial \lambda}\right)^2 \frac{1}{<A>^2} \ , \tag{8.17}$$

and take its value for $\lambda = 0$. By using the relation (5.13) [$\delta \mathcal{D} \propto \mathcal{D}$ is allowed], we obtain

$$\frac{\partial <A>}{\partial \lambda} = \text{Tr} \frac{\partial A(t)}{\partial \lambda} \mathcal{D}(t) = \text{Tr} \frac{\partial A(t_0)}{\partial \lambda} D \ , \tag{8.18}$$

and hence

$$\frac{\partial^2 <A>}{\partial \lambda^2} = \text{Tr} \frac{\partial^2 A(t_0)}{\partial \lambda^2} D. \tag{8.19}$$

The relations (8.8) and (8.17-19) yield

$$\Delta \mathbf{Q}^2\bigg|_{t_1} = \frac{\partial^2 \varphi}{\partial \lambda^2}\bigg|_{\lambda=0} = \left(\text{Tr} \mathcal{D}(t_0)\frac{\partial^2 A(t_0)}{\partial \lambda^2}\right)_{\lambda=0} - \left(\text{Tr} \mathcal{D}(t_0)\frac{\partial A(t_0)}{\partial \lambda}\right)^2_{\lambda=0} \ , \tag{8.20}$$

since $<A>\big|_{\lambda=0} = 1$, and from (8.6) we get

$$\frac{\partial A(t)}{\partial \lambda}\bigg|_{\lambda=0} = -\mathbf{L}^{(1)}, \tag{8.21a}$$

$$\frac{\partial^2 A(t)}{\partial \lambda^2}\bigg|_{\lambda=0} = (\mathbf{L}^{(1)}\,\mathbf{L}^{(1)} - \mathbf{L}^{(2)}). \tag{8.21b}$$

The formula (8.21b) seems to indicate that the knowledge of $\mathbf{L}^{(2)}$ is necessary in order to evaluate the first term of (8.20).

Fortunately, the (difficult) evaluation of $\mathbf{L}^{(2)}$ can be avoided by taking advantage of the relation

$$\frac{d}{dt}\,\mathrm{Tr}\,\mathcal{D}^{(0)}(t)\left(\frac{\partial^2 A(t)}{\partial \lambda^2} - \frac{\partial A(t)}{\partial \lambda}\,\frac{\partial A(t)}{\partial \lambda}\right)_{\lambda=0} = 0\ . \tag{8.22}$$

The proof of (8.22) is not straightforward. It relies on the fact that the operator

$$A\,\frac{\partial}{\partial \lambda}\,(A^{-1}\frac{\partial A}{\partial \lambda}) = \frac{\partial^2 A}{\partial \lambda^2} - \frac{\partial A}{\partial \lambda}\,A^{-1}\,\frac{\partial A}{\partial \lambda}\ ,$$

is an allowed variation δA (indeed, it is the product of A by a single particle operator), which can therefore be inserted in the equations (5.1) and (5.2).

With the help of (8.22) and (8.21a) the relation (8.20) takes the more convenient form

$$\Delta \mathbf{Q}^2\bigg|_{t_1} = \mathrm{Tr}\,\mathcal{D}^{(0)}(t_0)\mathbf{L}^{(1)}\mathbf{L}^{(1)}(t_0) - \left(\mathrm{Tr}\mathcal{D}^{(0)}(t_0)\mathbf{L}^{(1)}(t_0)\right)^2, \tag{8.23}$$

where $\mathbf{L}^{(1)}(t_0)$ is the first order part [see eq.(8.3)] of the single particle operator (7.4). A straightforward application of the generalized Wick theorem [7] to (8.33) gives for the <u>fluctuation at the time t1</u> the formula

$$\Delta \mathbf{Q}^2\bigg|_{t_1} = \mathrm{tr}\,B(t_0)\rho(t_0)B(t_0)[1-\rho(t_0)] \equiv \Delta^2 B(t_0), \tag{8.24}$$

where we have set again $\mathbf{L}^{(1)}(t) \equiv B(t)$ since equations (8.16) and (6.20) are identical, with the same boundary condition $B_{\alpha\beta}(t_1) = Q_{\alpha\beta}$.

The expression (8.24), which is the main result of the present Section, is definitely different (except in the trivial case of a non-interacting Hamiltonian) from the conventional value

$$\Delta Q^2 \big|_{t_1} = \text{tr } Q\rho(t_1)Q[1-\rho(t_1)] \ , \tag{8.25}$$

resulting from a naive use of Wick's theorem and of the TDHF equation (6.16).

Notice that, according to (8.24), the observable of interest **Q** is brought back to the time t_0 by the equation (6.20). Notice also that in our variational derivation of TDHF proper (Section 6), equation (6.20) appeared as a useless by-product, since its solution was not needed to evaluate $\langle Q \rangle_{t_1}$. In contrast, this solution is now required for a consistent evaluation of $\Delta Q\big|_{t_1}$ in the mean-field approach. The relationship between (6.20) and the time-dependent RPA has already been pointed out in Section 6.3 [see eqs. (6.20') and (6.21')]. The idea of dealing with particle-hole excitations, in some form or another, in order to enhance the predicted values for the fluctuations is not new [12]. What we witness here however is the automatic emergence of the RPA within a pure mean-field framework [the state $\mathcal{D}$ retains its uncorrelated form (6.3-4)] and, maybe more surprisingly, from a time-dependent variational principle.

A further simplification of practical interest arises in the result (8.24) when the initial state is a pure state [$\rho^2(t_0) = \rho(t_0)$]. In this case, the fluctuation (8.24) can be written as

$$\Delta \mathbf{Q}^2 \big|_{t_1} = \frac{1}{2} \lim_{\varepsilon \to 0} \left[\frac{\rho(t_0) - \sigma(t_0,\varepsilon)}{\varepsilon} \right]^2 , \tag{8.26}$$

where $\sigma(t,\varepsilon)$ is the solution of the TDHF equation (6.16) with the final boundary condition

$$\sigma(t_1,\varepsilon) = e^{i\varepsilon Q} \rho(t_1) e^{-i\varepsilon Q} \ . \tag{8.27}$$

To derive (8.26) from (8.24), we note first the relation

$$\text{tr } B\rho B(1-\rho) = -\frac{1}{2} \text{tr}[B,\rho][B,\rho] \ , \tag{8.28}$$

valid when $\rho^2 = \rho$. Next we remark that the quantity

$$\sigma(t,\varepsilon) = e^{i\varepsilon B(t)} \rho(t) e^{-i\varepsilon B(t)} \simeq \rho + i\varepsilon[B,\rho] , \tag{8.29}$$

is also a determinant which obeys, in first order in ε, the TDHF equation

$$i \frac{d\sigma}{dt} = [W(\sigma), \sigma] . \tag{8.30}$$

This can be seen from (6.16) and from

$$i \frac{d}{dt}[B, \rho] = [W(\rho), [B, \rho]] + [tr_2 V[B, \rho], \rho] , \tag{8.31}$$

an equation which results from (6.16) and (6.20). Using a TDHF code, one can obtain $\sigma(t_0, \varepsilon)$ by solving (8.30) with the boundary condition (8.27). The final formula (8.26) is obtained from

$$[B(t_0), \rho(t_0)] = -i \frac{\sigma(t_0, \varepsilon) - \rho_0}{\varepsilon} , \text{ by taking the limit } \varepsilon \to 0. \tag{8.32}$$

The numerical evaluation of (8.26) requires only the <u>use of</u> <u>the existing TDHF codes</u>. The steps are the following : a) <u>run first</u> <u>the usual TDHF equation</u> (6.16) from t_0 to t_1 in order to obtain $\rho(t_1)$ from the initial condition $\rho(t_0)$; b) perform at the time t_1 the unitary transformation (8.27) generated by the observable **Q** ; c) run backward from t_1 to t_0 the TDHF equation (8.30) for σ, with a set of smaller and smaller values of ε in order to take the limit (8.26). The formula (8.26) is used in the numerical applications discussed in Section 8.4.

8.2 <u>Some theoretical features of the approximate expression (8.24)</u>

<u>for fluctuations</u>

Using a variational criterion to *<u>shape the evolution upon the</u>* *<u>question being posed</u>* has produced a formula (8.24) for the fluctuations of single particle observables which differs from the trivial formula (8.25). By looking for the characteristic function, we have obtained simultaneously the expression (8.14) for <**Q**> and (8.24) for Δ**Q**, which are not related to each other by Wick's theorem. This illustrates our considerations upon the relative nature of an approximate state: although the uncorrelated density operator (6.3) underlies the approximation (8.14), (8.24), the dependence of this operator upon the source entering the characteristic function *<u>simulates some correlations produced during the evolution</u>* under the influence of the interactions.

In order to check that the deviations from Wick's theorem introduced by (8.24) bring real improvement, we shall present some numerical tests in Section 8.3 and 8.4. Beforehand, let us review some rewarding theoretical features [8] of the results obtained in Section 8.1 .

Consider first the problem of *<u>conservation laws</u>*. Let a single particle observable **Q** (like a component of the total momentum **P** or of the total angular momentum) commute with H. In the exact evolution, all its moments are time-independent. While the TDHF result (6.16)(8.14) is satifactory in this respect for the average $\langle \mathbf{Q} \rangle$, the naive expression (8.25) provided by Wick's theorem for the fluctuation $\Delta \mathbf{Q}$ *does not remain constant* with the time t_1, as it should, when ρ evolves according to (6.16). In the characteristic function approach, this theoretical difficulty disappears. Indeed, the stationarity of (3.1) together with the boundary condition at t_1 lead to the (exact) constant solution $A(t) = \exp(-\lambda \mathbf{Q})$. The resulting stationary value of I yields for the characteristic function the expression

$$\varphi(\lambda) = \ln \, \mathrm{Tr} \, e^{-\lambda \mathbf{Q}} \, D \quad , \tag{8.33}$$

which does not depend on t_1. All the moments of the conserved s.p. observable $\mathbf{Q}$ remain constant with time.

Such an improvement may have implications of practical interest. In the calculation of the momentum dispersion of the final fragments produced by a heavy-ion collision, the observable $\mathbf{Q}$, which is the total momentum in some region of space, commutes with the channel Hamiltonian after fragmentation. In principle, the time at which its fluctuation is evaluated is irrelevant (once the reaction has taken place, and in the absence of evaporation). This requirement is violated by the usual TDHF treatment. In contrast, eq.(8.24) should give a momentum dispersion without spurious dependence on the final time t_1.

Another theoretical improvement concerns the *spreading of wave packets*. Consider a Galilean invariant system, localized in space at the initial time t_0, which evolves according to a Schrödinger equation. As is well known, the fluctuation at the time t_1 of its center of mass position $\mathbf{R}$ varies according to

$$\Delta \mathbf{R}^2(t_1) = \Delta \mathbf{R}^2(t_0) + (t_1-t_0)^2 \Delta \mathbf{P}^2/M^2 \quad , \tag{8.34}$$

where $\Delta \mathbf{P}$ is the momentum fluctuation (constant with time) and M the total mass. However, the approximate TDHF equation admits time-dependent solutions, obtained by boosting the localized HF static solutions through Galilean transformations, for which $\Delta \mathbf{R}^2$ remains (unphysically) constant with time. Here again, our modified mean-field approach removes the difficulty, and provides for the spreading of the wave packet the exact expression (8.34).

This point also helps to clarify the meaning of the static HF approximation. The absence of spreading appears from this viewpoint as a pathology of this approximation, which obviously violates the exact equation (8.34) since the s.p. density ρ (a solution of the static HF equation) does not evolve. If we choose a static HF solution as our initial state $D(t_0)$, it would evolve under the action of the exact Hamiltonian H (of which it is not an eigenstate).

In the approximation considered in Section 8.1, the average (8.14) of a s.p. observable **Q** remains constant with time, but the expression (8.24) for its fluctuation depends on the time difference $t_1 - t_0$. For the center of mass position, the exact time dependence (8.34) is even recovered. We are thus led to interpret a static HF solution as an approximation for an exact time-dependent wave packet rather than for a stationary solution of the Schrödinger equation.

8.3 The Lipkin-Meshkov-Glick model

In Ref.[13], Bonche and Flocard have calculated the approximation (8.24) for the fluctuation in the LMG model [14]. Although its physical relevance to nuclear structure is questionable, this model provides useful preliminary tests by allowing comparison with exact solutions. Let us recall that it involves two levels having the same degeneracy N, and populated by N fermions. The s.p. states are labelled by the quantum numbers $\sigma \pm 1$ and $p = 1, 2, \ldots N$. The Hamiltonian,

$$ H = \frac{\varepsilon}{2} \sum_{\sigma p} \sigma c^{\dagger}_{\sigma p} c_{\sigma p} - \frac{V}{2} \sum_{\sigma p p'} c^{\dagger}_{\sigma p} c^{\dagger}_{\sigma p'} c_{-\sigma p'} c_{-\sigma p} \quad , \tag{8.35} $$

is exactly soluble in terms of the *s.p. operators*

$$ J_{\pm} = \sum_{p} c^{\dagger}_{\pm p} c_{\mp p} \quad , \quad J_z = \frac{1}{2} \sum_{\sigma p} \sigma c^{\dagger}_{\sigma p} c_{\sigma p} \quad , \tag{8.36} $$

which generate an SU(2) algebra. In particular, the multiplet containing the ground state (to which the discussion is restricted) is obtained by diagonalizing

$$ H = \varepsilon J_z + V(J^2_y - J^2_x) \quad , \tag{8.37} $$

in the representation $j = N/2$ of the quasi-spin $\vec{J}$.

Various examples of evolution have been worked out, exactly and in the TDHF approximation for the averages $<J_i>$ (i = x, y, z) of s.p. observables, exactly and in the approximation (8.24) for their fluctuations ΔJ_i. This investigation has led to the following general conclusion, valid even for a strong interaction. Whenever the time-dependent averages $<J_i>$ are well reproduced by the TDHF approximation, a comparable agreement is reached by means of (8.24) for the dispersions ΔJ_i, even though the TDHF answer (8.25) for ΔJ_i may badly fail. When the exact evolution of $<J_i>$ and its TDHF approximation diverge from each other, exact and approximate dispersions diverge as well. The similar quality of fit for the expectation values and for the fluctuations of s.p. observables is likely to stem from the unified treatment of Section 8.1, based on the optimization of a characteristic function.

A detailed analysis of the results can be found in Ref.[13] . We shall only discuss here the instructive situation in which the *initial state* D is a *static* Hartree-Fock solution. The exact evolutions of the $\langle J_i \rangle$ and ΔJ_i are non-trivial, whereas the TDHF approximation predicts constant values for them. In constrast, the variational principle for a state and an observable leads to a time-dependence for the ΔJ_i. An example is displayed on Fig.1, corresponding to the parameters $N = 40$, $\varepsilon = 1$ MeV, and $\chi \equiv V(N-1)/\varepsilon = 0.5$. The time unit is 6.10^{-22} s. The static HF solution corresponds to the eigenvalue $-N/2$ of J_z. A naive HF estimate of the fluctuations gives the time-independent values $(\Delta J_x)^2 = (\Delta J_y)^2 = 10$, $\Delta J_z = 0$.

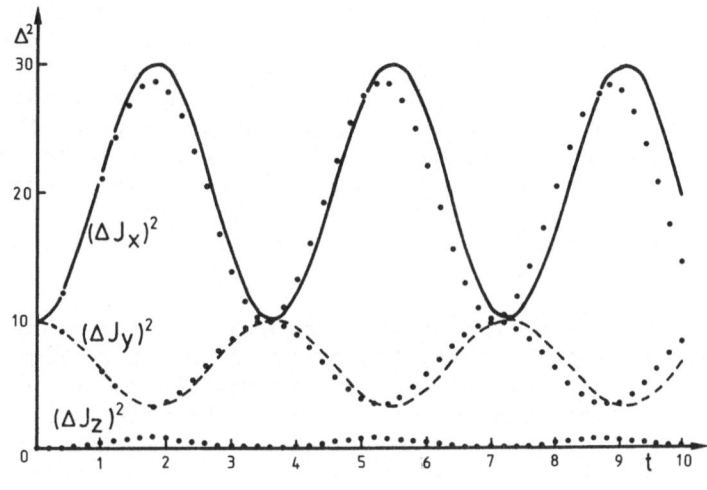

FIGURE 1

The exact time-dependences of $(\Delta J_x)^2$, $(\Delta J_y)^2$ and $(\Delta J_z)^2$ are represented by the points ; the predictions of (8.24) are represented by the solid line for $(\Delta J_x)^2$, by the dotted line for $(\Delta J_y)^2$, and by the horizontal line $\Delta J_z = 0$. The agreement is quite satisfactory. The exact time-dependence of the s.p. averages $\langle J_i \rangle$ have not been drawn. While $\langle J_x \rangle$ and $\langle J_y \rangle$ remain 0, $\langle J_z \rangle$ varies, but its value does not deviate from the initial value -20 by more than 0.33 ; the agreement with the constant HF values parallels the agreement for the fluctuations illustrated by the figure. These behaviours confirm the interpretation (given at the end of Section 8.2) of static HF solutions as approximations for exact wave-packets yielding constant averages $\langle J_i \rangle$ and time-dependent fluctuations ΔJ_x and ΔJ_y.

8.4 A mass transfer calculation

A more realistic calculation, directly relevant to the problem of the mass distribution widths in heavy ion collision, has been performed by Bonche and Flocard in the same Ref.[13]. For computational reasons they have chosen a simple situation : the collision between two ^{16}O ions at the energy of 80 MeV in the center of mass, with an angular momentum of $30\hbar$. This choice corresponds to an impact parameter just above fusion, a situation expected to favour the fluctuations in the final products. For the sake of simplicity, use has been made of the BKN force [15] as in many early calculations.

At the initial time t_0, the two ions are set apart at a distance of 10 fm. The TDHF evolution was followed up to the time $t_1 = t_0 + 18 \ 10^{-22} s$, with a time step $\Delta t = 0.04 \ 10^{-22} s$. The separation between the two fragments is then 9.0 fm, the final kinetic energy 22.0 MeV, and the Coulomb repulsion 10.2 MeV. At this distance, it can be safely assumed that the fragments no longer interact through nuclear forces, so that the final kinetic energy at the time $+\infty$ will be 32.2 MeV. This particular choice of initial conditions leads therefore to a deeply inelastic collision which exhibits a significant orbiting.

In order to define the observable $\mathbb{Q} = N$, which counts the number of nucleons in one outgoing fragment, a binary final state is assumed. The operator N is chosen diagonal in the $\mathbf{r}$ representation ; if x denotes the coordinate along the principal inertia axis of the total system at time t_1, with its origin at the center of mass, N is equal to $\Theta(x) = 1$ in the region $x > 0$ where one ion lies, and $\Theta(x) = 0$ in the region $x < 0$ corresponding to the other ion.

A set of backward TDHF calculations has been performed from t_1 to t_0 for several small values of ε (including $\varepsilon = 0$). The convergence of the quantity (8.26) toward $\Delta^2 N$ has been checked by letting ε decrease down to 10^{-4} and by verifying the stability of the results for values of ε less than 10^{-3}.

The value of the dispersion obtained by this method is

$$\Delta N = 1.42 \ ,$$

to be compared with

$$\Delta N_{TDHF} = 0.81 .$$

The use of the formula (8.24) results therefore in a 75% increase in the width of the fragment mass distribution over the TDHF prediction. This result looks encouraging but unfortunately this single calculation (selected for its feasibility with the available computational facilities) does not allow a direct comparison with the experimental data. Similar calculations should be performed over a sufficient range of the impact parameter and of the incident energy. They should also be extended to heavier systems (including asymmetric situations), for which the discrepancy between TDHF and experiment is known to reach one order of magnitude. Calculations for the same ^{16}O- ^{16}O system at the same energy but for a head-on collision have given comparable results [16].

More realistic interactions than the BKN force should also be used. This is likely to produce a further increase in the mass dispersion predicted by (8.26). Indeed, the spin-isospin degeneracy of the BKN force and the symmetry between the two ^{16}O fragments, have allowed the authors of Ref.[13] to restrict the calculation to four orbitals. This means a great saving in computational time but also a drastic reduction of the degrees of freedom. Since one may expect the dispersion to increase with their number, the use of a more realistic force could lead to a larger value for ΔN.

The formula (8.26) has also been applied by Troudet and Vautherin [17] to the calculation of the mass dispersion, through particle emission, of a ^{40}Ca nucleus which experiences monopole oscillations. The emission of particles occurs because of the coupling to the continuum states of the self-consistent field. The single-particle observable $\hat{N} = Q$ is the number of particles inside a bounded region of space. The fluctuation ΔN^2 measures the mass dispersion of the residual nuclei after particle emission. The value $\Delta \hat{N}^2$ calculated from (8.26) is 19,6 whereas it is 5.27 for the conventional expression (8.25), the expectation value $<\hat{N}>$ being 33.10.

Before closing this Section about fluctuations, let us remark that there is another mechanism which, within the mean-field approach, might affect their values. Two types of correlations are relevant for a proper evaluation of $\Delta Q(t_1)$: those generated by the evolution, and those existing in the initial state. The former are approximately accounted for by the formalism fo Section 8.1 which has optimized, in the sense of the variational principle (3.1), the approximate equations of motion with respect to $\Delta Q(t_1)$. The approximate initial state however is supposed to be given. In Ref.[13] for instance, as in all TDHF calculations, each initial fragment is described by a boosted HF solution. Although this choice may seem natural, it is not fully consistent with our viewpoint. Indeed, the present context suggests looking also for an optimization of the approximate initial state with respect to the quantity $\Delta Q(t_1)$. The optimization of the evolution has already carried back the evaluation of $\Delta Q(t_1)$ to the initial time t_0 through the formula (8.24). One might try to adapt the uncorrelated initial state to the evaluation of the fluctuation $\Delta \mathbf{B}(t_0)$. Although we do not here elaborate this point further, the best choice (which should simulate the correlations existing in the initial state) is not expected to be the static HF solution [8].

9. CONCLUSION

In order to determine the "best" time-dependent mean-field equations suited to our purpose, we took advantage of a variational formulation, in which not only the state of the system, but also a time-dependent observable enter as trial quantities. The observable of interest appears as the boundary condition at the final time while the boundary condition at the initial time is provided by the knowledge about the preparation of the system. When an approximation is sought by restricting the trial spaces, the optimized evolution of an approximate state depends usually on the observable considered. This feature may seen surprising, since the exact evolution of a state is independent of the observation to be performed. It has however the same origin as the familiar appearance of non-linearities or the breaking of invariances in self-consistent approximations. It is the price to be paid for keeping the variational space small enough to render the equations of motion tractable. The density operator most appropriate for predicting some average <A> may well be badly suited for some other average <B>. An approximation for a state should be introduced with the specification of the observables for which it is appropriate.

In our variational scheme, TDHF appears as the best mean-field equation of motion for predicting averages of s.p. observables. When the optimization is performed for other quantities, different mean-field theories come out.

In particular, the optimization of the characteristic function of a s.p. observable led, whithin the mean-field framework, to a definite set of coupled equations. Since the characteristic function encompasses both $<\mathbb{Q}>$ and $\Delta\mathbf{Q}$, these equations underlie in a consistent fashion the determination of $<\mathbf{Q}>$ by TDHF and of the fluctuation $<\mathbb{Q}>$ by the formula (8.24). On the theoretical side, the formalism brings in real improvements upon a straightforward use of TDHF (see Section 8.2). On the pratical side, solving the equation (8.26) does not require more work than running a TDHF code several times. The numerical tests discussed in Section 8.4 should pave the road toward more realistic calculations directly comparable to the experimental data. (Notice that no adjustable residual interaction is needed). We do not claim that these calculations will explain all the data. (We have seen in Section 8.4 that a proper evaluation of $\mathbf{Q}(t_1)$ may also require the additional

206

optimization of the initial uncorrelated state). But we feel that, before leaving the mean-field framework, one should explore all its potentialities if one wants to disentangle the physical mechanisms at work.

Let us finally return, as in Section 3, to the case of unrestricted trial spaces. There, the variational principle provided the exact dynamics of the many-body system both in the Schrödinger and in the (backward) Heisenberg pictures. As it was mentioned in Section 3.4, the formalism may be viewed as an extension of the Lippmann-Schwinger theory to the Liouville representation. We surmise therefore that the unrestricted version of the variational principle for a state and an observable may turn out to be useful in topics such as response problems (see Section 4) or perturbation and scattering theories.

REFERENCES

[1]   R. Balian and M. Vénéroni, to be published in Annals of Physics ; a first account is given in Phys. Rev. Lett. 47(1981), 1353, 1765 (E).

[2]   K. Gottfried, "Quantum mechanics", p. 240, Benjamin, New York, 1966.

[3]   See Balian's lectures in the present volume.

[4]   B.A. Lippmann and J. Schwinger, Phys. Rev. 79(1950), 469.

[5]   J.P. Blaizot and G. Ripka, Phys. Lett. 105B(1981), 1.

[6]   N.G. Van Kampen, Physica Norvegica 5(1971), 279.

[7]   M. Gaudin, Nucl. Phys. 15(1960), 89.

[8]   R. Balian and M. Vénéroni, Phys. Lett. 136B(1984), 301, and in preparation ; R. Balian, P. Bonche, H. Flocard and M. Vénéroni, Nucl. Phys. A 428(1984), 79c-94c.

[9]   R. Balian and E. Brézin, Nuovo Cimento B64(1969), 37.

[10]   S. Levit, Phys. Rev. C21(1980), 1594.
       Y. Alhassid and S.E. Koonin, Phys. Rev. C23(1981), 1590.
       See also references in H. Reinhardt, Nucl. Phys. A390 (1982), 70.

[11]   J.W. Negele, Rev. Mod. Phys. 54(1982), 913.
       K.T.R. Davies, K.R. Sandhya Devi, S.E. Koonin and M.R. Strayer, in "Heavy ion science" (D.A. Bromley, Ed.), Plenum, New York, 1983.

[12]   See for instance D. Brink, Nucl. Phys. A409(1983) p.220c ; S. Yamaji and M. Tohyama, Phys. Lett. 147B(1984), 399.

[13]   P. Bonche and H. Flocard, to be published in Nucl. Phys.A.

[14]   H.J. Lipkin, N. Meshkov and A.J. Glick, Nucl. Phys. 62
         (1965), 188, 199, 211.

[15]   P. Bonche, S.E. Koonin and J.W. Negele,
         Phys. Rev. C13(1976) 1226.

[16]   J.B. Marston, Senior Thesis, Kellog Radiation Laboratory,
         Caltech (1984), unpublished.

[17]   T. Troudet and D. Vautherin, Phys. Rev. C31(1985), 278.

DYNAMICS OF THE RELATIVISTIC HEAVY ION COLLISIONS

J. Cugnon

Physics Department B5
University of Liège, Sart Tilman
B-4000 Liège 1, Belgium

1. INTRODUCTION

1.1. The domain[1]

The boundaries of the relativistic heavy ion collision domain are loosely defined. Obviously, the lower limit is certainly at an incident beam energy (per nucleon) larger than the highest energy discussed by Prof. M. Lefort in his lectures,[2] about 44 MeV/A. The upper limit separating relativistic and ultrarelativistic collisions (discussed elsewhere by Prof. G. Baym[3]) is even more diffuse. As a convenient but perhaps not very meaningful guide, we will use the present operating accelerators. The Bevalac has accelerated for now twelve years ions up to Ar from 250 MeV/A to 2.1 GeV/A (which corresponds to a Lorentz factor in the lab frame from $\gamma_{inc} = 1.26$ to 3.23). Very recently Nb and La beams have been accelerated and Au and U beams are going to be exploited in the very near future. The Synchrophasotron at Dubna has the more energetic ions available nowadays : up to 4 GeV/A, but unfortunately, only light ions (up to ^{12}C) are available. In the last months or so, C and O beams have been accelerated in the Saturn machine in Saclay at an energy around 1 GeV/A. The Ganil machine in France and the MSU machine should cover the range between 100 MeV and 400 MeV per nucleon, but they are just starting now. As most of the now available data have been obtained in Berkeley, our purpose here is to discuss the physics in the Bevalac range, with some excursions in the Ganil regime.

1.2. The motivation

The fascinating aspect of the Bevalac experiments is the fact
that they offer the opportunity to probe highly excited, dense
nuclear matter. The hope was to extract from experimental data the
nuclear matter equation of state, i.e. the energy per nucleon of a
nuclear extended system at density ρ and at temperature T. Up to now,
this relation is known with accuracy for only one point, namely
$T = 0$, and $\rho = \rho_0 = 0.17$ fm^{-3}. Properties in a small domain around
this point are carried by the compression modulus K, which is known
with a reasonable accuracy. Even more fascinating appear the theoretical speculations of the last few years, which suggest novel
phases of hadronic matter. Fig. 1 summarizes the main features of
this new development.

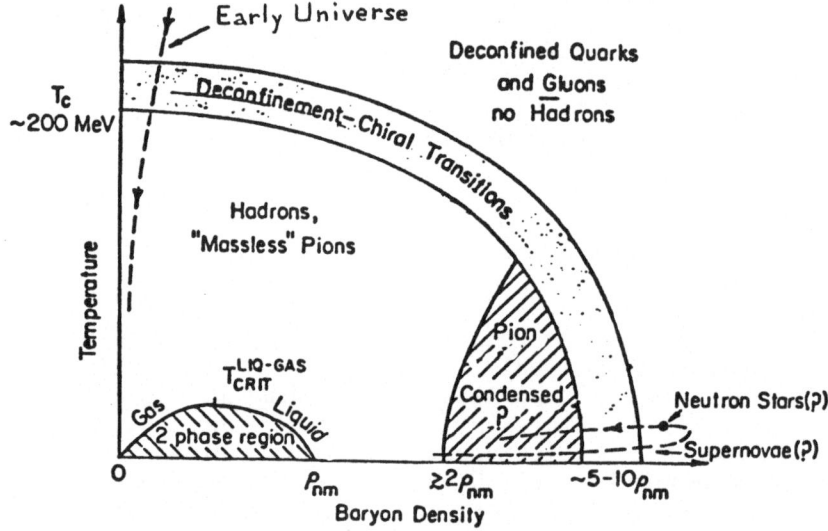

Fig. 1. Phase diagram for hadronic matter, with possible probes of
cosmological and astrophysical nature. Adapted from Ref. 3.

At high temperatures and high densities, quarks become deconfined
and propagate freely and perhaps loose their mass. Because of spin-
isospin long range nuclear forces a pion condensed phase may exist
at low temperatures and for $\rho \gtrsim 2 \rho_0$. At low temperature and for
$\rho < \rho_0$, the nuclear matter could exist as a mixture of a gas and a
liquid phases. The central part of the diagram would correspond to
the ordinary liquid phase of nuclear matter. Some authors make a
distinction between nuclear matter as a fluid of neutrons and
protons and the hadronic matter which is composed of nucleons,
pions (and some other mesons) and Δ's (and some heavier resonances).
However, it is likely that mesons and resonances appear progress-
ively and therefore, there is no serious reason to speak of two
distinct phases.

The detection of the new phases of matter as well as the determination of the hadronic matter equation of state has appeared as a
kind of Holy Grail that the experimentalists around the Bevalac had
to quest (this was the only tool to do such a thing if exception is
made of some cosmological and astrophysical indications). However,
even if gross estimates of densities and temperatures reached during
the collision process indicate that exotic phases may be reached,
several severe conditions must be fulfilled in such a way that one
can speak of a phase transition : (1) the surface effects must not
be important ; (2) thermal equilibrium must be realized, at least
locally (this term will be discussed later on) ; (3) the compressed
matter must live long enough for the transition to take place. None
of these requirements are clearly fulfilled in the systems studied
up to now, as we will see. Therefore, one can expect to see the
onset of critical phenomena only.

From the experimental point of view, only the numerous debris
of the often violent reaction are detected. It appears like a
formidable challenge to extract from all the complicated patterns
of these events (in a streamer chamber experiment, f.i.) any property of the dense system, because (1) the signal is not very well
known ; (2) it may take the form of a tiny bump superimposed on a
huge background ; (3) final state interaction or experimental bias
may distort the signal. Clearly, careful and systematic studies are
required as well as a good theoretical guide which provides the
basic features of the reaction mechanism. The search for signals
has acted as a driving force for both experimentalists and theorists,
who in this way have accumulated an impressive amount of new physics,
as well as its accompanying amount of deception. We are going to
present these new developments in the following chapters.

2. SIMPLE PICTURE: PARTICIPANTS AND SPECTATORS

2.1. Simple classification of events

Global detectors (like streamer chamber) reveal that events
can be grossly classified into three types :
(a) peripheral events, which are expected to occur for large
impact parameters. In these events, the momentum transfer between
the two partners is small and the projectile simply fragments and
produces a jet of particles in the forward direction ;
(b) central events, which presumably occur at small impact
parameters. These events are violent and a total disintegration of
the system results from a large momentum transfer between the two
partners ;
(c) events, where the two features coexist : a forward jet
is to be seen along with many tracks at large angles.

Counter experiments looking at the emission of protons
corroborate these features, as shown in Fig. 2.

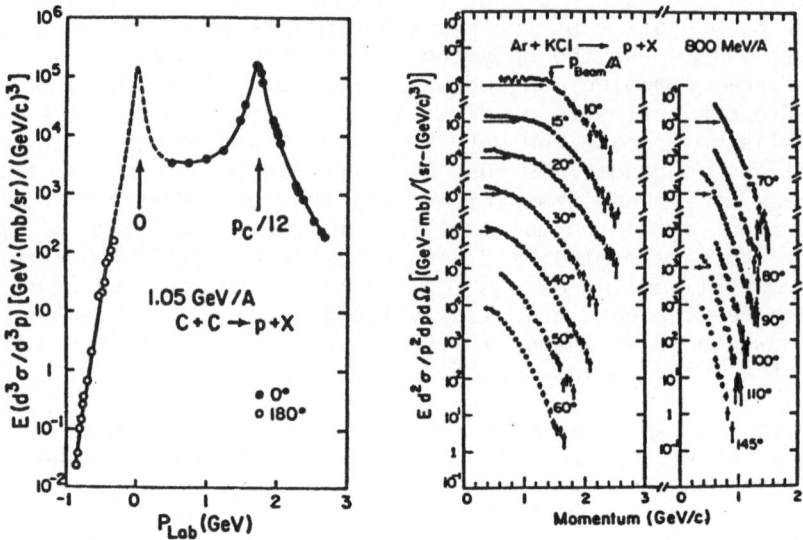

Fig. 2. Left : proton spectrum at 0° and at 180° showing the spectator peaks and the participants in between. Right : proton
spectrum at various angles. Adapted from Ref. 4.

Clearly, at 0° protons are detected with an appreciable rate around
the velocity of the projectile. A corresponding peak for the target
is expected. The particles detected in these peaks are suggestively
called spectator particles. The region in momentum space where these
particles are detected is named the fragmentation region. At 0°
protons with intermediate velocities are detected. They are produced
by large momentum transfers. When one gets off the beam direction,
they are more and more visible. Those particles are called participants. These observations suggest that in a first approximation the
nucleons are behaving independently, like in a gas.

2.2. Characteristic lengths

It is instructive, when tackling a new domain, to look at the
characteristic lengths (or times) of the system under study. Often,
this helps to build a general frame, which may serve to generate a
simplified picture, that is to be improved further and further. In
Fig. 3, we have indicated several characteristic lengths and
energies. In the Bevalac region, the incident energy per particle
is notably larger than both the Fermi energy and the binding energy
per particle. Therefore, binding effects as well as Fermi statistics
are not expected to play an important role. Because this region is
above the threshold, pion production will be an important aspect of
the collision process. Along the ordinate axis, we exhibit some
characteristic lengths. As can be seen, the range of the nucleon-
nucleon (NN) force r_s is smaller than the average inter-nucleon

distance d, which in turn is smaller than the radius R of the nuclei.

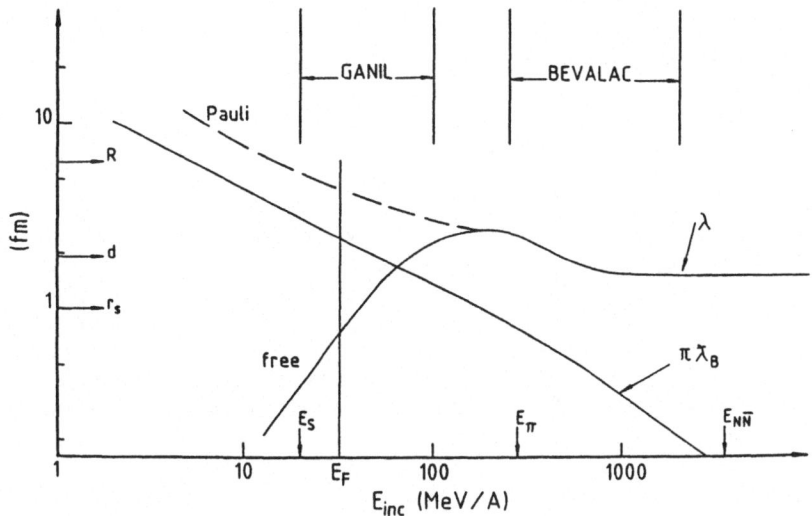

Fig. 3. Typical characteristic lengths and energies for a colliding
nucleus-nucleus system. E_π is the threshold for pion production in the NN system. See text for detail.

Another interesting parameter is the de Broglie wave length $\lambdabar_B$
for the relative NN motion. It is smaller than d in the Bevalac
domain. Those parameters are such that the collision processes can
be viewed (in a kind of roughly approximate picture) as a succession
of independent binary collisions between classical nucleons, although
it should be stressed that the strict validity of this picture would
rather require $\lambdabar_B$, $r_s \ll d \ll R$ and not simply "smaller" signs.
Also Fig. 3 gives the mean free path λ as calculated in the gas
limit (with and without Pauli blocking correction). We will come
back on the role of this parameter. Finally, Fig. 3 shows that the
Ganil physics could be quite complicated because in this energy
domain, many parameters are comparable. In the remaining part of
this section, we are going to review all the information that can
be extracted from experiment using only this simple picture of
successive collisions.

2.3. Role of the geometry

It is now a popular idea that the geometry itself determines
the way nucleons are separated into participants and spectators :
If a nucleon is in the part of the projectile (target) intercepted
by the target (projectile), it will be a projectile (target) participant. If, on the other hand, the nucleon belongs to the outer
parts of the system, it will be a spectator. Such a geometrical
picture leads to a proton participant cross-section

213

$$\sigma_p = \pi r_o^2 (Z_P A_T^{2/3} + Z_T A_P^{2/3}) \qquad (2.1)$$

and to a projectile spectator cross-section

$$\sigma_{sp}^P = \pi r_o^2 (Z_T A_T^{2/3} + 2 A_P^{1/3} A_T^{1/3}) \qquad . \qquad (2.2)$$

The mass dependence has been checked experimentally by Nagamiya[4,5] in light to medium-heavy systems. The agreement with Eqs. (2.1-2) is quite good if he used $r_o = 1.2$ fm for the participants and 0.95 fm for the spectators. The smaller value for the spectators is already indicative of a correction to the simple geometrical picture, because of the finite angle NN scattering. Another expected modification comes from the (limited) transparency of the nuclei, due to finite mean free path.

2.4. Evidence for multiple collisions

The best way to study the properties of the participant region is to look at 90°-c.m. for equal-mass systems, because one is then far from the spectator region. Proton energy spectra at 90° c.m. in several systems is given in Fig. 4. Typical features are observed.

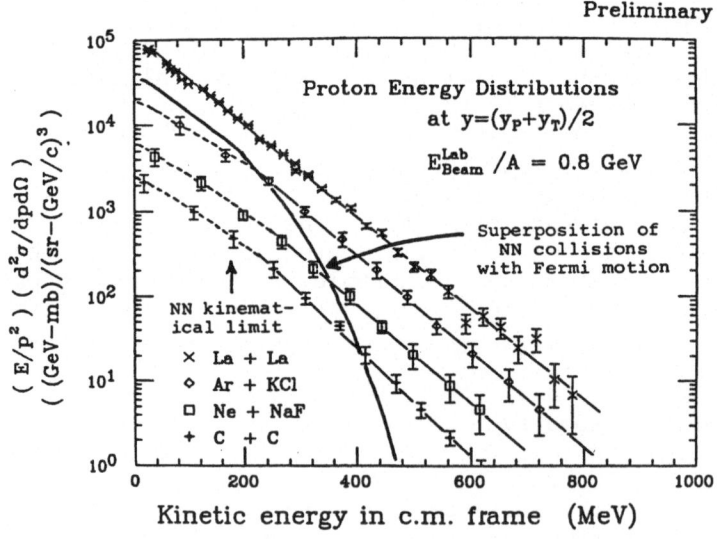

Fig. 4. Proton energy spectrum at 90° in the c.m. frame for various symmetric systems. The full line indicates the kinematical limit obtained by taking account of a reasonable Fermi distribution. Adapted from Refs. 6-7.

First, the spectrum is rather the same for all the systems. This suggests that the relevant quantity is the beam energy per nucleon rather than the total energy. This is in keeping with the idea that the basic mechanism is a superposition of NN collisions. Second, high energy protons are emitted abundantly well beyond the kinematical limit, even if Fermi motion is taken into account. Third, the high energy domain approaches an exponential, with deviations on the low energy side. Fourth, the slope parameter E_0 characterizing the exponential shape is quite large : 70-80 MeV. The slope parameter is also increasing with the mass of the system indicating that high energy protons are produced via multiple collisions.

These very simple observations suggest that through multiple collisions, the participant nucleons are spread over the whole momentum space that they have at their disposal (the so-called phase space or thermal limit). This is not true, however, because of two observational facts, which indicate that some nucleons are just making a small number of collisions.

2.5. Evidence for hard scattering

If the thermal limit was attained, the angular distribution would be isotropic in the c.m. of equal mass systems. The experiment shows that it is forward (and backward) peaked, suggesting that some nucleons are making just one or two collisions, since the NN differential cross-section is forward peaked. The presence of such a hard component, as it is named, has been nicely demonstrated in correlation experiments by Nagamiya et al.[8-10]. The idea is to detect a proton at one angle and to look for coincidence with the proton detected in another counter, whose direction may be varied. They observed that the coincidence yield is enhanced when the position of the second counter and the energy of the second proton is such that the two outgoing protons correspond to a possible final state of a proton-proton system with the same velocity as the incoming nuclei. Those protons thus do not interact with the other nucleons. It is understandable that the frequency of such pairs provides a measure of the mean free path λ of nucleons since the probability for a nucleon to do not suffer an additional collision is roughly $\exp(-R/\lambda)$, where R is the radius of the interaction zone. In proton-nucleus interactions, R is well defined, and the experiments lead to a value of $\lambda \simeq 2.4$ fm around 0.8 Gev.[10-11] With this value, one can use the correlation experiments to determine R. The extracted values are consistent with two-pion and two-proton interferometry.

2.6. Pion energy spectra

An important piece of data is the 90° c.m. spectrum of (negative) pions in equal mass systems. They look pretty much exponential (without any shouldering like in the proton case). The

extracted values of the slope parameter E_0 are given for the
Ne + NaF system in Fig. 5, along with the proton slope parameter.

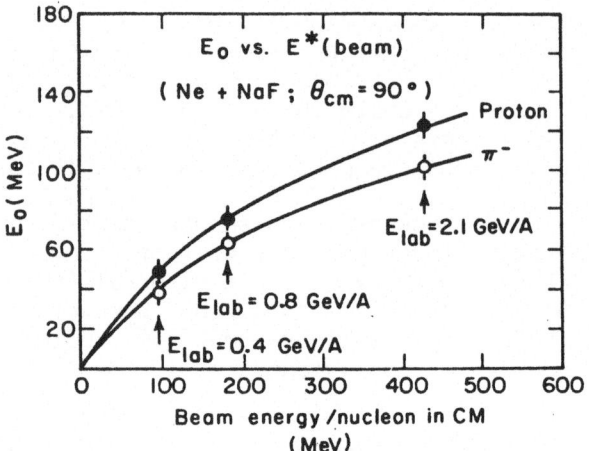

Fig. 5. Values of E_0 for both pions and protons as a function of
the beam energy per nucleon in the c.m. frame. Figure taken
from Ref. 12.

As expected, E_0 is increasing with the incident energy. In addition,
the values for the pions are systematically smaller than the values
for the protons. There are several proposed explanations for
$E_0(\pi^-) < E_0(p)$, in terms of mean free path,[4] of the importance of
the Δ-resonances as a pion source,[13,14] or of blast wave.[15] We will
come back to these possibilities. Let us point out here the last
explanation, which implies the rapid expansion of the participant
system, an expected feature in view of the pressure that can be
built if the matter is coming to stop in the c.m. system.

2.7. Summary

All these considerations lead to the following picture, which
can be adopted as an organization principle in classifying data and
models. The nucleons are separated into participants and spectators
in a geometrical way. The spectators are little affected, whereas
the participants are spread over the momentum space by the virtue
of the two-body collisions. Some are scattered many times, some only
once. They are very likely compressed in an early stage and afterwards expand very quickly under the action of the pressure.

3. NON EQUILIBRIUM DYNAMICS

3.1. Quantal and classical aspects

Ideally, the heavy-ion collision should be described by a complete quantum theory of strong interactions, QCD as it is generally believed. Needless to say that this is out of scope for the time being. Even a field theoretical description using point-like nucleons and mesons is hopeless, except in very simplified one-dimensional models. The next less fundamental step is to consider point-like nucleons (we disregard pion production for the presentation) whose interaction can be described in terms of potentials or perhaps in terms of scattering amplitudes or cross-sections. With this starting point of view, various many-body theories are at our disposal, both in the classical and in the quantum regimes. Fig. 6 gives the relationship between these theories.

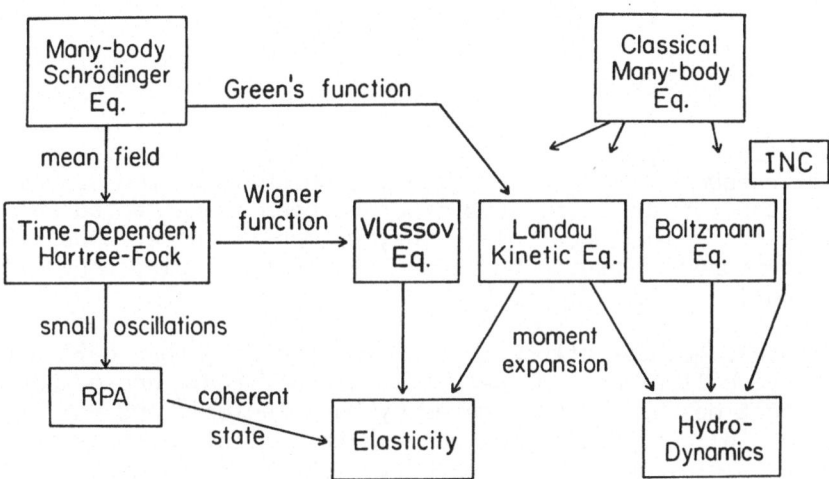

Fig. 6. The relationship between many-body theories. Adapted from Ref. 16.

Although, as we shall see, the most appropriate and tractable theory is the Boltzmann kinetic equation (or rather the closely related cascade model), it is worth to see whether one should start from a classical or a quantal point of view. This question cannot be answered globally and the choice may vary according to the observables and the kinematic domain one is looking at. For instance, the elastic scattering, which accounts for nearly one half of the total cross-section, has a diffractive quantal nature. However, one is much more interested in the reaction cross section, inclusive or partially exclusive measurements. In these cases, the interference between partial waves is expected to play a secondary role, because of the very many final states. Therefore (and also because the nucleus-nucleus wave number is very large), the classical concept of impact parameter may be used and one may simply sum over impact parameters.

Looking a little bit further at quantum features of the collision path, one realizes that in a nucleon-nucleon (NN) collision, the maximum angular momentum ℓ_{max} is not large : for a relative momentum of 1 GeV/c, $\ell_{max} \approx 3$, if one considers the range of nucleon-nucleon forces. Thence, the NN scattering should in principle be described at a quantum level.

Let us now turn to the possible interference between NN scatterings. Has the scattering wave function reached its asymptotic (far zone) behaviour, before hitting another nucleon ? Two successive scatterings occur *in the average* after a mean free path λ

$$\lambda = \frac{1}{\rho \ \sigma_{NN}}$$ (3.1)

The near zone limit is at least the range of the NN force r_s (see Fig. 3). Around 1 GeV/A, $\lambda \approx 1.6$ fm, which is of the same order as the range of the force, and interferences are expected. However, in nucleus-nucleus collisions, so many interferences will occur (compared to proton-nucleus) that, very likely, they will cancel out.

Another quantum effect arises from the fact that during the successive collisions, the particles are off their energy shell. This aspect has been investigated by Danielewicz[17] and can be stated as follows. During two successive collisions the energy of the particle cannot be fixed better than with an uncertainty of $\delta\epsilon = \hbar/\delta t$, where δt is the time interval separating the collisions. One readily has

$$\frac{\delta\epsilon}{\epsilon} \approx \frac{2}{\lambda \ k} \approx \frac{\lambdabar_B}{\lambda}$$ (3.2)

which is of the order of 25% or less. According to ref. 17, this has profound implications on the equilibration process. It is not clear yet whether this effect differs from the previous one.

Let us finally mention quantum effects linked with quantum statistics, in the depopulation of the Fermi spheres and also in the final state interaction. The latter leads to interference between two protons or two pions, if they have small relative momentum.

3.2. From many-body theory to a kinetic equation

The very first starting point is the many-body Schrödinger equation, if we adopt the quantum point of view. Although, as we will see, a classical approach is largely sufficient around 1 GeV/A, we nevertheless choose this presentation, because it may give contact with the low-energy side. We closely follow the presentation of Ref. 16. The approximation scheme is better described on the A-body density matrix

$$\tilde{\rho}(\vec{r}_1,\ldots\vec{r}_A,\vec{r}_1',\ldots\vec{r}_A',t) = \psi^*(\vec{r}_1,\ldots\vec{r}_A,t)\psi(\vec{r}_1',\ldots\vec{r}_A',t) \quad (3.3)$$

from which one may define the one-body density matrix

$$\rho^{(1)}(\vec{r},\vec{r}',t) = A \int d^3r_2 \ldots d^3r_A \; \tilde{\rho}(\vec{r},\vec{r}_2\ldots\vec{r}_A,\vec{r}',\vec{r}_2,\ldots\vec{r}_A,t) \quad (3.4)$$

and the two-body density matrix

$$\rho^{(2)}(\vec{r}_1,\vec{r}_2,\vec{r}_1',\vec{r}_2',t) = A(A-1) \int d^3r_3 \ldots d^3r_A$$
$$\tilde{\rho}(\vec{r}_1,\vec{r}_2,\ldots\vec{r}_A,\vec{r}_1',\vec{r}_2',\vec{r}_3\ldots\vec{r}_A,t) \; . \quad (3.5)$$

The equation of motion for $\rho^{(1)}$ is obtained by integrating the Von Neumann equation

$$\frac{\partial\tilde{\rho}}{\partial t} = \frac{i}{\hbar} [H,\tilde{\rho}] \quad (3.6)$$

over all but one coordinates:

$$\frac{\partial\rho^{(1)}}{\partial t} (\vec{r},\vec{r}',t) = -i \frac{\nabla^2 - \nabla'^2}{2m} \rho^{(1)}$$

$$+ i \int d^3r_2 \; (v(\vec{r},\vec{r}_2) - v(\vec{r}',\vec{r}_2)) \; \rho^{(2)}(\vec{r},\vec{r}_2,\vec{r}',\vec{r}_2,t) \quad . \quad (3.7)$$

The evolution of $\rho^{(1)}$ involves the quantity $\rho^{(2)}$. Analogously, the evolution of $\rho^{(2)}$ involves $\rho^{(3)}$, etc. One so generates a hierarchy of equations, similar to the BBGKY equations[18] for the classical distribution functions, and also to the Martin-Schwinger[19] formalism for the Green functions. The idea is to truncate the hierarchy somewhere to close the set of equations.

It is interesting at this stage to introduce the Wigner transform[20] of the one-body density

$$f_1(\vec{r},\vec{p},t) = \int \frac{d^3s}{(2\pi\hbar)^3} \; e^{\frac{i}{\hbar}\vec{p}\cdot\vec{s}} \; \rho^{(1)}(\vec{r} + \frac{\vec{s}}{2} , \vec{r} - \frac{\vec{s}}{2}, t) \quad . \quad (3.8)$$

This Wigner function behaves in many ways as the classical one-body distribution function (we will use the same symbol to designate the two quantities). In particular the particle density ρ, the particle current $\vec{j}$ and the momentum tensor are given by the same integrals

$$\rho(\vec{r},t) = \rho^{(1)}(\vec{r},\vec{r},t) = \int d^3p \, f_1(\vec{r},\vec{p},t) \tag{3.9}$$

$$\vec{j}(\vec{r},t) = \frac{1}{m} \left\{ \left(\frac{\vec{\nabla} - \vec{\nabla}'}{2i} \right) \rho^{(1)}(\vec{r},\vec{r}') \right\}_{\vec{r}=\vec{r}'} = \int d^3p \, \frac{\vec{p}}{m} \, f_1(\vec{r},\vec{p},t) \tag{3.10}$$

$$\tau_{ij}(\vec{r},t) = \int d^3p \, p_i \, p_j \, f_1(\vec{r},\vec{p},t) \quad . \tag{3.11}$$

The main difference is that the Wigner function is not necessarily non-negative. Similarly, one can define the analog of the two-body distribution function

$$f_2(\vec{r}_1,\vec{r}_2,\vec{p}_1,\vec{p}_2,t) = \int \frac{d^3s_1}{(2\pi\hbar)^3} \frac{d^3s_2}{(2\pi\hbar)^3} \, e^{\frac{i}{\hbar}(\vec{p}_1 \cdot \vec{s}_1 + \vec{p}_2 \cdot \vec{s}_2)}$$

$$\rho^{(2)}(\vec{r}_1 + \frac{\vec{s}_1}{2}, \vec{r}_2 + \frac{\vec{s}_2}{2}, \vec{r}_1 - \frac{\vec{s}_1}{2}, \vec{r}_2 - \frac{\vec{s}_2}{2}, t) \quad . \tag{3.12}$$

Using Eq. (3.8), Eq. (3.7) can be put the form

$$\frac{\partial}{\partial t} f_1 + \frac{\vec{p}}{m} \cdot \vec{\nabla} f_1 = i \int \frac{d^3\vec{x}}{(2\pi)^3} \, e^{\frac{i}{\hbar} \vec{p} \cdot \vec{x}} \int d^3r'' \times$$

$$[v(\vec{r} + \frac{\vec{x}}{2}, \vec{r}'') - v(\vec{r} - \frac{\vec{x}}{2}, \vec{r}'')] \, \rho^{(2)}(\vec{r} + \frac{\vec{x}}{2}, \vec{r}'', \vec{r} - \frac{\vec{x}}{2}, \vec{r}'', t)$$

$$\tag{3.13}$$

or

$$(\frac{\partial}{\partial t} + \frac{\vec{p}}{m} \cdot \vec{\nabla}) f_1 = i \int \frac{d^3\vec{x}}{(2\pi)^3} \, e^{i\vec{p} \cdot \vec{x}} \int d^3r'' \times$$

$$[v(\vec{r} + \frac{\vec{x}}{2}, \vec{r}'') - v(\vec{r} - \frac{\vec{x}}{2}, \vec{r}'')] \int d^3p_1 \, e^{-i\vec{p}_1 \cdot \vec{x}} \int d^3p_2 \, f_2(\vec{r},\vec{r}'',\vec{p}_1,\vec{p}_2,t)$$

$$\tag{3.14}$$

This equation is not very useful since it cannot be solved exactly. So we turn to its usual approximations. The factorization of $\rho^{(2)}$ into two one-body density matrices

$$\rho^{(2)}(\vec{r}_1,\vec{r}_2,\vec{r'}_1,\vec{r'}_2,t) = \rho^{(1)}(\vec{r}_1,\vec{r'}_1,t)\ \rho^{(1)}(\vec{r}_2,\vec{r'}_2,t)$$

$$- \rho^{(1)}(\vec{r}_1,\vec{r'}_2,t)\ \rho^{(1)}(\vec{r}_2,\vec{r'}_1,t) \qquad (3.15)$$

closes the equation (3.13) on itself as an equation for f_1. Retaining the first term in Eq. (3.15) leads to the Hartree approximation which may be written as

$$[\frac{\partial}{\partial t} + \frac{\vec{p}}{m}.\vec{\nabla} - (\vec{\nabla}U).\vec{\nabla}_{\vec{p}}]\ f_1(\vec{r},\vec{p},t) = 0 \qquad , \qquad (3.16)$$

where U is the Hartree potential

$$U(\vec{r}) = \int v(\vec{r} - \vec{r'})\ \rho^{(1)}(\vec{r'},\vec{r'})d^3r' \qquad , \qquad (3.17a)$$

or

$$U(\vec{r}) = \int d^3r'\ d^3p\ v(\vec{r} - \vec{r'})\ f_1(\vec{r'},\vec{p},t) \qquad . \qquad (3.17b)$$

This equation, obtained by retaining the first order term in the expansion of the bracket in (3.13) as a power series of $\vec{x}$, (this amounts to assume that the potential is a smooth function) is formally the same as the Liouville equation for free particles in an external potential. The new feature here is that U is generated by the particles themselves and is density dependent : it is known as the *Vlassov* equation. Except for this density dependence, the equation conserves density in phase space, and does not contain expected non-equilibrium effects.

The Hartree-Fock approximation, which leads to the well-known TDHF formalism gives rise to a complicated equation in phase space. The dynamics implied by this equation is essentially the same as the one of the Vlassov equation once a Slater determinant has been chosen as a starting point for the latter. It is generally designated as the one-body dynamics and is extensively covered in Refs. 21.

The one-body distribution function is influenced by collisions also. The effect of the latter is introduced by higher order approximation. Such a program has been carried out within the formalism of the Green's functions by Kadanoff and Baym[22]. We merely quote the result here, which is an equation of the form

$$[\frac{\partial}{\partial t} + \frac{\vec{p}}{m}.\vec{\nabla} - (\vec{\nabla}U).\vec{\nabla}_{p}]\ f_1(\vec{r},\vec{p},t) =$$

$$\int \frac{d^3 p_2}{2\pi} \frac{d^3 p_3}{2\pi} \frac{d p_4}{2\pi} \, W(\vec{p}_3\vec{p}_4 \to \vec{p}\vec{p}_2) \, \delta^3(\vec{p}) \, \delta(E) \; \times$$

$$[\tilde{f}_3\tilde{f}_4(1 - \tilde{f})(1 - \tilde{f}_2) - \tilde{f}\tilde{f}_2(1 - \tilde{f}_3)(1 - \tilde{f}_4)] \qquad (3.18)$$

where

$$\tilde{f}_i = f_1(\vec{r},\vec{p}_i,t)/(2\pi\hbar)^3 \qquad (3.19)$$

and where W is a transition rate for collisions between particles of initial momenta $\vec{p}_3$, $\vec{p}_4$ and final momenta $\vec{p},\vec{p}_2$. The delta functions in (3.18) stand symbolically for the conservation laws. This equation is known as the *Landau* equation. In Ref. 22, this result is obtained in a power series expansion in term of the interacting potential v. In this case

$$U = \int d^3 r' \, d^3 p' \, v(\vec{r} - \vec{r}')f_1(\vec{r}',\vec{p}',t) \qquad (3.20a)$$

$$W(p_3p_4 \to pp_2) = \frac{1}{2} \, [v(\vec{p} - \vec{p}_3) - v(\vec{p}_2 - \vec{p}_4)]^2 \; . \qquad (3.20b)$$

We have to say that we considerably simplify the presentation here, neglecting aspects like off-shell behaviour, local approximation, etc. We think however that the basic features are preserved.

In the nuclear case, the interacting potential is so strong that, even in a dilute system, the expansion in terms of v breaks down. The usual argument is that, in that particular case, two colliding particles interact repeatedly with each other like in a usual free scattering case. Therefore, it is then more plausible to use

$$U = 0 \qquad (3.21a)$$

$$W(p_3p_4 \to pp_2) = |\langle p_3p_4 \, |T| \, pp_2\rangle|^2 \; , \qquad (3.21b)$$

where T is the usual T-matrix. Whether the two heavy ion system is sufficiently dilute is still an open question. Eq. (3.18) with the choice (3.21) is denoted as the Boltzmann equation. More precisely, this is the (non-relativistic) Uehling-Uhlenbeck equation which becomes the Boltzmann equation when the $(1 - \tilde{f})$ factors are removed in Eq. (3.18).

3.3. The intranuclear cascade model (INC)

The intranuclear cascade model is based on a simulation of the collision events. The method presents the great advantage of being capable of handling problems with a large number of degrees of

222

freedom (up to several hundreds) at a microscopic level.

The model describes the collision process as a succession of classical *binary* collisions between *on-shell* particles proceeding *as in free space*. We will not describe it here into detail (see Ref. 23 for that and for a comparison between the various approaches). It is sufficient to say that the particles are given randomly initial positions and momenta consistent with the nuclear density, Fermi motion and incident beam energy. The particles are travelling along straight line trajectories until a pair of them reaches its minimum distance of approach d_{min}. If

$$d_{min} \leq (\frac{\sigma_{tot}}{\pi})^{\frac{1}{2}} \quad , \qquad (3.22)$$

where σ_{tot} is the total cross-section, the particles are forced to scatter. The final momenta of the particles are determined by energy-momentum conservation laws and by choosing the scattering angle at random according to a distribution law derived from the experimental differential cross-section. If several channels are open, the final state is chosen at random according to the weights of the respective cross-sections. The straightline motion is then resumed until another pair reaches its minimum distance of approach, etc. The procedure is repeated to obtain a sufficient statistic . Observables and other physical quantities are obtained by ensemble averages. Several features are added like pion production (see later), relativistic kinematics, Pauli blocking, etc.

A widely spread opinion is that INC calculations are just a numerical trick to solve the Boltzmann equation. To support this statement, one invokes the conditions of validity which are roughly the same for the two approaches.

The conditions of validity of the Boltzmann equation are not a trivial matter. They are, however, established heuristically in many textbooks. We follow here Ref. 24. First of all, the basic assumption is that there exists an interval Δt such that

$$\tau_c \ll \Delta t \ll \tau_r \quad , \qquad (3.23a)$$

where τ_c is the (elementary) collision time and τ_r the relaxation time. This condition is easy to understand : one cannot look for variations over characteristic times which are too short since one has neglected the change in the system when the trajectories of the particles are different from a straight line. On the other hand the relaxation of the system, i.e. the evolution of strong local inhomogeneities, cannot be described by derivatives if the relaxation time is too short. The condition (3.23a) is often stated as a low density condition (which may not be strictly equivalent)

$$r_s^3 \, \rho \ll 1 \quad . \tag{3.23b}$$

Physically, the latter condition ensures that the successive collisions are well separated in space-time.

In deriving the Boltzmann equation, one implicitly assumes that the distribution function does not change significantly over the time interval Δt. In other words, one assumes that the Boltzmann equation is local in time. Similarly, one also neglects space variation of the distribution function over lengths $\Delta \ell$, typical of the distance travelled by the particles during the time Δt : the equation is local in space.

Finally, the factorization

$$f_2 = f_1 f_1 \tag{3.24}$$

is required : this is the celebrated "Stosszahlansatz" or "molecular chaos" assumption, which introduces irreversibility as is well known.

Much less stringent assumptions are required for the INC picture to be correct. Since one does not deal with time derivative, one needs not worry about a characteristic interval Δt smaller than the relaxation time. Only condition (3.23b) is required. The very structure of the INC calculation is not subject to conditions on locality in space-time. Finally, whether a two-body collision occurs or not depends in the INC picture on the probability of finding two nucleons at sufficiently close points with an appropriate relative momentum, a quantity which has something to do with the two-body distribution function f_2. Therefore, two-body correlations effects are really taken into account. Three-body and higher correlations are neglected, as in Boltzmann equation. In conclusion, it can be said that the INC model goes beyond the Boltzmann equation and beyond the Landau equation as far as the collision term is concerned. Furthermore, the INC is a theory for the evolution of the one-body distribution function, but also of the two, three, ... A-body distribution functions. This is equivalent to say that the INC can predict event-to-event fluctuations

3.4. Classical many-body equations

The heavy ion case has been studied through the so-called equations of motion approach.[25-30] The latter consists in solving the Hamilton equations for A nucleons interacting through central repulsive short-ranged plus attractive long-ranged potentials. This method, which seems to be more fundamental than a kinetic equation (Landau, Boltzmann, INC, ...) presents some technical difficulties : non-physical solutions at low energy, proper choice of the nucleon-nucleon potential, introduction of pion production and large computation times compared to INC. Moreover, the relative motion of two colliding nucleons is treated completely classically, which seems very doubtful in view of what we have said in Section 3.1. This may be important for the evolution in phase space. On the other hand, the equation of motion approach can incorporate two-body interaction energy, contrarily to a kinetic equation approach.

3.5. Moment expansion of the Landau equation

For the Boltzmann and Landau equations, there are five and only five conserved quantities (called the collisional invariants) :

$$I_n = \int d^3r \ d^3p \ \gamma_n \ f_1(\vec{r},\vec{p},t) \quad , \tag{3.25}$$

with

$$\gamma_n = 1, \ \vec{p}, \ p^2 \quad . \tag{3.26}$$

The conservation laws may be written in a local form by taking the lowest moments in $\vec{p}$ of the Landau equation (3.18). The zeroth moment yields (with (3.9-10), and assuming a finite system) the equation of continuity

$$\frac{\partial \rho}{\partial t} + \vec{\nabla}.\vec{j} = 0 \quad , \tag{3.27}$$

or

$$\frac{\partial \rho}{\partial t} + \vec{\nabla}.(\rho\vec{u}) = 0 \quad . \tag{3.28}$$

The last equation defines $\vec{u}$, the average macroscopic velocity. The first moment gives (with Eq. (3.11))

$$m \frac{\partial \vec{j}}{\partial t} + \vec{\nabla}.\underset{\approx}{\tau} + \rho \ \vec{\nabla} \ U = 0 \tag{3.29}$$

or

$$m \frac{\partial \vec{j}}{\partial t} + \vec{\nabla}.\underset{\approx}{\tau} + \vec{\nabla}(\rho U - V) = 0 \quad , \tag{3.30}$$

where

$$V[\rho] = \int_0^\rho U(\rho') \ d\rho' \quad . \tag{3.31}$$

Equation (3.30) may be rewritten in terms of $\vec{u}$. One readily gets

$$\rho[\frac{\partial \vec{u}}{\partial t} + \vec{u}.\vec{\nabla} \ \vec{u}] + \vec{\nabla}.\underset{\approx}{\Pi} = 0 \tag{3.32}$$

where

$$\Pi_{ij} = S_{ij} + \delta_{ij}(\rho U - V) \quad . \tag{3.33}$$

The quantity

$$S_{ij} = m^2 \int d^3p \ \delta v_i \ \delta v_j \ f_1(\vec{r},\vec{p},t) \qquad (3.34)$$

where $\delta\vec{v}$ is defined by

$$\vec{p} = m(\vec{u} + \delta\vec{v}) \quad , \qquad (3.35)$$

is the momentum tensor in the matter (local) rest frame. The quantity Π is the momentum flux tensor.

The second moment equation is obtained, after some tedious algebra. It reads

$$(\frac{\partial}{\partial t} + \vec{u}.\vec{\nabla}) \ S_{ij} = -\sum_k (S_{ik} \ \nabla_k \ u_j + S_{jk} \ \nabla_k \ u_i) - S_{ij} \ \vec{\nabla}.\vec{u}$$

$$- m \ \vec{\nabla}.\vec{J} + m \int I \ \delta v_i \ \delta v_j \ d^3p \qquad . \qquad (3.36)$$

In this equation, $\vec{J}$ is the internal energy current

$$\vec{J} = \int d^3p \ \frac{(\vec{p} - m \ \vec{u})^2}{2m} (\frac{\vec{p}}{m} - \vec{u}) \ f_1(\vec{r},\vec{p},t) \quad , \qquad (3.37)$$

and I stands symbolically for the collision term of the Landau equation. Defining the energy density by

$$\rho \ e(\vec{r},t) = \int d^3p \ \frac{p^2}{2m} \ f_1(\vec{r},\vec{p},t) - \frac{1}{2} \ \rho \ u^2 = \frac{1}{2} \ \text{tr} \ \underset{\approx}{S} \quad , \qquad (3.38)$$

one obtains

$$\rho[\frac{\partial}{\partial t} + \vec{u}.\vec{\nabla}]e + \vec{\nabla}.\vec{J} = -\sum_i \sum_k S_{ik} \ \nabla_k \ u_i \qquad . \qquad (3.39)$$

The collision term contribution in (3.36) vanishes when the trace is taken.

One generally wants to close the moment equations. This is usually done phenomenologically, as in hydrodynamics. We postpone the discussion of this case to Section 5. We will show here that the long relaxation time limit gives the elasticity limit. Indeed, if a uniform damping due to the collision term is assumed,

$$m \int I \ \delta v_i \ \delta v_j \ d^3p \approx - \frac{(S_{ij} - \frac{1}{3} \delta_{ij} \ \text{tr} \ \underset{\approx}{S})}{\tau} \quad , \qquad (3.40)$$

where τ is the relaxation time, Eq. (3.36) can be written, assuming no energy current ($\vec{J} = 0$),

$$(\frac{\partial}{\partial t} + \vec{u}.\vec{\nabla})S_{ij} + \tau^{-1}(S_{ij} - \frac{1}{3}\delta_{ij} \, tr \, \underset{\approx}{S}) =$$

$$-\sum_{k}(S_{ik} \nabla_k u_j + S_{jk} \nabla_k u_i) - S_{ij} \vec{\nabla}.\vec{u} \quad . \quad (3.41)$$

In the long relaxation time limit, the second term may be neglected. By using the notation

$$\frac{D}{Dt} = \frac{\partial}{\partial t} + \vec{u}.\vec{\nabla} \quad , \quad (3.42)$$

the derivative of (3.32) gives

$$\frac{D}{Dt}(\rho \frac{D}{Dt}\vec{u}) + \frac{D}{Dt} \vec{\nabla}.\{S_{ij} + \delta_{ij}(\rho U - V)\} = 0 \quad . \quad (3.43)$$

Using (3.41), we may write

$$\frac{D}{Dt}(\rho \frac{D}{Dt}\vec{u}) - \vec{\nabla}.\{\sum_{k}(S_{ik} \nabla_k u_j + S_{jk} \nabla_k u_i) + S_{ij} \vec{\nabla}.\vec{u}\}$$

$$+ \vec{\nabla} \frac{D}{Dt}(\rho U - V) = 0 \quad (3.44)$$

Now, if we assume that S_{ij} is close to its equilibrium value

$$S_{ij} = \delta_{ij} \rho \frac{\langle v^2 \rangle}{3} \quad , \quad (3.45)$$

one has

$$\frac{D}{Dt}(\rho \frac{D}{Dt} u_j) - \sum_{i} \nabla_i \{\rho \frac{\langle v^2 \rangle}{3}(\nabla_i u_j + \nabla_j u_i) + \vec{\nabla}.\vec{u}\}$$

$$- \nabla_j \{\rho^2 \frac{\partial U}{\partial \rho} \vec{\nabla}.\vec{u}\} = 0 \quad . \quad (3.46)$$

This is the classical equation for vibrations in a solid state which is ordinarily written in terms of displacements

$$\rho \frac{\partial^2 d_i}{\partial t^2} + \lambda \nabla_i(\vec{\nabla}.\vec{d}) + \mu \sum_{j} \nabla_j(\nabla_i d_j + \nabla_j d_i) = 0 \quad . \quad (3.47)$$

The Lamé coefficients are

$$\mu = \frac{1}{3} \rho \langle v^2 \rangle \quad , \quad \lambda = \frac{1}{3} \rho \langle v^2 \rangle + \rho^2 \frac{\partial U}{\partial \rho} \quad . \quad (3.48)$$

This limit is not useful for the heavy ion case. We just give it for the sake of completeness. In heavy ion physics, one is probably closer to the short relaxation time limit, which leads to usual hydrodynamics, as indicated in Fig. 6.

4. EVOLUTION IN PHASE SPACE

4.1. Introduction

The kinetic theories that we have studied in the previous chapter describe the evolution of the system in phase space. However, only the asymptotic properties of the system are measured, f.i. the momentum of the particles. One thus may be interested in the evolution of the system in momentum space only. In Section 4.2, we will investigate this question in some detail. In Section 4.3, we will discuss several aspects of the evolution of the system in phase space. Finally, Section 4.4 will contain a small discussion of the concept of freeze-out as seen from a kinetic point of view.

4.2. Evolution in momentum space

Many models have been proposed to study this aspect. Randrup[31] has studied the Uehling-Uhlenbeck equation (Eq. (3.18)) without convection and potential terms in an homogeneous medium. The r-dependence may be integrated out and the model is applied to two uniform infinite media in relative motion. This may help to look at the equilibration process. However, as Fig. 7 shows, the latter is very much influenced by geometrical properties.

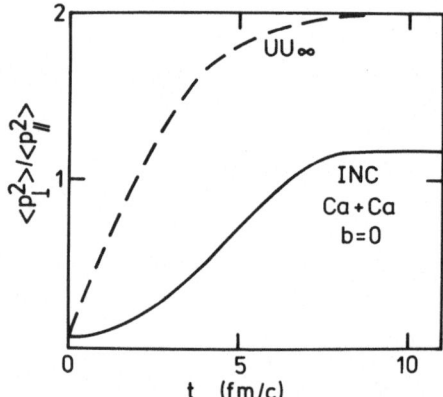

Fig. 7. Comparison of the equilibrium process as predicted by a uniform medium calculation[31] (dotted line) and a 3-dimensional INC calculation[13] (full line) : time evolution of the ratio between average squared longitudinal and perpendicular particle momentum.

The equilibration is much less rapid with real finite nuclei. In the latter case, binary collisions are drastically reduced because of surface and finite size effects. Furthermore, thermalization is not achieved because the system desintegrates rapidly. Let us remind that experimental data do indicate a lack of thermalization. The work of Ref. 31 also shows that the thermalization process is rendered much more efficient by the possible Δ-production. The latter transforms the longitudinal kinetic energy into mass energy and helps to populate the so-called midrapidity region,[13,31] whereas elastic scattering starts to populate the elastic circle (see Fig. 8).

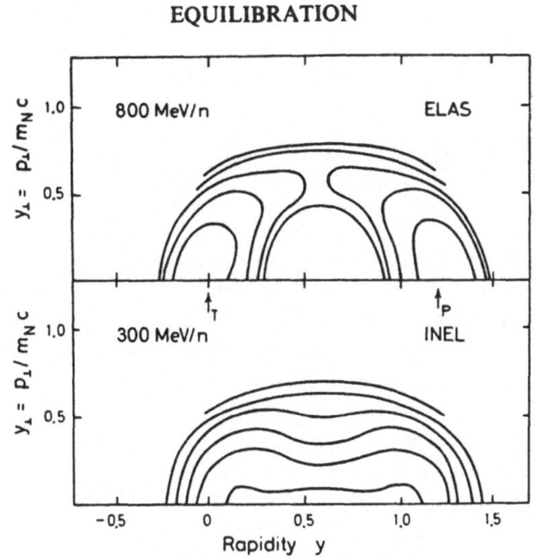

EQUILIBRATION

Fig. 8. Contour plots of the distribution of nucleons in rapidity space after their first collision when only elastic NN collisions are considered (upper part, maxima close to the arrows) and when Δ-production is introduced (lower part, maximum close to the $y_\perp = 0$ axis). From Ref. 31.

The features of the final momentum distribution are also interesting to analyze in terms of the number of collisions n undergone by the nucleons. There are numerous models which are dedicated to this study. Let us only mention here the diffusion model of Pirner and Schrürmann[32,33,34] and the closely related approach of Malfliet based on a linear approximation of the Boltzmann equation in momentum space.[35,36] Before entering into the discussion, we make the connection with the experimental data. The invariant nucleon production cross-section can be written as

$$E \frac{d^3 \sigma}{d p^3} = \int_0^{b_{max}} 2 \pi \, b \, d b \int d^3 r \, E \, f_1(\vec{r}, \vec{p}, b, t) \qquad , \qquad (4.1)$$

where we have made explicit the impact parameter dependence of f_1. The quantity b_{max} is the maximum impact parameter ($\approx R_1 + R_2$). The integral (4.1) is time-independent, since for freely moving particles, one has

229

$$f_1(\vec{r}, \vec{p}, t) \approx f_0(\vec{r} - \frac{\vec{p}}{m} t) \qquad . \tag{4.3}$$

We may write the following expansions in terms of n, disregarding the spectators $(n \geq 1)$:

$$\nu = \int d^3r \ d^3p \ f_1(\vec{r}, \vec{p}, b, t) = \sum_{n \geq 1} f(n) \tag{4.4}$$

where ν is the number of participants, and

$$(E \frac{d^3 \sigma}{d p^3}) = \sum_{n \geq 1} (E \frac{d^3 \sigma}{d p^3})_n \quad , \quad \sigma_n = \int \frac{d^3 p}{E} (E \frac{d^3 \sigma}{d p^3})_n \quad . \tag{4.5}$$

Of course, the terms in Eq. (4.5) are not observable quantities, but it is instructive to look at them in the frame of a specific model. Figure 9 shows the quantities entering into Eqs. (4.4-5) as calculated by an INC calculation.[37] The striking feature is the wide distribution of the collision frequencies for central as well as inclusive (integrated over b) collisions. In the average, for central collisions, the nucleons are making 2.78 and 3.24 collisions, for Ne + Ne and Ca + Ca, respectively (~ 4 for Nb + Nb at 400 MeV/A). However, a non negligible number of them are making one collision only.

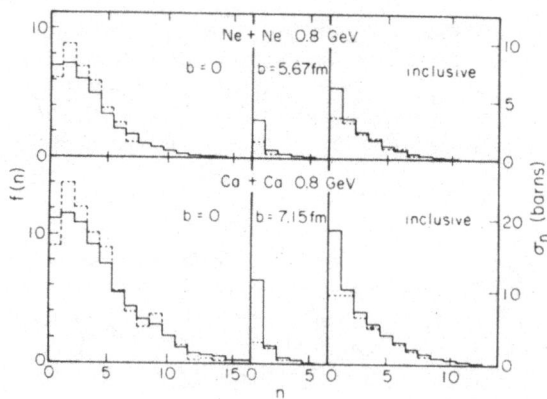

Fig. 9. Quantities $f(n)$ for two impact parameters (left scale) and σ_n (right scale) (Eqs. (4.4-5)), as calculated in Ref. 37. The full lines refer to the final nucleons whereas the dotted lines refer to Δ-particle ($\times$ 8).

The spectra are very different according to the number n. They are displayed in Fig. 10, for the Ca + Ca system. For n = 1, the so-called single-scattering component,[38,39] the spectrum is largely dominated by the differential NN cross-section, which is rather well

forward-peaked (see also Fig. 8). As n is increasing the central
rapidity region fills up whereas the projectile rapidity region
is depleted. For n larger than 3, the distribution is close to an
isotropic exponentially decreasing (in energy) distribution.

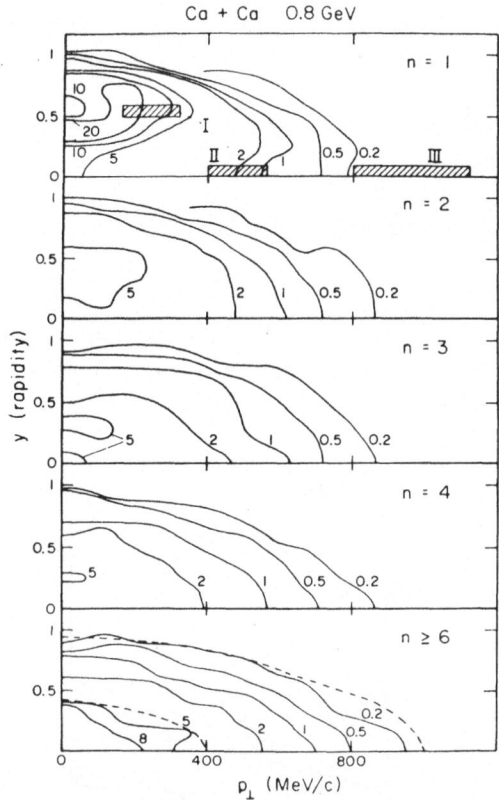

Fig. 10. Invariant cross sections to produce a proton after n
collisions as a function of the c.m. perpendicular momentum
$p_\perp$ and c.m. rapidity y. They are given in units of 0.462×10^4 mb c^3 GeV^{-2}. Only one half of the rapidity space is
shown. To give an idea of the isotropy n $\geq$ 6, curves of
equal total energy of 1020 and 1371 MeV (dotted curves)
have been drawn. Taken from Ref. 37.

As a consequence, only a part of the system may be considered as
thermalized (or thermal-like), whereas the remaining nucleons are
making few collisions. The importance of the latter part is directly
related to the size of the system (and also to the energy) and is
decreasing with increasing mass. Let us also mention that the
analysis based on the interaction cluster model[40-43] or the so-
called phase space model[44-46] leads to the same conclusion, namely
that only a part of the system may be considered as equilibrated.

4.3. Evolution in phase space

It is customary to show matter density distribution. We will not do it here because, as we have seen , it is very likely that the equilibrium is not reached. The interested reader may look at Refs. 13, 47, 48, 49. It is however interesting to look at the evolution of the gross properties of the system in phase space. The latter may be conveniently characterized by the second moments of the one-body distribution functions

$$M_{ij}(t) = \int d^3r \, d^3p \, \xi_i \, \xi_j \, f_1(\vec{r},\vec{p},t) \quad , \qquad (4.6)$$

where the six-dimensional vector ξ is

$$\xi = \{\vec{r},\vec{p}\} \quad . \qquad (4.7)$$

They are 21 of them describing the spatial and momentum extensions of the system, as well as the possible r-p correlations. Some of them are given in Fig. 11.

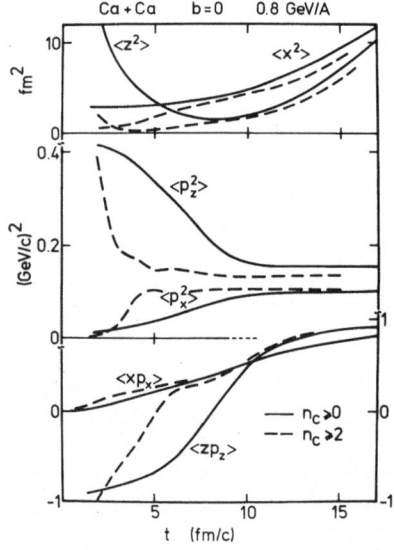

Fig. 11. Time evolution of various moments of the one-particle distribution function f_1. The non diagonal moments $\langle xp_x \rangle$ and $\langle zp_z \rangle$ have been normalized by dividing by $(\langle x^2 \rangle \langle p_x^2 \rangle)^{\frac{1}{2}}$ $(\langle z^2 \rangle \langle p_z^2 \rangle)^{\frac{1}{2}}$, respectively. z stands along the beam axis. The full lines correspond to taking all the particles, whereas the dotted lines amount to selecting those that made two collisions at least. Adapted from Ref. 50.

The latter, as well as Fig. 12, gives the result of an INC calculation. It shows that the collision process may be roughly divided

into two stages : a compression stage and an expansion stage. At
the end of the compression stage, the system is longitudinally
compressed into a pancake shape. The transformation of the longitudinal motion into random motion (central part of Fig. 11) goes on a
little bit further, although it is not complete, as we have repeatedly
said. As time proceeds, the shape of the system becomes more and more

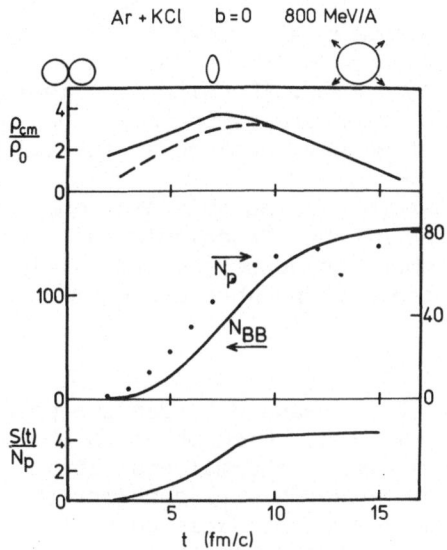

Fig. 12. Time evolution of the baryon density (dashed line : participants only, full line : all baryons included) normalized
to the normal density of nuclear matter (top), of the number
of baryon-baryon collisions (centre, scale on the left),
of the number of participants (centre, scale on the right),
and of the entropy of the participant system divided by
the final number of participants (bottom). Taken from
Ref. 50.

The evolution of the system reveals interesting and complicated
features. Figure 12 shows that the system (more precisely the participants in the system) gains entropy during the compression stage.
We recall that the entropy per baryon is related to the one-particle
distribution function by

$$\frac{S}{A} = 1 - A^{-1} \int d^3r \, d^3p \, f_1(\vec{r},\vec{p},t) \, \ln \, [f_1(\vec{r},\vec{p},t)(2\pi\hbar)^3] + 4 \ln 2$$

(4.8)

or, roughly, by

$$\frac{S}{A} \approx 1 - \ln \, [\bar{f}_1(2\pi\hbar)^3] + 4 \ln 2$$

(4.9)

if the distribution function is sufficiently close to a constant
over the available phase space. The term $4 \ln 2$ enters into (4.8)

because we have assumed spin-isospin degeneracy. The bracket is in some sense the average number of nucleons per natural unit in phase space. After t ≈ 8 fm/c, the system expands at constant entropy. The gain in entropy from configuration space is compensated in momentum space. But this compensation lasts a small time span. After ~ 10-11 fm/c, the system is still expanding, but the momentum distribution does not change anymore (see Fig. 11). We face then a free expansion characterized by a strong correlation between position and momentum. In fact, the behaviour of the xp_x and zp_z correlations allows to consider the system as a more or less globally equilibrated system, i.e. with little correlations, during a very short time interval, say from 7 to 9 fm/c. The situation looks a little bit better for the baryons which are making more than one collision. This is in keeping with what we have noticed at several places.

Let us mention before closing this section, that entropy of the order 4.4 as calculated by the INC calculation corresponds to a fermion-occupancy in phase space of the order 1/10 per unit volume. We will come back to the entropy when discussing deuteron production.

4.4. The freeze-out

By this, one generally means a sudden disappearance of the interaction between particles, which makes an abrupt transition between an expanding system in thermal equilibrium and a freely expanding system. It is reasonable to consider that the freeze-out takes place when the mean freepath λ is equal to the mean distance d between particles. The latter quantity is related to the density ρ by

$$d \approx \rho^{-1/3} \quad . \tag{4.10}$$

Therefore, the freeze-out occurs for

$$\rho = \rho_{fo} \approx (\frac{1}{\sigma_{NN}})^{3/2} \quad . \tag{4.11}$$

After the maximum compression (see Fig. 11), the average energy is such that $\sigma_{NN} \sim 20$ mb, which makes $\rho_{fo} \approx 2 \rho_0$. This seems quite a large value. However, we see in Fig. 11 that it is approximately the point beyond which the $\vec{r}.\vec{p}$ correlations are growing considerably. This is just what we expect in a simple-minded picture of the freeze-out as examplified in Fig. 13 for the one-dimensional case. During the isoentropic expansion, the system expands in staying in a thermal equilibrium state : without any $\vec{r}.\vec{p}$ correlations, and keeping the same phase space extension. The box in Fig. 13 increases along the r-axis and shrinks correlatively along the p-axis. After the freeze-out, the box is distorted by the free expansion of the system and $\vec{r}.\vec{p}$ correlations are building up. The calculation shown in Fig. 11 seems to follow this pattern with some deviations however. The binary collisions seem to persist after the $\vec{r}.\vec{p}$ correlations start to grow, i.e. after t ≈ 10 fm/c in this case. This presumably comes from fluctuations in the internucleon distance and the non-uniformity of the density. In addition, the expansion prior to freeze-out (from t ≈ 7 to t ≈ 10 fm/c) seems to produce some (limited) amount of entropy : about half a unit per baryon. This obviously comes to collisions, which act as a source of viscosity.

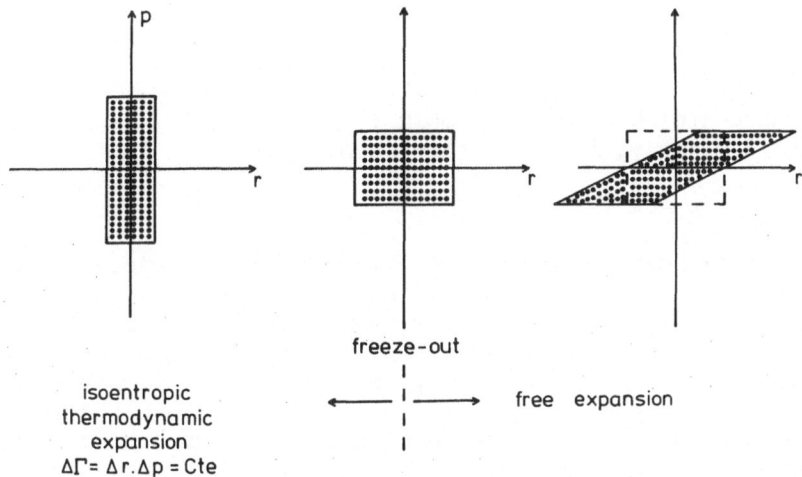

Fig. 13. Schematic illustration of the freeze-out transition showing the building of $\vec{r}.\vec{p}$ correlations.

Recently,[51] it has been shown that such an amount of entropy can be generated by a viscosity of a free nucleon gas. In summary, it seems that the concept of freeze-out is consistent with the cascade dynamics, although some $\vec{r}.\vec{p}$ correlations are built prior to the moment at which NN collisions cease.

5. HYDRODYNAMICS

The conservation equations obtained from the Landau kinetic equation (3.18) (we however drop here the $(1 - \tilde{f})$ factors) are Eqs. (3.27), (3.32) and (3.39). They however imply unknown quantities and are not closed on themselves. The usual way to close them is phenomenological. The starting point is to observe that the collision term vanishes for

$$f_1(\vec{r},\vec{p},t) = \rho(\vec{r},t)(2\pi m T)^{-3/2} \exp(- \frac{(\vec{p} - m \vec{u}(\vec{r},t))^2}{2m\,T}) \quad , \quad (5.1)$$

which corresponds to a local equilibrium. The Boltzmann constant is set equal to one. In this case,

$$S_{ij} = \frac{1}{3} p \, \delta_{ij} \quad , \quad p = \rho T \quad , \quad \vec{J} = 0 \quad . \qquad (5.2)$$

235

In general, (5.1) does not provide a stationary solution, even if we remove the t-dependence. The reason is that the convection term in the l.h.s. of the Landau equation drives particles from one point to the other. Obviously, corrections to (5.1) will come from gradients of the intensive macroscopic quantities ρ , $\vec{u}$ and T. Because S is a second rank tensor, it should be corrected as

$$S_{ij} = \frac{1}{3} p\, \delta_{ij} - \eta(\nabla_j u_i + \nabla_i u_j - \frac{2}{3} \delta_{ij}\, \vec{\nabla}.\vec{u}) - \zeta\, \delta_{ij}\, \vec{\nabla}.\vec{u} \quad (5.3)$$

where η and ζ are the shear and bulk viscosity coefficients, respectively. Similarly, the heat flow should be of the form

$$\vec{J} = -\lambda\, \vec{\nabla}\, \rho - \varkappa\, \vec{\nabla}\, T \quad , \quad\quad\quad\quad\quad\quad (5.4)$$

where $\varkappa$ is the thermal conductivity. Therefore, the basic equations are the continuity equation (3.27), Eq. (3.32) with S_{ij} given by (5.3) and the energy equation (3.39) with the expressions (5.3) and (5.4). They are known as the Navier-Stokes equations describing the behaviour of a real fluid. They still contain too many unknowns. The situation is cured by eliminating the pressure and the internal energy in terms of ρ and T. The functional relationship is assumed to be the same as in equilibrium and is known as the equation of state

$$p = p(\rho, T) \quad\quad\quad\quad\quad\quad\quad\quad\quad\quad (5.4a)$$

$$e = e(\rho, T) \quad . \quad\quad\quad\quad\quad\quad\quad\quad (5.4b)$$

One thus ends with five scalar equations for five variables $(\rho, \vec{u}, T)$.

It should be realized that the hydrodynamical approach *basically* assumes a local equilibrium. In other words, it averages over all the local fluctuations inherent to the microscopic structure of the system. In particular, all the fluctuations with characteristic length smaller than the thermalization mean free path λ_{th} are averaged out. In principle, hydrodynamics can describe disturbances with a scale not smaller than λ_{th}. The latter quantity is not necessarily the usual mean free path λ, because in a collision, a particle can lose a very small amount of energy. This is precisely the case in the GeV/A range, where a particle needs at least two or three collisions to be thermalized.[52,53] The quantity λ_{th} is certainly not small compared to the dimension of the system and hydrodynamics is hardly expected to be valid. On the other hand, just to close the discussion about hydrodynamics in general, it should be mentioned that this approach is not necessarily restricted to weakly interacting gases. Actually, the Euler equation (i.e. the Navier-Stokes without viscosity) may be derived in very general grounds, once one has decided to forget about microscopic fluctuations.

236

Even though, hydrodynamics has been applied very soon in this field because one is dealing with supersonic velocities. Actually, the hydrodynamical equations admit shock wave solutions (due to their non-linearity). The shock compression and heating are linked (for a plane shock) by the Rankine-Hugoniot equation, (that we write here in its relativistic form)

$$(e^2 - e_o^2) + p(\frac{e}{\rho} - \frac{e_o}{\rho_o}) = 0 \tag{5.5}$$

In this equation, which applies to the rest frame of the compressed matter, the indexed quantities correspond to the before-shock zone. The interesting feature of this equation is that one can study the collision between identical nuclei, if we idealize them as two slabs (infinite in the perpendicular direction). The after shock energy density e must be equal to the initial energy density as seen in the c.m. frame :

$$e = \gamma_{c.m.} \; e_o \quad , \tag{5.6}$$

with

$$\gamma_{c.m.} = (1 + \frac{T_{lab}}{2W_o})^{\frac{1}{2}} \quad , \qquad W_o = m_N c^2 - B \quad , \tag{5.7}$$

T_{lab} being the lab incident kinetic energy. For a given equation of state $e = e(\rho,s)$, (s = entropy density), Eqs. (5.5) and (5.6) can yield a solution for ρ, s and p for any energy T_{lab}. In particular, if we take an ideal gas ($e = \frac{3}{2} p \rho^{-1}$), one readily obtains

$$\frac{\rho}{\rho_o} = \frac{5}{2} (\gamma_{c.m.} (E) - \frac{3}{2} \gamma_{c.m.}^{-1} (E)) \quad . \tag{5.8}$$

It has been shown that shocks can generate high entropy ($S/A \approx 1-4$), high compressions ($\rho/\rho_o \approx 2-6$) and high temperatures.[56]

Solving the Navier-Stokes equations in three dimensions is a formidable task and has not been carried out up to now. But solving the Euler equations is feasible and has been done. Numerical cell-size effects introduce the so-called "numerical" viscosity,[48,57] whose control is not complete and which seems to be small.[58]

Since one is measuring particles, the microscopic structure has to be reintroduced at some place making comparison with data possible. This is generally done by assuming the fluid cells "freeze out" when the density comes below a certain value ρ_f. The most sophisticated method is to determine the yield and the spectra of different particles by a chemical equilibrium calculation using the parameters of the cells : ρ_f, T and its flow velocity. The cross-sections for a specific species seem to be sensitive to the freeze-out procedure,[59,60] but the sum charge cross-sections are roughly insensitive.

Figure 14 illustrates the results of an hydrodynamical calculation[62] when the transport coefficients are set equal to zero.

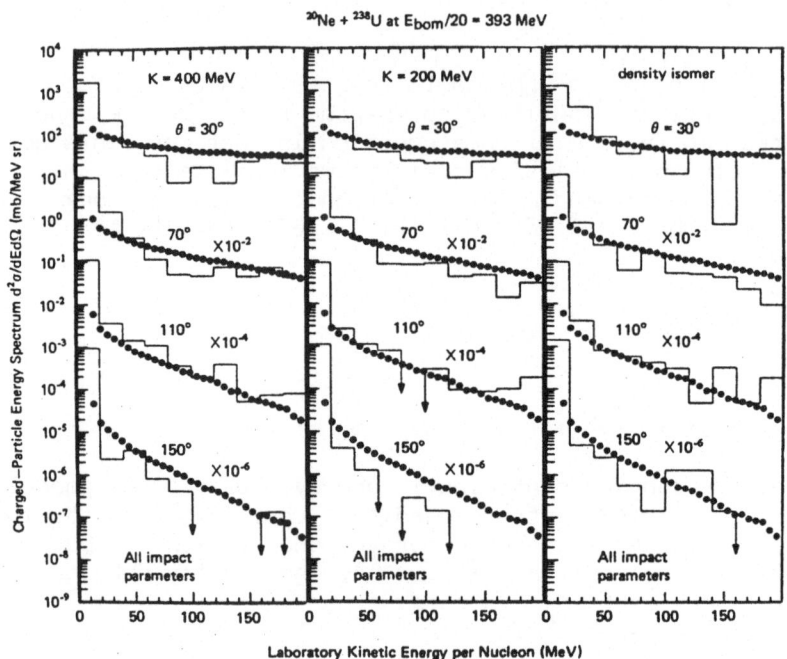

Fig. 14. Comparison of charge-inclusive data[61] (dots) with nonviscous hydrodynamical calculations[62] (histogram). Results for three equations of state are shown.

The results are good on the qualitative level only. Figure 14 touches also at the sensitivity of the results on the equation of state. Due to the large uncertainty of the calculation, it is hard to tell anything about a possible sensitivity. Recent calculations[63] seem to indicate that some observables like the flow angle is sensitive to detail of the equations of state (see later).

So far, the full Navier-Stokes equation could be solved only in the two-dimensional approximation.[60,64,65] The viscosity effects make the entropy increase by only ≈ 20 %.

Definite evidence for the validity of hydrodynamics or even for the manifestation of some specific features has not yet been obtained. Two candidates for such an evidence are the sidewards proton

emission in Ne + U at 400 MeV/A[66] and the recent Nb + Nb data at
400 MeV/A.[65] We will discuss these two points in the next section.

6. TOWARD HOT DENSE MATTER PHYSICS ?

6.1. Introduction

As said in the introduction, the difficult task is to relate
observables to the properties of the intermediate excited, compressed
system, and hopefully to connect these ones to bulk properties of
nuclear matter. The inclusive measurements are not very useful in this
respect as they are largely dominated by phase space (more exactly,
the volume of the available momentum space). In the huge amount of
data accumulated, three of them (at least) seem or have seemed
promising in this perspective. We are going to discuss them successi-
vely. The following procedure will be adopted. The INC model, which
embodies a trivial equation of state will serve as a reference for
"conventional" physics. Hydrodynamics will be looked for indication
of eventual equation of state effects.

6.2. Deuteron formation and the entropy puzzle

It is an experimental law[12,67-69] that the production cross-
section for a fragment A in the participant region is related to
proton cross-section by

$$E_A \frac{d^3 \sigma_A}{d p_A^3} = C_A \left(E_p \frac{d^3 \sigma_p}{d p^3}\right)^A \quad , \tag{6.1}$$

where $E_A = A E_p$, $\vec{p}_A = A\vec{p}$, and where C_A is a constant independent of
energy and angle. This law at first sight suggested the so-called
coalescence idea, namely that composite particles are just formed
because nucleons can be close to each other in phase space at the
end of the collision process. The latter prescription may be made
a little bit more precise for the deuteron case in the following
equations

$$E \frac{d^3 \sigma}{d P^3} = E \int 2 \pi b \, d b \int d^3 R \, f_d^{(b)}(\vec{R},\vec{P}) \tag{6.2}$$

with

$$f_d(\vec{R},\vec{P}) = \frac{3}{4} \int d^3r \, d^3p \, f_2^{np}(\vec{R}+\frac{\vec{r}}{2}, \vec{R}-\frac{\vec{r}}{2}, \frac{\vec{P}}{2}+\vec{p}, \frac{\vec{P}}{2}-\vec{p}) g_d(\vec{r},\vec{p}) \, , \tag{6.3}$$

where we have left out the indice b. In the latter equation, f_2^{np}
is the two-body distribution function specialized to neutron-
proton pairs and $g_d(\vec{r},\vec{p})$ is the distribution function of the relative distance and momentum in the deuteron, normalized to unity. It
should be stressed[70-72] that Eq. (6.3) expresses the building of
correlations like in a deuteron but does not necessarily exclude
that they are carried by heavier composites. The asymptotic emission

of definite composite particle requires knowledge of higher order distribution functions. Therefore (6.2-3) give the cross-section for the formation of ("d") deuteron-like objects. Assuming that the n-p correlations are roughly the same in any light composite, it can be estimated how many "d"'s are present in ^{3}H, ^{3}He,... nuclei. For instance, counting clusters up to ^{4}He, one has

$$\sigma_{"d"} = \sigma_d + \frac{3}{2}(\sigma_{3_H} + \sigma_{3_{He}}) + 3\,\sigma_{4_{He}} \quad . \tag{6.4}$$

Similarly, one may define proton-like objects "p" :

$$\sigma_{"p"} = \sigma_p + \sigma_d + \sigma_{3_H} + 2(\sigma_{3_{He}} + \sigma_{4_{He}}) + \dots \quad , \tag{6.5}$$

which is nothing but the charge production cross-section. Formula (6.2-3) are a generalization of the old coalescence model.[67-73] Indeed, if g_d is very sharply peaked in phase space and if $f_2 \approx f_1 f_1$, one readily recovers the power law (6.1) for deuteron emission. The coefficient C_A may even be computed if one takes a gaussian form for g_d

$$g_d(\vec{r},\vec{p}) = \frac{1}{(r_o\sqrt{\pi})^3} \frac{1}{(p_o\sqrt{\pi})^3} \exp\left(-\left(\frac{r}{r_o}\right)^2 - \left(\frac{p}{p_o}\right)^2\right) \quad , \tag{6.6}$$

and for f_1 ($j = n,p$)

$$f_1^{(j)}(r,p) = \frac{N_j}{(R_o\sqrt{\pi})^3} e^{-\frac{r^2}{R_o^2}} \frac{1}{(\sqrt{2\pi\,m\,T})^3} e^{-\frac{p^2}{2\,m\,T}} \tag{6.7}$$

where N_j is either the neutron or proton number. In this case, one gets

$$f_d(\vec{R},\vec{P}) = \frac{3}{4} \frac{1}{\left(1 + \frac{r_o^2}{2R_o^2}\right)^{3/2} \left(1 + \frac{p_o^2}{m\,T}\right)^{3/2}} f_1^{(n)}\left(\vec{R},\frac{\vec{P}}{2}\right) f_1^{(p)}\left(\vec{R},\frac{\vec{P}}{2}\right) \tag{6.8}$$

If one assumes that only N_j is b-dependent in (6.7), one readily gets (for N = Z systems)

$$\left(\frac{d^3\sigma}{d\,P^3}\right)_d = C_d \left[\left(\frac{d^3\sigma_p}{d\,p^3}\right)_{p=\frac{P}{2}}\right]^2 \quad , \tag{6.9a}$$

with

$$C_d = 48 \times \frac{1}{\sigma_o} \frac{\hbar^3}{(\sqrt{2\pi} \, R_o)^3} \frac{1}{(1 + \dfrac{r_o^2}{2R_o^2})^{3/2} \, (1 + \dfrac{p_o^2}{mT})^{3/2}} \quad , \qquad (6.9b)$$

and

$$\frac{1}{\sigma_o} = \frac{\int 2\pi b \, db \, N_p^2(b)}{(\int 2\pi b \, db \, N_p(b))^2} \approx \frac{3}{8} \frac{1}{\pi \, b_{max}^2} \qquad . \qquad (6.9c)$$

The above argumentation leads to a power law between deuteron-like and proton-like cross-sections, whereas the experimental law (6.1) holds for free particles. Whether this difference is meaningful or not has never been investigated. The correction term in r_o/R_o was introduced for the first time in Ref. 74. In general, r_o is small compared to R_o and Eq. (6.9b) reduces to

$$C_d \approx \frac{48}{\sigma_o} \frac{1}{(\sqrt{2\pi} \, R_o)^3} \frac{1}{(1 + \dfrac{p_o^2}{mT})^{3/2}} \qquad . \qquad (6.9d)$$

The coefficient $p_o \approx 200$ MeV/c ($\approx \hbar/r_o$ because of the normalization of g_d) and therefore the temperature dependence (which reflects the bombarding energy dependence) is not likely to be observed. In passing, we notice that the same calculation can be carried out if there is some $\vec{r}.\vec{p}$ correlations, like ($-1 \leq y \leq 1$)

$$f_1(\vec{r},\vec{p}) \sim e^{-\dfrac{r^2}{R_o^2}} \, e^{-\dfrac{p^2}{2mT}} \, e^{-2y \dfrac{\vec{r}.\vec{p}}{R_1 \sqrt{2mT}}} \qquad . \qquad (6.10a)$$

In this case, the r.h.s. of Eq. (6.9b) must be multiplied by

$$(1 + y^2)^{-3} \, (1 + y^2 \frac{r_o^2}{(r_o^2 + 2 R_o^2)} \frac{p_o^2}{p_o^2 + mT}) \qquad . \qquad (6.10b)$$

In view of the uncertainty on the parameters entering in (6.9b) and (6.10b), it is hardly believable that one can extract the value of y from experiment, but the analysis has never been undertaken.

A very puzzling aspect of the deuteron production is that a

thermal model[83] (see also refs. 84-87) can also reproduce the power law (6.1). Such a model yields

$$\frac{N_d}{N_n N_p} = \frac{3}{4V} \exp(B/T) \qquad , \qquad (6.11)$$

where V is the volume of the thermalized system and B is the deuteron binding energy. For the heavy ion case, the argument of the exponential is very small. One has an expression similar to (6.9a) with this time

$$C_d = \frac{48}{\sigma_o} \frac{1}{\sqrt{2}} \frac{\hbar^3}{V} \qquad , \qquad (6.12)$$

which as we said is very similar to (6.9c). Thus the power law may come from two very different dynamical schemes. Either the deuterons are formed by chance at the freeze-out or they are regularly created and destroyed during the collision process. Let us however mention that the deuteron cross-section has been calculated by Eqs. (6.2)-(6.3) using the INC model to evaluate the two-body distribution function, in Refs. 70,74. The results of Ref. 74 are presented in Fig. 15.

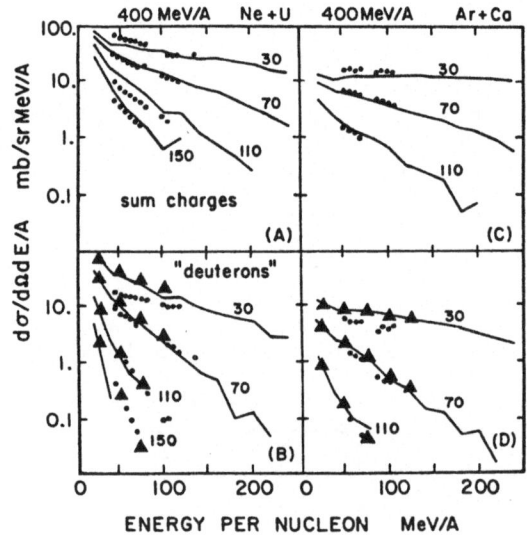

Fig. 15. Comparison between the INC calculation of Ref. 74 and the proton-like and deuteron-like (triangles) experimental cross-section. For the deuterons, the dots indicate the free deuteron yield. The experimental data are from Ref. 75.

They are obtained by taking f_2 at the time where both nucleons have their last interaction. The agreement is within a factor 2, which is remarkably good. The latter result is quite interesting from another point of view, since it may explain the peak at non-zero angle in the proton cross-section (see Section 5). Actually the sum charge (primordial proton) cross-section would be maximum at $0°$, but in this region, the nucleons would be so numerous that they would be "eaten up" by the deuteron formation. For more detail, see Ref. 4.

The fact that coalescence and thermal models give similar results may be accidental or due to the deuteron formation mode in a thermal model. Indeed, in a kinetic model (applicable to dilute gases), the deuteron balance would be given by (we forget the convection term)

$$\frac{\partial}{\partial t} f_d(\vec{P}_{23}) = \int d^3p_1 \, d^3p_2 \, d^3p_3 \, d^3p_1' \, \{f_1(\vec{p}_1) \, f_1(\vec{p}_2) \, f_1(\vec{p}_3) \, f_1(\vec{p}_1')$$
$$W(\vec{p}_1,\vec{p}_2,\vec{p}_3 \rightarrow \vec{p}_1' \, \vec{P}_{23}) - f_1(\vec{p}_1') \, f_d(\vec{P}_{23}) \, W(\vec{p}_1',\vec{P}_{23} \rightarrow \vec{p}_1\vec{p}_2\vec{p}_3)\} \quad ,$$

$$(6.13)$$

since the energy-momentum conservation requires a third partner $(n+p+N \rightarrow N+d)$. However, because the deuteron is almost unbound, the transition probability is close to

$$W(\vec{p}_1,\vec{p}_2,\vec{p}_3 \rightarrow \vec{p}_1',\vec{P}_{23}) \approx \delta(\vec{p}_1-\vec{p}_1') \, \delta(\vec{p}_2+\vec{p}_3 - \vec{P}_{23}) \, w(\vec{p}_2,\vec{p}_3,\vec{P}_{23}) \quad .$$

$$(6.14)$$

If the dynamics is such that the nucleons close in phase space inevitably makes a deuteron, one may write

$$W(\vec{p}_1,\vec{p}_2,\vec{p}_3 \rightarrow \vec{p}_1',\vec{P}_{23}) \approx \delta(\vec{p}_1-\vec{p}_1') \, \delta(\vec{p}_2 - \frac{\vec{P}_{23}}{2}) \, \delta(\vec{p}_3 - \frac{\vec{P}_{23}}{2}) \quad (6.15)$$

Therefore

$$\frac{\partial}{\partial t} f_d \propto f_1(\frac{\vec{P}_{23}}{2}) \, f_1(\frac{\vec{P}_{23}}{2}) - f_d(P_{23}) \quad . \tag{6.16}$$

The equilibrium condition requires

$$f_d(\vec{P}) \approx f_1(\frac{\vec{P}}{2}) \, f_1(\frac{\vec{P}}{2}) \quad , \tag{6.17}$$

which is the basic coalescence relation.

Whatever the deuteron source is, there is a relationship between the deuteron yield and the entropy gained by the system. Originally, this has been established by Siemens and Kapusta[76] using the thermal model. The same relation can be derived on the basis of relation (6.3). We follow the presentation of Ref. 70. The number of deuteron-like particles is given by

$$N_{"d"} = \frac{3}{4} \int d^3R \, d^3P \int d^3r \, d^3p \, f_2^{np}(\vec{R}+\frac{\vec{r}}{2}, \, \vec{R}-\frac{\vec{r}}{2}, \, \frac{\vec{P}}{2}+\vec{p}, \, \frac{\vec{P}}{2}-\vec{p}) \, g_d(\vec{r},\vec{p}) \quad .$$

$$(6.18)$$

If we assume g_d to be very narrow and $f_2 \approx f_1 f_1$, we obtain (for a N = Z) system

$$N_{"d"} = \frac{3}{16} \int d^3R \ d^3P \ [f_1(\vec{R},\frac{\vec{P}}{2})]^2 \quad , \tag{6.19}$$

where f_1 refers to nucleon, either proton or neutron. We may write (6.19) as

$$N_{"d"} = \frac{3}{2} \langle f_1 \rangle \int d^3r \ d^3p \ f_1(\vec{r},\vec{p}) = \frac{3}{2} \langle f_1 \rangle N_{"p"} \quad . \tag{6.20}$$

In this equation $\langle f_1 \rangle$ is the average of f_1 on its own distribution

$$\langle f_1 \rangle = \frac{\int d^3r \ d^3p \ [f_1(\vec{r},\vec{p})]^2}{\int d^3r \ d^3p \ f_1(\vec{r},\vec{p})} \quad . \tag{6.21}$$

Now, the entropy (Eq. (4.8)) is given by

$$\frac{S}{A} = 1 + 4 \ln 2 - \langle \ln f_1 \rangle \quad . \tag{6.22}$$

By the Sackur-Tetrode formula, one may relate $\langle \ln f_1 \rangle$ to $\ln \langle f_1 \rangle$. Using numerical values, one gets

$$\frac{S}{A} = 3.95 - \ln \frac{N_{"d"}}{N_{"p"}} \quad . \tag{6.23}$$

In Ref. 76, the same relation is obtained, but, since the dilute gas limit is used, the formula involves free deuterons and protons. The physical content of Eq. (6.23) is rather transparent. The larger the entropy is, the farther from each other the nucleons are in phase space, the smaller the chance of building deuterons is.

Originally, the entropy was extracted (through (6.23)) from *integrated* deuteron cross-section and compared to ordinary equation of state and to INC evaluations (Eq. (4.8)) for central collisions. The "experimental" entropy was found substantially larger than the theoretical estimates, as shown in Fig. 16. The full curve is a one-dimensional hydrodynamical calculation with two adjusted parameters (η and ρ_f , see Section 5). It contains however an interesting aspect. At low incident energy, few entropy is created and many heavy clusters are expectedly formed. At the freeze-out, many of them decay into smaller ones increasing the observed deuteron yield. Apart from this aspect, which is effective at low energy only, there is a large discrepancy between theory and experiment : this consti-tutes the so-called *entropy puzzle*. There has been several proposi-tions to remove the puzzle, like the excitation of new degrees of freedom[76] or the reduction of the deuteron binding force inside a medium,[78] a very similar effect to the Mott transition in a plasma.

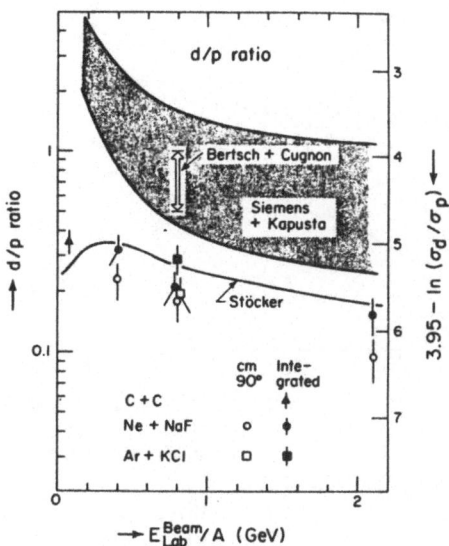

Fig. 16. Observed d/p ratios and entropy. Data for equal-mass collisions are compared with various theoretical predictions by
Siemens and Kapusta (thermal),[76] Bertsch and Cugnon (cascade),[70] and Stöcker (hydrodynamics + thermal break-up).[77]
Adapted from Ref. 12.

The explanation comes very likely from a strong impact parameter or
multiplicity dependence. The charged multiplicity dependence of the
"d"/"p" ratio has been demonstrated by Gutbrod's group[80] and later
by Nagamiya's group.[81] Their results are contained in Fig. 17, along
with INC calculations for Ar + KCl.[23,50,82] Interestingly enough,
the "d"/"p" increases with the charge multiplicity (both experimentally and theoretically), suggesting that the occupation in phase
space is substantially larger in high multiplicity (central) collisions. The question which naturally arises is whether a saturation
really occurs and why it occurs. According to Refs. 23, 70, 82, the
bulk dynamics can occur for small impact parameters only, because
particles which are within a mean free path from the surface do
not participate to the phase space spreading in the same way, even if
in the momentum space, there is perhaps no real difference. They are
not confined by outer parts of the system. The situation is illustrated in Fig. 18, which shows a steady decrease of the calculated
entropy as b decreases leading to a sort of saturation for small
impact parameters. The content of Fig. 19 goes in the same direction
showing that relation (6.23) holds quite well for small impact
parameters. For all these reasons, it is more reasonable to compare
saturation (or maximum) values of Fig. 17 with theory (if the latter
cannot predict surface effects), rather than integrated quantities,
like in Fig. 16.

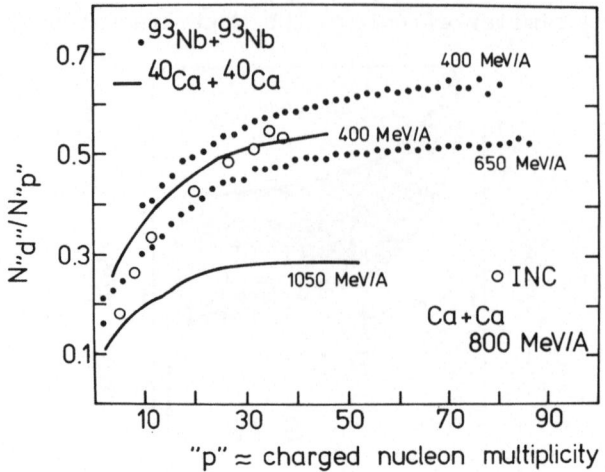

Fig. 17. Experimental "d"/"p" ratio versus charged multiplicity for
four systems. The full dots are data from Ref. 81, the lines
represent the data of Ref. 80 and the open dots are calculations from Ref. 82. For the latter the abscissa represents
half the average participant number for several impact parameters ranging from zero to 7.15 fm.

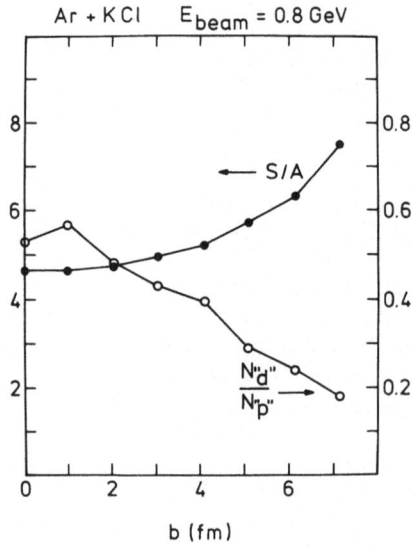

Fig. 18. INC calculation of the entropy per baryon and the deuteron
proton ratio as a function of impact parameter. Adapted
from Ref. 82.

246

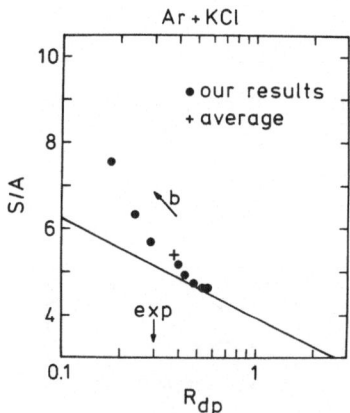

Fig. 19. INC calculated entropy and "d" over "p" ratio compared with
Eq. (6.23) (full line). The dots correspond to regularly
spaced impact parameters ranging from 0 to about 7.5 fm. The
arrows indicate the experimental value of $R_{dp} = N_{"d"}/N_{"p"}$
when integrated cross-sections are used. Adapted from Ref.82.

One then obtains Fig. 20, where the experimental bulk entropy is
somewhat between 4 and 5 units, whereas the INC model predicts a
value a little too small. The original "puzzle" has thus substantially
deflated, but the remaining discrepancy should not be underestimated,
since it corresponds to a phase space occupancy, wrong by ~ 40 %.

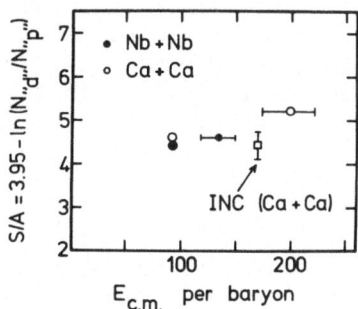

Fig. 20. Experimental values of the "d"/"p" ratio (or the entropy)
for large multiplicities compared to the INC calculations
of Ref. 70.

247

6.3. Pion multiplicity and compression energy

A very important piece of data was obtained by the Stock's streamer chamber group[88] and is shown in Fig. 21.

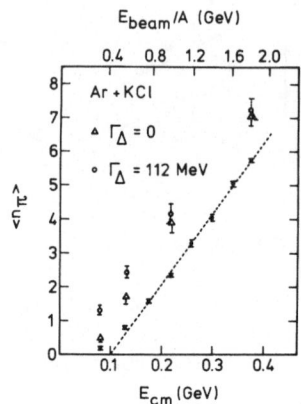

Fig. 21. The negative pion yield measured by a central trigger for Ar + KCl collisions (data around the dotted line) is compared to the prediction of INC (circles). The triangles are the results obtained in a calculation with frozen Δ-isobars. Taken from Ref. 14.

Here is presented the (negative) pion multiplicity from the central trigger, which, according to Ref. 88, collects all the small impact parameters up to b = 2.2 fm. To understand fully the importance of these data, it is useful to remind of two complementary experimental results. (1) The pion multiplicity distribution is consistent with incoherent production,[88,89] which, for a given impact parameter, implies a Poisson law.[90] (2) There is a strong correlation between the pion and the charge multiplicity,[88,89] heavily suggesting that central collisions produce most of the pions. Figure 21 has been the starting point of a long-standing debate. The INC calculations of Ref. 14, which was the most reliable calculation at that time, was not able to reproduce the data. If the following reactions were considered

$$NN \overset{\rightarrow}{\leftarrow} N\Delta \quad , \quad N\Delta \rightarrow N\Delta \quad , \quad \Delta\Delta \rightarrow \Delta\Delta \tag{6.24a}$$

with pions coming from the decay of the Δ's at the end of the cascade, the multiplicities (triangles in Fig. 21) came out rather well, but the spectrum was really bad (see Ref. 14 for the detail). If the following reactions

$$\Delta \overset{\rightarrow}{\leftarrow} \pi N \tag{6.24b}$$

were introduced during the cascade, on the basis of the natural Δ lifetime, the spectrum is very much improved, but the pion multiplicity is overestimated (circles in Fig. 21), especially close to the effective threshold. The increase of the pion yield simply reflects the higher energy which is required to make a pion if only (6.24a) is retained, because of the large mass of the Δ-resonance.

The discrepancy between INC predictions (with (6.24b)) and experiment has been interpreted by the authors of Ref. 91 as a manifestation of the compression energy. The idea is the following. A part of the available energy is necessary to compress the matter and therefore less kinetic energy is available to produce pions. This demands two subsidiary conditions. First, the number of pions is just determined by the kinetic (or thermal) energy. Second, the number of pions is fixed at (or by) the maximum density. The latter observation is corroborated by the INC calculation of Ref. 13,14. At the maximum compression, there are almost only Δ particles. During the expansion, the number of Δ's diminishes, but the number of free pions is compensating this decrease. Globally, the total number of pions (free + those hidden in Δ's) seems to stay constant after the moment of maximum compression. Therefore, it seems that one may write

$$n_\pi^{INC} (E_o) = f^{INC}(E_o, \overline{\rho}) \quad , \tag{6.25}$$

which expresses that, within the frame of the INC model, the number of pions is a function (whose analytical form is unknown) of the initial energy (which is the same as the kinetic energy) and of the average maximum density $\overline{\rho}$. In Ref. 91, it is assumed that, in the real world, the function is the *same* as in (6.25), but that the available energy is reduced by the compression energy :

$$n_\pi^{exp} (E_o) = f^{INC}(E_o - E^c(\overline{\rho}), \overline{\rho}) \quad . \tag{6.26}$$

Comparison between (6.25) and (6.26) is then used to extract $E^c(\overline{\rho})$, when $\overline{\rho}$ is taken from the INC evaluation. The results are given in Fig. 22, and are remarkable in several respects. Let us just mention that the compression energy is obtained for a wide range of densities. The procedure of Ref. 91 has been very much disputed, because it looks like taking some good aspects from a model and rejecting some bad aspects at the same time. However, it is fair to say that the authors of Ref. 91 have been very careful to eliminate all subsidiary sources of uncertainty, as by making an independent analysis in terms of multiplicity rather than in energy. Moreover, Harris and Stock[92] have made a similar analysis, abandonning the hypothesis (6.25) in favour of

$$n_\pi^{exp}(E_o) = f^{th}(E_o - E^c(\overline{\rho}), \overline{\rho}) \quad , \tag{6.27}$$

i.e. using a thermal model. The only reference to the INC model is the value of $\overline{\rho}$, which is borrowed from it. They arrived at the same result. (Incidentally, it means that chemical equilibrium is more or less realized in the INC). This additional result makes the analysis more convincing, but the whole procedure leaves still some skepticism. Nevertheless, it seems that the discrepancy is so large at small energy that something should be lacking in the INC model.

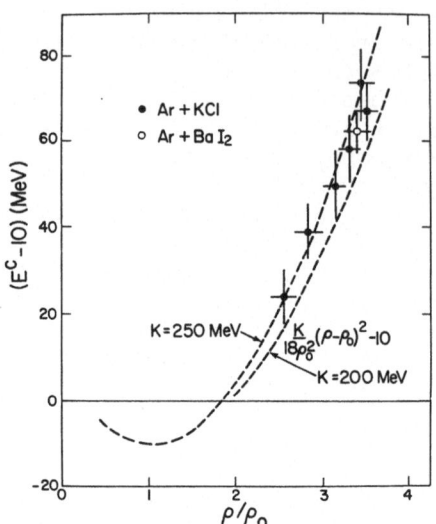

Fig. 22. Compression energy versus the density as determined from the comparison between Eqs. (6.25)-(6.26). See text for detail. Taken from Ref. 91.

Let us remind that the basic premise of the INC is the assumption that the collision process is a sequence of binary collisions between on-shell particles proceeding as in free space. If the model fails somewhere, either the whole model should be rejected (this would be unlikely in view of the numerous successes of the model) or one (or several) of the assumptions (indicated by the underlined terms) have to be relaxed. For the last choice, one may think of several possibilities. Most of them have concentrated on the second assumption. In most cases, the idea is to introduce in one way or the other an average potential for the nucleons.

In Ref. 93, the particles are put off their mass shell to simulate the initial binding effects and they are allowed to come on their mass shell once they make a collision. This procedure is inspired by the observation that particles are spreading very fast in phase space. Presumably, the average field is destroyed at the same rate.

In Ref. 94, the standard INC model is run to calculate the average density and by a simple folding method the average field. Then the cascade is run again with an average field (a model close to the Landau equation). In Ref. 95, the influence of a hard core

250

is added through the Enskog procedure.[96,97] Let also mention that the effect of interaction on the pion multiplicity has also been studied recently[30] in a kind of equation of motion approach. All these approaches aimed to introduce interaction energy effect. It is not clear however whether an average potential, a one-body property, can account for the interaction energy, a two-body property. In addition, one is facing in all these approaches a "double-counting" problem. If a part of the dynamics is put into an average field, the free collision term should be corrected for that. We will not discuss this topic any further, because the situation is far from clear for the time being.

Let us finally notice that the relaxation of the third assumption has not been investigated up to now. One may very well conceive that the production mechanism of the pions is very much altered by the presence of the surrounding matter. Also, nobody has looked at the influence of the pion propagation.

In conclusion, it is very likely that interaction energy effects are manifestating themselves in the excitation function of the pion multiplicity, but one is still far from extracting the function $E^c(\rho)$ without any ambiguity.

6.4. Collective flow

One usually characterizes as collective a phenomenon by which several particles behave cooperatively to the construction of a physical quantity. Expectedly, a collective behaviour arises from the bulk properties of the system (although nuclear physics has accustomed us with collective vibrations which involve single-particle properties coupled to surface degrees of freedom). At least, there exists a good example : the giant isoscalar monopole state. Its energy is given (asymptotically, i.e. as $A^{-1} \to 0$) by the compression modulus of nuclear matter.

A collective flow would thus correspond to a correlated emission of particles (or momentum or energy) in a preferential direction. The hydrodynamics basically relying on the bulk properties of matter, it was originally believed that the observation of a collective flow would provide an evidence of hydrodynamical behaviour. Now, it seems to us that the issue can be formulated in two questions : (1) How to put in evidence a collective flow ? (2) Are there different predictions of the flow in the frame of INC and hydrodynamics, respectively ?

In the relativistic heavy ion case, we have already indicated that signals of a collective flow are possibly given by the proton angular distribution in Ne + U at 400 MeV/A [59] and by the comparison between proton and pion temperatures, which could indicate an explosive radial expansion.[15] But it is generally believed that the

collective flow will be brought to light more easily by studying the
so-called global variables.[98-105] They have been used successfully
in high-energy physics to put jets in evidence in high multiplicity
events. The idea is to characterize with a few numbers the topology
of an event with a large number of ejectiles, whose exact description
would require many and many variables. The most common of the global
variables is the sphericity tensor

$$Q_{ij} = \sum_{\nu} \gamma(p^{\nu}) \, p_i^{\nu} \, p_j^{\nu} \quad , \tag{6.28}$$

where p_i^{ν} is the ith Cartesian coordinate (in the c.m. system) of the
momentum of the νth ejectile. The quantity $\gamma(p^{\nu})$ is a (scalar)
weighting factor. The most interesting choices are

$$\gamma = 1/|p^{\nu}| \quad , \quad 1/2m_{\nu} \quad , \quad 1/(p^{\nu})^2 \quad , \tag{6.29}$$

well suited to study the flow of momentum, energy and number of particles, respectively. For $\gamma = 1$, the sphericity tensor is a part of
the phase space tensor (4.6), when proper average over events has
been carried out. The choice $\gamma = 1/2m_{\nu}$ deserves particular attention,
since it is coalescence-invariant. This means that a deuteron has the
same weight as two nucleons having the same velocity. Such a choice
is very well suited to models which do not include composite formation
explicitly.

Geometrically, the tensor (6.28) represents an ellipsoid, which
has three axes and three eigenvalues λ_i , $i = 1,2,3$, that we label
in decreasing order. We call $\vec{e}_i$ the corresponding normalized eigenvectors. The eigenvalues are given by (this time, not necessarily in
descending order)

$$\lambda_i = \frac{1}{3} \, tr \, \underset{\approx}{Q} + 2p \, \cos \left[\frac{1}{3} \cos^{-1} \left(\frac{r}{p^3} \right) + (n-1) \frac{\pi}{3} \right] \quad , \tag{6.30}$$

with

$$p = \frac{1}{3} \left[(tr \, \underset{\approx}{Q})^2 - 3 \, A \right]^{\frac{1}{2}} \quad , \tag{6.31}$$

$$r = \frac{1}{6} (A \, tr \, \underset{\approx}{Q} + 3 \, B) + \frac{1}{27} (tr \, \underset{\approx}{Q})^3 \quad , \tag{6.32}$$

$$A = \sum_{\{ij\}} (Q_{ii} Q_{jj} - Q_{ij}^2) \quad , \tag{6.33}$$

$$B = \sum_{[ij\,k]} \varepsilon_{ijk} Q_{1i} Q_{2j} Q_{3k} \quad . \tag{6.34}$$

In Eq. (6.33), the summation extends over the combinations of the
three indices two by two. In Eq. (6.34), the summation runs over the

permutations and ε_{ijk} is the usual antisymmetric tensor. Since Q_{ij} contains six independent numbers, three additional quantities are needed to complete the description of the ellipsoid, namely to give its orientation. They are usually taken as the polar angles (ϑ,φ) of the largest axis with respect to the x,y,z axes (let us remind that the z-axis coincides with the beam axis), and χ, the third Euler angle (i.e. the angle between the second largest axis and the intersection between the $(\vec{e}_x, \vec{e}_y)$ plane and the $(\vec{e}_2, \vec{e}_3)$ plane. Since the trace of $\underset{\sim}{Q}$ is a conserved quantity (for usual γ's), one of the variables is useless. It is customary to consider the aspect ratios

$$q_1 = \frac{\lambda_1}{\lambda_3} \quad , \qquad q_2 = \frac{\lambda_2}{\lambda_3} \quad , \tag{6.35}$$

instead of the three eigenvalues. The physically accessible values of q_1 and q_2 are given in Fig. 23.

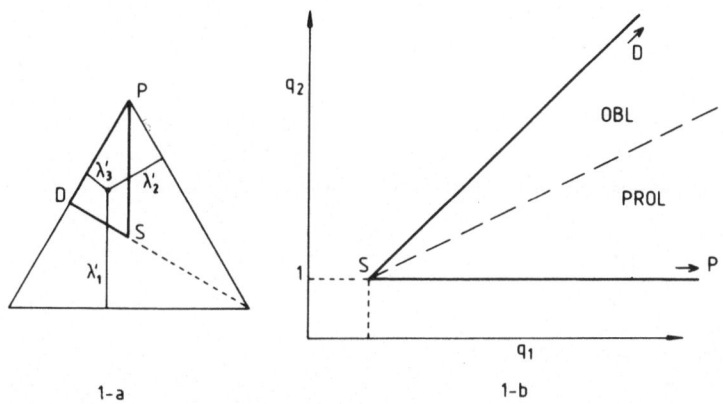

1-a   1-b

Fig. 23. Two ways of characterizing the shape of an ellipsoid. See text for more detail. The heavy lines delineate the accessible area. P = pencil, S = sphere, D = disk. The primes indicate that the eigenvalues are normalized to $\lambda_1' + \lambda_2' + \lambda_3' = 1$. Adapted from Ref. 102.

It would be desirable, for the sake of convenience, to characterize events with the smallest possible number of global variables. Generally, one restricts oneself to one shape variable q_1 and one angular variable ϑ. Several other shape variables may be constructed with the eigenvalues λ_i, such as the sphericity

$$S = \frac{3}{2} \frac{\lambda_2 + \lambda_3}{\lambda_1 + \lambda_2 + \lambda_3} \quad , \tag{6.36a}$$

the flattness

253

$$F = \frac{\sqrt{3}}{2} \frac{\lambda_2 - \lambda_3}{\lambda_1 + \lambda_2 + \lambda_3} \qquad , \qquad (6.36b)$$

the "jettiness"

$$j = \frac{\lambda_1 - \lambda_2}{\lambda_1 + \lambda_2 + \lambda_3} \qquad , \qquad (6.36c)$$

the "prolateness"

$$\Phi = \frac{\lambda_2 - \lambda_3}{\lambda_1 - \lambda_3} \qquad , \qquad (6.36d)$$

or the eccentricity

$$\epsilon = \frac{\lambda_s - \frac{1}{2}(\lambda_2 + \lambda_r)}{\lambda_1 + \lambda_2 + \lambda_3} \qquad . \qquad (6.36e)$$

In the last relation, λ_s is the eigenvalue for the axis of quasi-symmetry. It is λ_1 if $\lambda_1 - \lambda_2 \geq \lambda_2 - \lambda_3$ and λ_3 if not ; λ_r is the remaining eigenvalue. Another commonly used variable is the thrust, given by

$$T = \max_{\vec{n}} \frac{\sum_{\nu} |\vec{n}.\vec{p}^{\nu}|}{\sum_{\nu} |\vec{p}^{\nu}|} \qquad . \qquad (6.37)$$

The direction $\vec{n}_T$ which gives the maximum is the thrust direction. Finally, several simple variables, like the longitudinal energy fraction

$$f_3 = \frac{\sum_{\nu} (p_3^{\nu})^2 / 2m_{\nu}}{\sum_{\nu} (p^{\nu})^2 / 2m_{\nu}} \qquad , \qquad (6.38)$$

or the fragment multiplicity N_{ν}, are considered. The last two variables are simply related to the sphericity tensor :

$$f_3 = \frac{Q_{33}}{\text{tr } \underset{\approx}{Q}} \qquad , \qquad N_{\nu} = \text{tr } \underset{\approx}{Q} \qquad , \qquad (6.39)$$

254

for $\gamma = 1/(2m_\nu)$ and $1/(p_\nu^2)$, respectively.

The values of typical shape variables are given in Table 1 for three special types of events : a spherical event, where the emission is isotropic, a disk-like event, where the emission is isotropic in a plane only and a pencil-like event, where particles are emitted in two opposite directions.

Table 1. Values of the global variables
for limiting events

Global variable	Sphere	Pencil	Disk
q_1	1	∞	∞
q_2	1	undefined	∞
S	1	0	$\dfrac{3}{4}$
F	0	0	$\sqrt{3}/4$
j	0	1	0
Φ	undefined	0	1
ε	0	1	$-\dfrac{1}{2}$
T	$\dfrac{1}{2}$	1	$2/\pi$

The analysis is made event by event. It is worthwhile to discuss what is actually implied by this procedure. The probability for producing an event with sphericity tensor elements Q_{ij} of value T_{ij} is given by

$$\Phi(T_{ij}) = \sum_N P_N \int d^3p_1 \ldots d^3p_N \; f_N(\vec{p}_1,\ldots,\vec{p}_N) \; \delta(Q_{ij}(\vec{p}_1,\ldots,\vec{p}_N) - T_{ij})$$

$$(6.40)$$

where Q_{ij} is the expression (6.28), f_N is the N-body distribution function (in momentum space) and P_N is the probability of having an event with N ejectiles. The average value of Q_{ij} is

$$\langle Q_{ij} \rangle = \int \Phi(T_{ij}) \; T_{ij} \; d \, T_{ij} =$$
$$\sum_N P_N \int d^3p_1 \ldots d^3p_N \; f_N(\vec{p}_1,\ldots,\vec{p}_N) \; Q_{ij}(\vec{p}_1,\ldots,\vec{p}_N) \, . \quad (6.41)$$

If the correlations are unimportant ($f_N \approx \prod_i f_1$), this expression reduces to

$$\langle Q_{ij} \rangle = \sum_N P_N \int d^3p \; \gamma(p) \; p_i \; p_j \; f_1 \; (\vec{p}) \qquad . \tag{6.42}$$

In this case, there is no other information in Q_{ij} than in f_1. (Let us notice that this is not necessarily the same for the λ_i's, since the relation between the λ_i's and the Q_{ij}'s is not linear). The fluctuation of the elements Q_{ij} is given by

$$\sigma^2_{ij} = \langle Q^2_{ij} \rangle - \langle Q_{ij} \rangle^2 \qquad . \tag{6.43}$$

With the help of (6.40), we have

$$\langle Q^2_{ij} \rangle = \int \Phi(T_{ij}) \; T^2_{ij} \; d \; T_{ij} =$$

$$\sum_N P_N \int d^3p_1 \ldots d^3p_N \; f_N(\vec{p}_1, \ldots, \vec{p}_N) \; [Q_{ij}(\vec{p}_2, \ldots, \vec{p}_N)]^2 \qquad . \tag{6.44}$$

If the correlations are negligible, we have

$$\langle Q^2_{ij} \rangle = \langle N(N-1) \rangle \; [\int d^3p \; \gamma \; p_i \; p_j \; f_1(\vec{p})]^2 + \langle N \rangle \int d^3p \; \gamma^2 \; p^2_i \; p^2_j \; f_1 \tag{6.45}$$

where $\langle N \rangle = \sum_N N \; P_N$; consequently, we have

$$\frac{\sigma_{ij}}{\langle Q_{ij} \rangle} = \frac{1}{\sqrt{\langle N \rangle}} \; \frac{\{\int d^3p \; \gamma^2 \; p^2_i \; p^2_j \; f_1 - [\int d^3p \; \gamma \; p_i \; p_j \; f_1(\vec{p})]^2\}^{\frac{1}{2}}}{\int d^3p \; \gamma \; p_i \; p_j \; f_1} \qquad . \tag{6.46}$$

This important result, surmised in ref. 102, was obtained in Ref. 105. As we have seen, the averages over f_1 are more or less the same for different (symmetric) systems at the same energy per particle. Therefore, if the correlations play no role, the fluctuations would *decrease* with increasing mass. This is noting but that an extension of the properties of a system described by a canonical ensemble.

Fluctuations on the global variables are not so easy to estimate because they imply averages over eigenvalues and similar quantities, which are non linear functions of the elements Q_{ij}. Furthermore, the fluctuations mix with the so-called Jacobian effects in a very intricate manner. To understand what it means, let us take two simple examples.

(1) Let us assume that the dynamics leads in the average to spherical events. But, because of fluctuations, the ellipsoid is sometimes prolate and sometimes oblate. Now, since $q_1 \geq 1$ by definition, it is evident that $\langle q_1 \rangle$ will be larger than unity and that the difference will be larger and larger when the fluctuations are increasing.

(2) Let us consider events with a fixed elongated shape, but with the major axis pointing in the average toward the beam (z) direction, with fluctuations. It is also clear that the average angle ϑ will be different from zero.

Both examples show the bias introduced by the mapping of a manifold (Q_{ij}) into another one $(\lambda_i, \vartheta, \varphi, \chi)$ in such a way that an interior point is transformed into a point on an edge. This is particularly clear for the point $q_1 = 1$ (see Fig. 23) and the point $\vartheta = 0$. This may be formalized in the following way. The differential cross-section for making a tensor of elements T_{ij} is

$$\frac{d^6 \sigma}{[d\, T_{ij}]} = \int d^2 b \sum_N P_b(N) \int d^3 p_1 \ldots d^3 p_N\ f_N^{(b)}(\vec{p}_1, \ldots, \vec{p}_N)\ \prod_{i \leq j} \delta(Q_{ij} - T_{ij})$$

$$(6.47)$$

where $[d\, T_{ij}]$ stands for the six-dimensional element. If $f_N^{(b)}$ does not depend on the impact parameter b, one has

$$\frac{d^6}{[d\, T_{ij}]} = \sum_N \sigma_N \int d^3 p_1 \ldots d^3 p_N\ f_N(\vec{p}_1, \ldots, \vec{p}_N)\ \prod_{i \leq j} \delta(Q_{ij} - T_{ij})$$

$$= \sum_N \sigma_N\ \Phi_N\ (T_{ij}) \quad . \qquad\qquad (6.48)$$

If we turn to the eigenvalues and angles, we can write

$$\frac{d^6 \sigma}{d\lambda_1\, d\lambda_2\, d\lambda_3\, d\vartheta\, d\varphi\, d\chi} = \frac{\partial(T_{ij})}{\partial(\lambda_1, \lambda_2, \lambda_3, \vartheta, \varphi, \chi)}\ \frac{d^6 \sigma}{[d\, T_{ij}]} \quad , \qquad (6.49)$$

where we have introduced the Jacobian of the transformation. It is given by[105]

$$\frac{\partial(T_{ij})}{\partial(\lambda_1, \lambda_2, \lambda_3, \vartheta, \varphi, \chi)} = J(\lambda_i)\ \sin \vartheta \quad , \qquad\qquad (6.50)$$

$$J(\lambda_i) = (\lambda_1 - \lambda_3)(\lambda_1 - \lambda_2)(\lambda_2 - \lambda_3) \quad . \qquad\qquad (6.51)$$

The two examples above can be easily formulated. One may start with (Eq. (6.48))

$$\Phi_N(T_{ij}) \, \alpha \, \exp \left[- \frac{N}{2} \, \text{tr} \, \underset{\approx}{T} \, \underset{\approx \text{th}}{T}^{-1} \right] \quad , \tag{6.52}$$

where T_{th} is the theoretical value matrix. For case (1) $\underset{\approx \text{th}}{T}$ is simply the unit matrix. In that case[105]

$$J(\lambda_i) \, \sin \vartheta \, \Phi_N(T_{ij}) \, \alpha \, \exp \left(- \frac{N}{2} \sum_{i=1}^{3} \, [\lambda_i - (1 - \frac{4}{N}) \, \ell n \, \lambda_i] \right) . \tag{6.53}$$

It can be checked numerically that this distribution (which has a maximum for $\lambda_1 = \lambda_2 = \lambda_3$) leads however to

$$\langle q_1 \rangle \approx 1 + \frac{3}{\sqrt{N}} + \frac{22}{N} \quad . \tag{6.54}$$

Similarly, case (2) corresponds to $\underset{\approx \text{th}}{T} = \text{diag}(1,1,r)$, with $r > 1$. Numerically, this case leads to a maximum in the distribution $dN/d\vartheta$ for the observed angle ϑ, which is away from zero degree. The false maximum may be avoided if one looks at the modified distribution

$$\frac{dN}{d \cos \vartheta} = \frac{1}{\sin \vartheta} \frac{dN}{d\vartheta} \quad . \tag{6.55}$$

A maximum for $\vartheta \neq 0$ in such a quantity can be considered as a manifestation of a true oriented flow. A modification similar to (6.55) can be introduced in the dN/dq_1 distribution to remove the distortion due to the Jacobian. See Ref. 105 for details.

What would be the differences between INC and hydrodynamics concerning the global variables ? Figure 24 gives the general trend of the variation of the average angle $\bar{\vartheta}$ (in the sense of the distribution $dN/d\vartheta$) and of the thrust (here the thrust has been preferred to the first aspect ratio q_1) as a function of the impact parameter for a symmetric system. There is a dramatic difference between the predictions of the two models. Hydrodynamics predicts a *true* flow angle (there is no fluctuation in an ordinary hydrodynamic calculation) at 90° for small impact parameter because of the short mean free path assumption which makes the matter opaque. The INC on the other hand, predicts for b = 0 smaller values of $\bar{\vartheta}$, but this does not correspond to a true flow in this case, but rather as explained above to the fluctuations. This is particularly well illustrated by Fig. 25. It shows the distribution of the projection of the extremity of the unit vector along the major axis on a plane perpendicular to the beam axis. This plot is free of Jacobian effects, at least around the beam axis. Clearly, for central collisions, the ellipsoid points in the average in the beam direction, with large fluctuations.

Hydrodynamics does predict roughly the same pattern for all (symmetric) systems and energies between, say 250 to 1000 MeV/A. If some energy dependence is predicted,[104] the large flow angles, linked to opacity, are always present. In the INC model, the situation is more subtle. First of all, as seen in Fig. 25, there is an average true flow for intermediate impact parameters. The flow angle increases with the size of the system. Also the fluctuations reduce with increasing mass system, in agreement with Eq. (6.46) (which would indicate the absence of strong many particle correlations in INC). In large systems, like U + U, even for b = 0 the flow angle is expected to be at finite angle.[103] As the energy increases, the flow angle is expected to decrease.[102] All these considerations are consistent with a mean free path of the order of a few fm and with the angular distribution of the NN scattering.

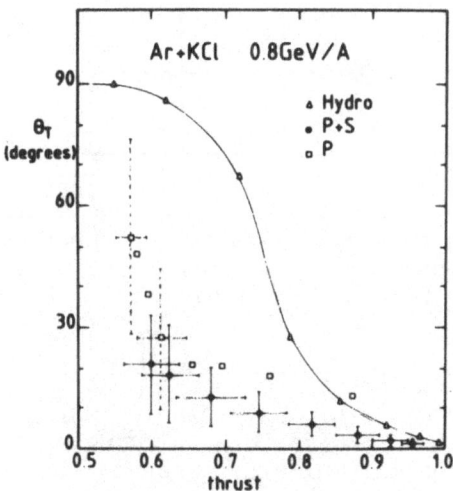

Fig. 24. Value of the thrust T and of the thrust angle ϑ_T for the Ar + KCl system at 800 MeV per nucleon and for impact parameters from 0 to 7.2 fm, as given by an INC calculation. The zero impact parameter points are on the left and the points for the largest one are on the right. The bars indicate two times the standard deviation due to the fluctuations. P means participants and S spectators. The full line gives the results of Ref. 100. The triangles correspond to the same eight impact parameters as those of the INC calculations. Taken from Ref. 102.

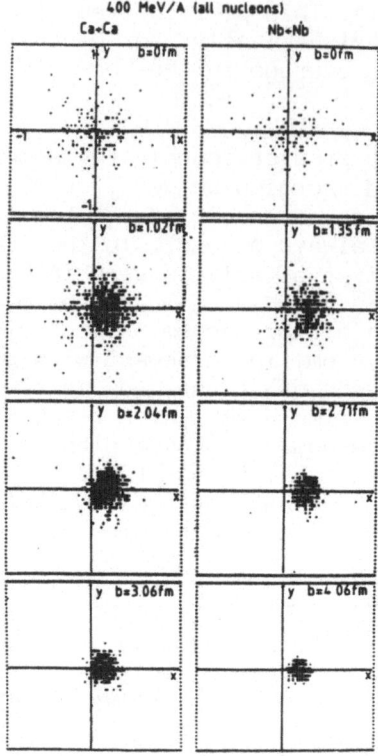

Fig. 25. Distribution of the projection of the extremity of the unit
vector along the major axis of the sphericity tensor on a
plane perpendicular to the beam axis, as a result of an INC
calculation. See text for more detail. From Ref. 106.

Recently,[66] the first measurements of the plastic ball/wall
system analyzed in terms of global variables have been presented.
The dN/d cos ϑ plot shows a maximum at zero degrees for Ca + Ca at
400 MeV/A, whatever the multiplicity observed in the system. For
Nb + Nb, a well-defined peak occurs in the dN/d cos ϑ plot. The
corresponding angle decreases with decreasing multiplicity, dis-
appearing for small multiplicities. The predictions of a recent INC
calculation[106] are given in Fig. 26. As can be seen from the bottom
of the figure, the agreement is rather good. As shown in Ref. 106,
no maximum happens in Ca + Ca because fluctuations are large enough
to wash out the small intrinsic flow angle. In Nb + Nb, the intrinsic
flow angle is larger and the fluctuations have decreased. Furthermore, as also shown in Ref. 106, the experimental acceptance of the
detector introduces a bias to large observed flow angles. Figure 27
shows the result of the hydrodynamical calculation of Ref. 63. The
latter produces a definite flow when a certain relationship is
assumed between the multiplicity and the impact parameter. Let us
however notice that results of Fig. 27 are not corrected for the
acceptance of the detector.

260

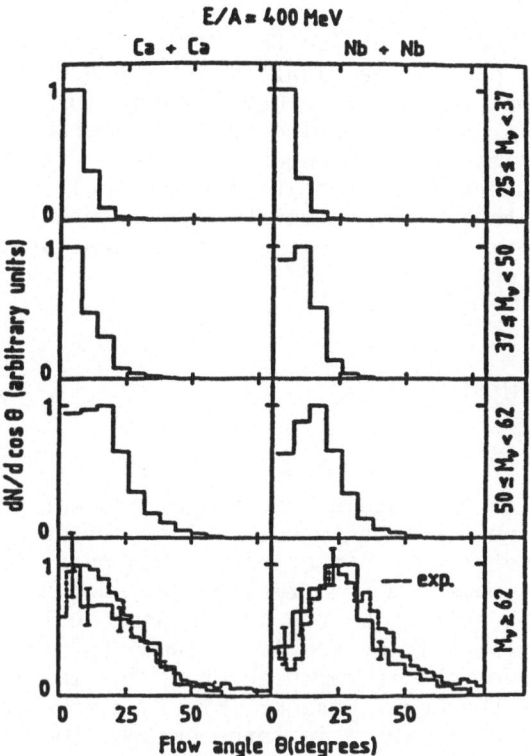

Fig. 26. Full lines : INC calculation of the dN/d cos ϑ as a function
of the multiplicity. Dotted lines : experimental data for
the largest multiplicity bin.[66] Taken from Ref. 106.

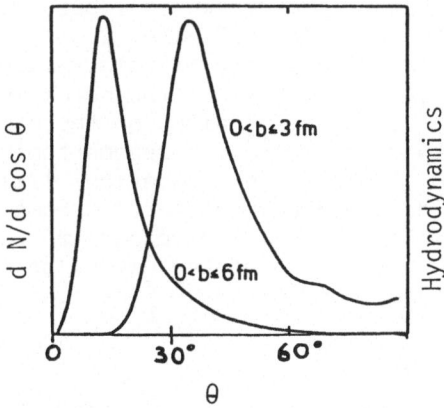

Fig. 27. dN/d cos ϑ plot for Nb + Nb at 400 MeV/A as generated by
an hydrodynamical calculation. The multiplicity M_c bins
are just reflecting selection on the impact parameter.
Adapted from Ref. 63.

There is some uncertainty in the INC calculation,[30,66] but it seems that the INC produces a flow. The flow angle is close to experiment, a little bit too small perhaps, and definitely smaller than the hydrodynamics prediction, which seems to overshoot the experimental value. The important consequence is that with heavy systems like Nb + Nb, we start to be sensitive to the bulk dynamics. Whether this is sensitive to equation of state or the detail of NN forces requires more experimental as well as theoretical study. Work in this direction is only starting.[30,95]

7. THE SPECTATOR PHYSICS

7.1. Introduction

Although not as spectacular as the participant physics, the spectator physics nevertheless presents interesting aspects. What happens to the spectators ? Essentially, this part of the system receives an energy-momentum transfer, by far much smaller than the one suffered in the average by the participant nucleons, but sufficient anyway to split this system into fragments. The momentum spectrum of the fragments is reminiscent of the initial state properties, namely of the Fermi motion. This is discussed in Section 7.2. When the spectator system is large enough, the mass spectrum of the fragments seems to follow a very simple power law. This feature is so disturbing that we postpone its discussion to the next chapter. Another interesting feature is the isospin properties. Many neutron rich isotopes can be formed in this way.[107-109] For the lack of space, we will not discuss this point. We have also to mention interesting topics, (which we will not discuss either), which are not properly pertaining to the spectator physics, but which are often discussed at the same time, presumably because experimentally they imply measurements at 0 and 180°. The first one is the backward and forward production of high-energy protons, which, as believed, are due either to a high energy tail of the Fermi motion[110] or to scattering by existing clusters,[111] although there may exist some relation between the two causes.[112] The second topic is the scaling of the pion production close to the kinematical limit. The latter was predicted by Schmidt and Blankenbecler.[113] Some deviations have been observed however.[114,115] It is not clear yet whether the scaling comes from the structure function of the nuclei or simply from phase space coming from interacting clusters.[116]

7.2. Fragment momentum spectra

One of the earliest experiment[117] performed at the Bevalac established that the fragments of ^{16}O and ^{12}C (observed at 0°) had the following distribution of parallel momentum $p_{//}$ in the projectile rest frame

$$\frac{d\sigma}{dp_{//}} \; \alpha \; \exp \; (\frac{p_{//}^2}{2 \; \sigma^2}) \qquad . \tag{7.1}$$

The width σ depends on the mass of the fragment, is maximum for half the initial mass. To a good approximation, the dependence is parabolic : $\sigma \sim F(A - F)$, if F and A are the fragment and projectile masses, respectively. Goldhaber[118] proposed a simple explanation of these observations. He assumed that the momentum of fragment F is obtained by summing F momenta picked up at random in the Fermi distribution. The Gaussian form (7.1) results from the central limit theorem. The values of σ can be calculated as follows. If $\vec{p}_i$ is the momentum of the ith nucleon, one has

$$\sum_{i=1}^{A} \vec{p}_i = 0 \qquad , \tag{7.2}$$

in the projectile rest frame. Squaring this expression, one obtains :

$$\sum_{i=1}^{A} \langle \vec{p}_i^2 \rangle + \sum_{i \neq j}^{A} \langle \vec{p}_i \cdot \vec{p}_j \rangle = A \langle \vec{p}_i^2 \rangle + A(A - 1) \langle \vec{p}_i \cdot \vec{p}_j \rangle = 0 \tag{7.3}$$

or

$$\langle \vec{p}_i \cdot \vec{p}_j \rangle = - \frac{1}{A - 1} \langle p_i^2 \rangle \qquad . \tag{7.4}$$

This last expression translates the correlation induced by the momentum conservation law. Even if the distribution of $\vec{p}_i$ is isotropic, there is some correlation between the momenta of two nucleons, because the total momentum must have a fixed value. Now, for a fragment F, one has

$$\langle (\sum_{i=1}^{F} \vec{p}_i)^2 \rangle = F \langle \vec{p}_i^2 \rangle + F(F - 1) \langle \vec{p}_i \cdot \vec{p}_j \rangle \qquad , \tag{7.5}$$

or because of (7.4)

$$\sigma^2 = \frac{1}{3} \langle (\sum_{i=1}^{F} \vec{p}_i)^2 \rangle = \frac{1}{3} \frac{F(A - F)}{A - 1} \langle \vec{p}_i^2 \rangle \qquad . \tag{7.6}$$

For the Fermi gas model, $\langle \vec{p}_i^2 \rangle = 0.6 \; p_F^2$, where p_F is the Fermi momentum, and thus, σ is related to p_F. The experiment suggests a value of $p_F \approx 220$ MeV/C, which seems smaller than the usual value.

The Goldhaber model implies momentum conservation, but does

not require energy conservation. This raises a kind of puzzle since apparently the spectator system receives an energy transfer (the energy sufficient to break up) but no momentum transfer. The transfer looks like a phonon exchange,[119] the nature of which is not understood well.[119,120] However, it is expected that the spectator system receives an impulse in the perpendicular direction.[4,121] As a result, the perpendicular momentum distribution of the fragments will be larger than the longitudinal momentum one (7.1). Noticing this difference, Hüfner et al.[122,123] have tried to extract the internal momentum distribution from the longitudinal distribution of the one-nucleon removal reactions, like $\alpha \to {}^{3}He$ or ${}^{16}O \to {}^{15}O$. The authors claim that the Hartree-Fock approximation gives a better momentum distribution, especially at high energy.[123] This looks very interesting since the high energy tail is not accessible by (e,e') or (e,e'p) experiments.

8. LIQUID-GAS PHASE TRANSITION ?

8.1. Introduction

The fragmentation of the projectile was studied in the early experiments at the Bevalac.[117] The mass yield was more or less described by the so-called abrasion-ablation model.[124] In the latter a certain number of nucleons are abraded from the projectile. The excited remnant subsequently decays by eventually emitting particles. But it was found in later experiments[125] that the mass yield in the target fragmentation region follows, for fragment masses $F \lesssim A/3$, a power law like

$$p(F) \propto F^{-\tau} \quad , \qquad\qquad (8.1)$$

where the exponent τ lies between 2 and 3. This yield is very much akin to a similar finding in nuclei bombarded with several GeV protons.[126] The authors of Ref. 126 realized that such a power law was predicted for the droplet size by Fisher's theory of condensation[127] and suggested that the observed fragmentation law signals a transition from a liquid phase to a gas phase in nuclei. Similar power laws were found[128,129] in the fragmentation resulting from heavy ion collisions between a few tens to about 200 MeV per nucleon. All these systems (spectators at high energy, nucleus after the passage of a fast proton, compound system in the Ganil regime) seem to share a common property. They are primarily excited to a few MeV per nucleon and possibly slightly compressed. Therefore, their evolution may be dominated by the nuclear equation of state for small temperatures and for densities close to ρ_o (Chapter 1).

In Section 8.2, we are going to describe briefly an approach which encompasses the abrasion-ablation model. It is essentially based on a two regime picture and phase space considerations. In Section 8.3, we will discuss the ideas which link the fragmentation

264

in many pieces (the so-called multifragmentation) to the equation of state.

8.2. A two-regime picture of the fragmentation

We here closely follow the presentation of Ref. 131, which gives a global view of the problem. The authors start with a detailed analysis of the mass yield of proton and light-ion induced reactions on relatively heavy targets. They arrive at the following observations : heavy fragments are produced in low-multiplicity events, whereas light (A $\lesssim$ 20) ones are produced in high multiplicity events. This suggests that two regimes are alternatively at work. The first one is an *evaporation* regime, in which the spectator system just looses a few very light particles. The second one is a *multifragmentation* regime, in which the system breaks up in many pieces of the size of light nuclei. The authors of Ref. 130 propose that the parameter which determines which one of the two regimes sets is the energy per particle E_0/A deposited in the spectator system by the participant nucleons which afterwards have escaped rapidly from the system. A very convincing analysis of the scaling properties of the mass yield[130] indicates that the critical value is $E_0/A \approx 3$ MeV, a value which seems consistent with the theoretical considerations of Ref. 132. If the energy per particle of a nucleus is less than this value, it stays bound, loosing its energy by evaporation. It cannot support a higher excitation energy without breaking up. The mass yield can be calculated after the introduction of a few reasonable assumptions: (1) the number of primary NN collisions done by the participants is given by the Glauber-Matthiae rule. (2) at every NN collision, an average amount ε_0 of energy is deposited in the spectator system. The quantity ε_0 is left as a free parameter. Comparison with experiment yields $\varepsilon_0 \approx 80$ MeV in agreement with the average transverse momentum in NN collisions. (3) In the multifragmentation regime, the number of fragments is equal to the number of primary NN collisions. (4) The fragmentation is assumed to be similar to a partition of a string of length A by n random cuts. The probability of finding a fragment of mass F is

$$P_n(F) = \frac{n(n-1)}{A} \left(1 - \frac{F}{A}\right)^{n-2} . \qquad (8.2)$$

An alternative statistical hypothesis is to be found in Ref. 133. The treatment of the evaporation process is quite standard.[134] With such a model, one can reproduce quite well the whole mass yield as illustrated in Fig. 28. The most important aspect of this result is that the yield in the F = 10-40 region cannot be accounted for by evaporation (at least in the present stage of its understanding) and demands the presence of another process. The detail for the region F $\lesssim$ 32 is displayed in Fig. 29. The model gives almost as good results as a fit by the power law (8.1). Let us also mention that this model, which goes beyond the abrasion-ablation model (essentially by the introduction of the multifragmentation), gives good results for the fragmentation of the projectile at high energy.

In conclusion, it seems that this model gives a reasonable description of the multifragmentation yield. It, however, does not provide any hint to the dynamical path followed by the system during the multifragmentation process.

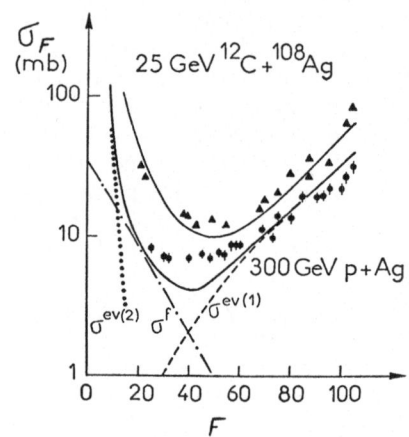

Fig. 28. Mass yield in p- and ^{12}C- Ag reactions. Data from Ref. 126. The prediction of Ref. 130 (full curve) for the p + Ag case is splitted into evaporation and multifragmentation components. Adapted from Ref. 131.

8.3. Multifragmentation and equation of state

It is known for a long time already that nuclear matter behaves like a Vanderwaals fluid,[135] i.e., that the isotherms display loops below a critical temperature T_c. This result, first obtained in the Hartree-Fock approximation[135-137] has been corroborated by more sophisticated calculations.[138,139] Without any doubt, it is due to

the attractive long-ranged plus repulsive short-ranged character
of the NN forces. Typically, the critical point occurs around
$\rho_c \approx 0.3\ \rho_0$ and $T_c \approx 20$ MeV, as shown in Figs. 30 and 31.

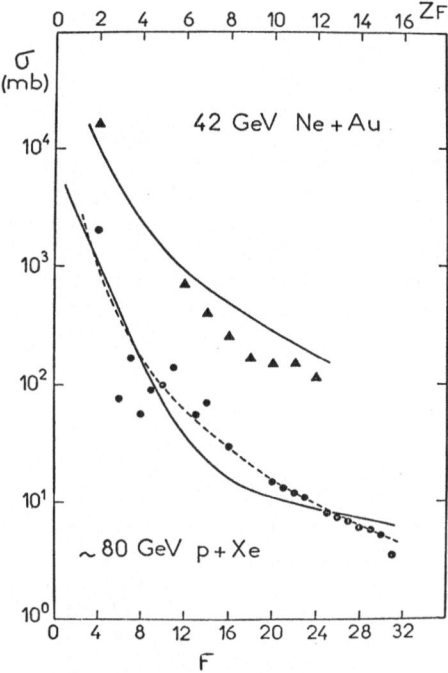

Fig. 29. Mass yield for the p + Xe and Ne-Au reactions. Data from
Refs. 125, 126. The full curves are the predictions of
Ref. 130 and the dashed curve is simply a fit by a power
law $A^{-\tau}$, with $\tau = 2.64$. Adapted from Ref. 131.

Are the physical aspects contained in Fig. 30 relevant to the
multifragmentation of a nuclear system ? For a positive answer,
several conditions should be met. First, the system must be large
enough to minimize surface effects, or, at least, corrections due
to these effects should be approximately known. The same remark

applies to the Coulomb forces. A second condition is that the system always stays in thermal equilibrium when it is expanding. In such a case, it can be represented by a point in the (p,ρ) (or (T,ρ)) plane. Although this may be less important, one would like the concept of freeze-out to be valid. If all these conditions are met, the system is prepared in a point on the (p,ρ) plane (f.i., after two ions have stopped each other in the Ganil machine), follows a trajectory in this plane and breaks up at the end of the trajectory (the freeze-out point). This point is obviously lying on the left of the co-existence region (see Fig. 30), i.e. in the area representing the gaseous phase.

Under certain circumstances, the trajectory will cross the co-existence zone and the system will undergo a phase transition. Before being sure that such an interesting phenomenon takes place, two questions must be answered. First, what are the characteristics of the trajectory ? Second, if a liquid-gas phase transition takes place, how to detect it ? We will first consider the second question, for which an answer seems to reside in Fisher's theory[127] of condensation.

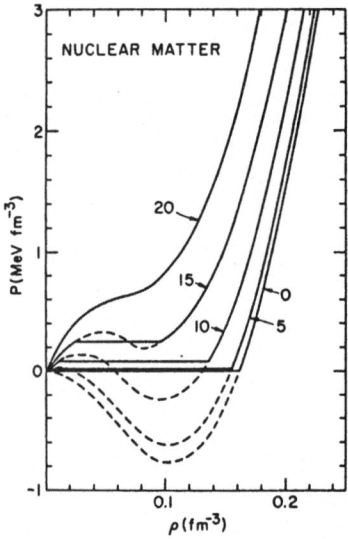

Fig. 30. Typical isotherms with the Maxwell construction (equal chemical potential) for symmetric nuclear matter (Ref. 139). The region enclosing the dotted lines is the coexistence region.

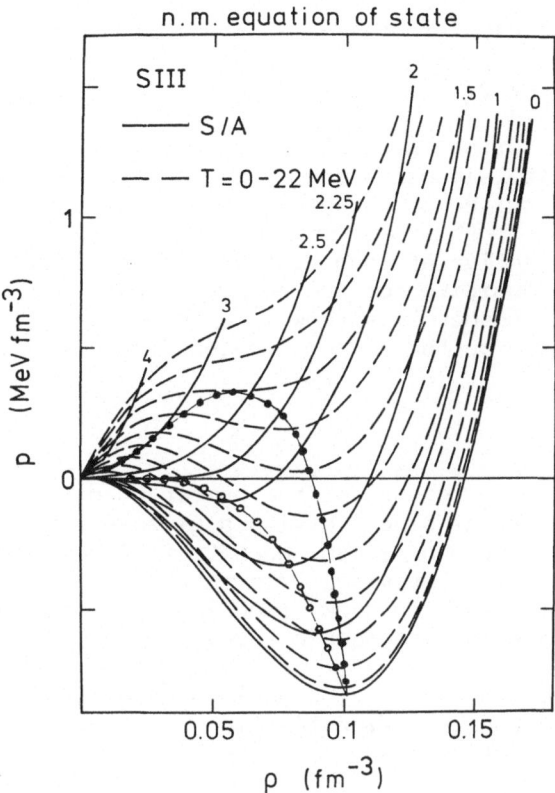

Fig. 31. Isotherms (dashed curves) and isoentropes (full curves) for
symmetric nuclear matter in the Hartree-Fock-Skyrme approximation. They give the pressure p as a function of the
baryon density ρ. The so-called Skyrme III force is used.
The value of the entropy per nucleon is indicated for each
of the isoentropes. The isotherms are given for temperatures
from T = 2 (bottom) to T = 22 MeV (top) by steps of 2 MeV.
The T = 0 isotherm coincides with the S/A = 0 isoentrope.
The heavy dots (the spinodal curve) enclose the instability
region at constant temperature whereas the open dots
delineate the instability zone at constant entropy.
Adapted from Ref. 144.

The basic idea of this theory, already formulated by Bijl,[140] is to
consider that in the gaseous phase particles can be bound in clusters,
whose average size is more and more important as the critical line
(the left borderline of the coexistence zone in Fig. 30) is
approached. The two main points of Fisher's theory are energy

and entropy of a cluster of size A. The energy is taken as

$$E_A = A E_0 + w \sigma_A \quad , \tag{8.3}$$

where σ_A is the surface area of the cluster and where $E_0 < 0$ and $w > 0$. This choice, inspired from the properties of molecular forces, is well suited to the nuclear case (which is not really surprising, since both molecular and nuclear forces are essentially attractive at long distances and repulsive at short distances). The positive surface energy favours (relatively) small surfaces. This tendancy to shrink is opposed by the entropy of the cluster which is expected to be composed of a bulk entropy and of a surface entropy

$$S_A = S_0 A + \omega \sigma_A \quad .$$

The second term is a measure of the number of different configurations for a cluster of size A having a given surface area σ_A. If the temperature is lowered or if the density is increased (both correspond to an increasing activity), the entropy will be less important, the clusters will start to grow (reducing their surface) and the droplets will go to a *macroscopic* size.

The above ideas can be clothed in a more mathematical form. We start with the grand partition function for a system containing any number of cluster of species i (here $\beta^{-1} = kT$)

$$Z(V,T,\mu) = \sum_{\substack{\{N_1,N_2,...\} \\ \{S_1,S_2,...\}}} e^{-\beta \sum_i N_i E_i(S_i)} \; e^{\beta \sum_i \mu_i N_i} \frac{(g_1(S_1))^{N_1}(g_2(S_2))^{N_2}...}{N_1! \; N_2! \; ...} \tag{8.5}$$

where S_i sums (for the moment) over all the possible states of the cluster i with a degeneracy $g_i(S_i)$. Equation (8.5) may be put in a compact form

$$Z(V,T,\mu) = \sum_{\{N_1,N_2,...\}} \prod_i \frac{(q_i z_i)^{N_i}}{N_i!} \quad , \tag{8.6}$$

where

$$q_i = \sum_{S_i} g_i(S_i) \; e^{-\beta E_i(S_i)} \tag{8.7}$$

is the partition function for the cluster of the ith type and where

$$z_i = e^{\beta \mu_i} \qquad (8.8)$$

is the corresponding activity. The summation and product symbols in (8.6) may be interchanged, as it can easily be checked (this is a general property of different species in the grand cannonical ensemble). One then gets

$$Z(V,T,\mu) = \exp \left\{ \sum_{i=1}^{\infty} q_i z_i \right\} \qquad . \qquad (8.9)$$

From this equation, one may derive the expression for the density of clusters of size A (or type i)

$$\langle N_A \rangle = \frac{\partial}{\partial \mu_A} (- \beta^{-1} \ln Z) \qquad , \qquad (8.10)$$

or, with the help of (8.9)

$$\langle N_A \rangle = q_A z_A \qquad . \qquad (8.11)$$

These results are quite general. The originality of Fisher's theory is the introduction of an intuitive form of the partition function q_i (8.7), which satisfies however precise physical requirements. The summation over S_i in (8.7) is considered as running over the possibilities of making a surface (this is not a restriction in the case of a lattice gas). It is further assumed that all the possible manners of making a surface are very well centered on an average surface area $\bar{\sigma}_A$. Therefore (8.7) is very close to

$$q_A \approx e^{-\beta(A E_o + w \bar{\sigma}_A)} e^{A S_o + \omega \bar{\sigma}_A} \qquad , \qquad (8.12)$$

since the entropy is nothing but the logarithm of the density of states. A correction factor should be applied to (8.12) to account for all the neglected possibilities of making a surface of area $\bar{\sigma}_A$ (called as "residual entropy"). Analytical considerations force this term to be of the form $A^{-\tau}$. This term is related to the "funny shapes" of a cluster (like a spaghetto) and to other combinatorial problems in self-avoiding random walks.[141] If one finally assumes that $\bar{\sigma}_A = a_o A^\sigma$, which is quite natural in nuclear physics, one finally gets

$$\langle N_i \rangle \alpha \; x^{A^\sigma} y^A A^{-\tau} \qquad , \qquad (8.13)$$

with

$$x = \exp [- a_o (\beta w - \omega)] \qquad , \qquad \qquad (8.14)$$

$$y = z \exp [- (\beta E_o - S_o)] \qquad . \qquad \qquad (8.15)$$

The expressions (8.13-15) are obtained by assuming $\mu_A = A\mu$, $z_A = z^A$, which amounts to assume that a cluster of size A is obtained by putting together A clusters of the smallest type (the nucleons). In this case, the nucleon density is

$$\langle N \rangle = \sum_A A \langle N_A \rangle \qquad . \qquad \qquad (8.16)$$

Now, it is clear that the convergence rate of (8.16) is governed (see (8.13)) by the value of y. On the left of the coexistence curve, x and y < 1 and $\langle N_A \rangle$ decreases rapidly with A. When the coexistence curve is approached (either because the temperature drops (Eq. (8.15)) or because the chemical potential increases (Eq. (8.3)), y → 1 and the $\langle N_A \rangle$ distribution decreases more slowly. If y > 1, the series (8.16) diverges, which physically means that very large clusters are favoured.

In Ref. 128, it is assumed that the freeze-out point is always lying on the coexistence curve as far as the freeze-out temperature T_{fo} is below the critical temperature T_c. Therefore, if $T_{fo} > T_c$, x > 1, y < 1 and the mass yield falls off more steeply than $A^{-\tau}$. At the critical point ($T_{fo} = T_c$), x = y = 1, the mass yield is exactly $A^{-\tau}$. Below the critical temperature, x < 1, y = 1, the mass yield falls off more steeply than $A^{-\tau}$ again. According to Ref. 128, it seems that this pattern is observed as the energy per particle increases steadily between 20 to 100 MeV/A. The authors of Ref. 128 extract the following parameters $T_c \approx 13$ MeV, $\tau \approx 1.8$ (the latter parameter is in principle related to the dimensionality of the system and to the so-called critical exponents). This success is encouraging despite of the many uncertainties that blur the analysis, namely those on the freeze-out density, on the temperature of the fragments,...

One may wonder whether the dynamical path followed by the system in the (p,ρ) plane has some influence on the multi-fragmentation itself. What is the evolution of the system once it is thermalized and possibly compressed ? The most important feature seems to be the *isoentropic* expansion of the system (Refs. 70, 142, 143 and Section 4). The system thus follows an isoentrope (full curves in Fig. 31) and not an isotherm. Along an isoentrope, the internal energy is transformed into flow motion, with a maximum efficiency : the work performed by the pressure is entirely transformed into macroscopic kinetic energy. Accordingly, the temperature of the system is decreasing.

In comparing with Figs. 20 and 31, it is already clear that the cluster formation in the participant system around 1 GeV per nucleon has little to do with the gas-liquid phase transition. The entropy per baryon of the system is too high and the isoentrope is always higher than the critical point in the (p,ρ) plane. In the Ganil regime (or in the spectator system at high energy), the system is "prepared" at a point around the upper right corner of Fig. 31. The associated isoentrope will cross the coexistence region, when followed toward small densities. Several scenarios are possible when the system reaches the coexistence curve (Fig. 30). If the evolution is sufficiently slow, the system may develop two phases (each at equal chemical potential) and evolves along the Maxwell construction. It eventually vaporizes and its composition is governed by Fisher's theory. On the other extreme, the evolution of the system may be so fast that the system does not have the time to develop two phases. It remains in the liquid phase, keeps on evolving on an iso-entrope : this is the *supercooling* phenomenon. The system is then in a metastable state. In usual isothermal fluid experiments, the super-cooling extends up to the spinodal curve (Fig. 31). However, here, as the expansion is at constant entropy, the supercooling may last until the encounter of the "isoentropic spinodal" (the open dots in Fig. 31), which joins all the points of vanishing $(\partial p/\partial \rho)_S$. Then the system enters the region of instability, characterized by a negative isoentropic "compressibility coefficient" $(\partial p/\partial \rho)_S < 0$. To understand the nature of this instability, let us consider an extended system with uniform particle density $\overline{\rho}$ at the top of which we add an oscillatory modification of the type

$$\rho(\vec{r}) = \overline{\rho} + a \sin (\vec{k}.\vec{r} - \omega t) \qquad . \tag{8.17}$$

At this modification is associated a velocity field

$$\vec{v}(\vec{r}) = \frac{a \omega \vec{k}}{\overline{\rho} \ k^2} \sin (\vec{k}.\vec{r} - \omega t) \qquad . \tag{8.18}$$

which guarantees the continuity equation. This introduces a modifi-cation in the kinetic (flow) energy

$$\delta K = \frac{M}{4 \ \overline{\rho}} \ V \ \frac{a^2 \ \omega^2}{k^2} \qquad , \tag{8.19}$$

where V is the volume of the system (surface effects are neglected) and M is the mass of the nucleon. If H is the bulk energy density, the modification in the internal energy U is

$$\delta U = \int \ [(\frac{\partial H}{\partial \rho})_S (\rho(\vec{r})-\overline{\rho}) + \frac{1}{2} \ (\frac{\partial^2 H}{\partial \rho^2})_S (\rho(\vec{r})-\overline{\rho})^2 + \ ...]d^3r \ . \tag{8.20}$$

Neglecting surface effects, one has in lowest (i.e. in second) order

$$\delta U = \frac{1}{4\bar{\rho}} V(\frac{\partial p}{\partial \rho})_S a^2 \quad . \tag{8.21}$$

The theory of small oscillations tells us that the frequency ω is given by

$$\omega^2 = \frac{k^2}{M} (\frac{\partial p}{\partial \rho})_S \quad . \tag{8.22}$$

When the derivative is negative, the frequency is imaginary. The fluctuations grow exponentially. It is suggested in Refs. 143, 144 that the multifragmentation may result from this kind of instability. However, no definite prediction has been done so far, because one is still far from having a complete dynamical theory. First, a model is needed for the expansion of a uniform hot sphere. Such a model already exists.[144,145,146] Second, the degrees of freedom associated with fluctuations like in (8.17) have to be introduced at the top of the global expansion of the system. This leads to a diffusion problem. This can be understood by looking at Fig. 32, which shows the typical variation of the internal energy of the system $U(\bar{\rho},a)$ generalized to nonuniform density by an expression like (8.17).

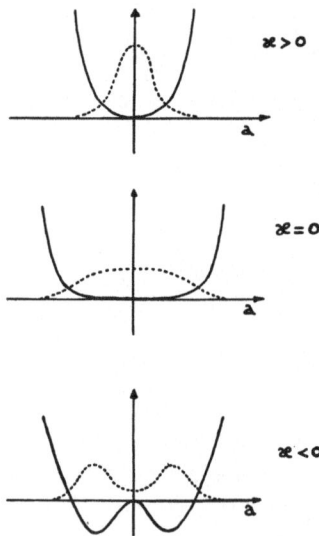

Fig. 32. Full curves : typical variation of the internal energy
$U(\bar{\rho},a)$ for different values of the isoentropic compressibility coefficient $\varkappa = (\partial p/\partial \rho)_S$. Dotted curves : typical
shapes of the distribution $P(a)$. See text for detail.

Considered as an analytic function of a, the series expansion of U begins as (see (8.21))

$$U(\overline{\rho},a) = U(\overline{\rho}) + c_2(\overline{\rho})a^2 + c_4(\overline{\rho})a^4 + \ldots, \qquad (8.23)$$

where $c_4 > 0$ and where

$$c_2(\overline{\rho}) = \frac{V}{4\overline{\rho}} \left(\frac{\partial p}{\partial \rho}\right)_S \qquad (8.24)$$

has the same sign as the isoentropic compressibility coefficient. Before reaching the "isoentropic spinodal", $c_2 > 0$ and $U(\overline{\rho},a)$ has a minimum at $a = 0$. If we consider the probability $P(a)$ of having a fluctuation of amplitude a, it will be centered at $a = 0$. Disregarding the fluctuations is probably a good approximation in this region. When $c_2 = 0$, the potential curve (8.23) opens up and as c_2 becomes more and more negative, the potential develops two pockets for two (physically equivalent) opposite values of a. As a consequence, the distribution $P(a)$ broadens and large values of a can be obtained. Physically, it means that the fluctuations grow spontaneously, leading to "cracks" in the system. The considerations above are a simplified description of the multifragmentation as a Landau-Ginzburg[147] model of a second order phase transition whose order parameter is the amplitude a of the fluctuations itself. A more elaborate description is however needed, which would include a more general form for the fluctuation (than (8.17)) and a diffusion term for the evolution of the distribution $P(a)$.

The most serious problem is to relate the observables, namely the mass yield to the fluctuation amplitude a. Other questions are related to the surface energy which may play a role (this is important in Fisher's theory) and to the Coulomb energy. These effects may obscure the manifestation of the bulk instability[148] (the most obvious modification is a lowering of the critical temperature to $T_c \approx 10$ MeV). However, despite the complications which soften the transition, it is very likely that multifragmentation may be due to the bulk instability. Indeed, the latter is due to attractive long ranged and repulsive short ranged forces. Even if thermodynamic concepts could not be strictly justified, it must happen situations where it is preferable for the system to exist either in a denser or in a more dilute form. This statement is supported by the recent two-dimensional Hartree calculation of Strack and Knoll.[149]

The issue is far from being settled, however. Another problem is the importance of the usual evaporation. Evidently, this process must be important when the system is excited at small pressure. Then, the system oscillates around a zero pressure state (see Fig. 31) and the only way for the system to loose energy is to emit particles.[144] However, there are indications that evaporation may be (at least partly) responsible for the multifragmentation itself.[150-152]

To conclude this section, it seems that multifragmentation happens when the system is not too much excited, but sufficiently excited for not loosing its energy by gentle evaporation. The relation between the properties of this phenomenon and the equation of state is not yet clearly established, but the actual situation seems promising.

9. CONCLUSION: OUTLOOK

From this account of the relativistic heavy ion physics, it is clear that this field is quite rich of diverse phenomena and in a phase of rapid development. Yet, the present review, supposed to be the basic material of a short series of lectures, is inevitably uncomplete. Among the topics that we did not cover, let us mention the production of strange particles, the two pion and two proton interferometry, the production of multibaryons, the production of secondary fragments with unusual mean free path (the anomalons). We did not treat these questions, either because of poor or uncertain experimental status, or because experimental data are really too scarce, or because of difficulties of interpretation. We would like however to mention the disturbing aspect of the strange particles production(K,Λ),[153-155] which seems to be produced with larger perpendicular momentum than expected,[156-160] especially the Λ's.[154] We did not discuss either the spin-isospin instability (the celebrated pion condensation) which seems to be ruled out by the present experiments.[161,162]

We are now in a situation, where the basic mechanisms of the reaction process are more or less understood. From the dozen or more models which were proposed a few years ago, only two have survived (when a simultaneous description of several aspects of the collision is needed) : the INC and hydrodynamics. Although it may not be clear from these notes, these denominations should be rather considered as generic names for two classes of closely related models. In the very recent times, the INC model is being refined to account for interaction energy. Also, there are a variety of hydrodynamic calculations (one fluid, two fluids, etc.).Both theories are designed to follow the evolution of the system in phase space. The main difference between the two approaches is twofold. First, hydrodynamics averages over local fluctuations, whereas INC keeps track of the microscopic structure all the time. Second, hydrodynamics can handle non trivial equation of state and is well suited to study the transport properties in a transparent way. Both approaches are however incomplete in the sense that they need some additional devices to treat the production of composites.

The basic mechanism can be outlined in this way. The system is divided in a more or less geometrical way into participant nucleons, which suffer large momentum transfers, and spectators which are much less perturbed. The participant system, whose mass decreases with increasing impact parameter is compressed substantially. It afterwards expands rapidly and decays into many energetic nucleons, pions and light composites. This dynamical path is largely dominated by a scheme of successive NN collisions. Due to the finite volume of the system, a few nucleons are making one or two collisions (a situation sometimes called the coronna effect). Therefore only a part of the system may be considered as in a more or less thermal equilibrium.

The ultimate goal is admittedly to extract from the observed participant properties the hadronic matter equation of state. Basically, two questions must be answered. Do we approach the bulk dynamics ? What are the observables sensitive to the equation of state ? It seems from the global analysis of the Nb + Nb data, that in this system and in heavier ones we are mainly watching the bulk dynamics at work. As for the second question, we have given in Section 6, three observables ("d"/"p" ratio, pion multiplicity, flow angle) which may be sensitive to detail of the equation of state. To our opinion, the flow angle is still too poorly determined (at least theoretically) to draw definite conclusion. The "d"/"p" ratio seems promising, but further investigation is needed. The pion multiplicity looks the most spectacular, although the interpretation of Ref. 91, which considers this quantity as a barometer, is still disputed. To our opinion, it constitutes however an obvious indication of interaction (two-body) effects. The present situation is really exciting because we feel that we are close to clearly isolate equation of state effects, even if the equation of state itself (and/or transport coefficients) would be difficult to extract because, presumably, thermodynamic concepts would be marginally applicable.

The future should probably concentrate on low energy because equation of state effects are expected to be more important in this regime. High energy (around 1 GeV/A) should not be forgotten anyway, because this region is suited to a detailed study of the mesonic properties of the matter, a topic which has not received the attention it deserves.

Paradoxically, the spectator system also provides an opportunity to study the nuclear matter equation of state, in another regime, of course. But, here too, no clear evidence of equation of state effects, a liquid-gas transition or instability in particular, is presently available and alternative (conventionally named) conventional models are also providing a good description of the data. Perhaps in this case, we do not yet have a very clear idea of the evolution of the system.

To summarize, in both cases we have indications but no evidence of equation of state effects. But we have to keep in mind that "the absence of evidence is not the evidence of absence".[163] This and the feeling that we are close to the goal will encourage us to pursue both the experimental and theoretical efforts.

ACKNOWLEDGEMENTS

I would like to express my sincere thanks to all the people with whom I had the pleasure to collaborate in this field. I would also like to thank Dr. D. Vautherin for discussions about phase transitions and Dr. X. Campi for providing me with a copy of his material before publication. I am also very grateful to Drs. B. Ascul and P. Atatras for an illuminating discussion about instability.

REFERENCES

1. The reader is assumed to be familiar with nucleus-nucleus and relativistic kinematics. If not, see, f.i. the Appendix of Ref. 4.
2. M. Lefort, this School.
3. G. Baym, "Quark Matter Conference 84", Helsinki, June 1984
4. S. Nagamiya and M. Gyulassy, Adv. Nucl. Phys. 13 (1984) 201.
5. V.I. Manko and S. Nagamiya, Nucl. Phys. A384 (1982) 475.
6. S. Nagamiya, "Quark Matter Conference 83", Brookhaven, Sept. 1983.
7. Y. Miake et al., University of Tokyo preprint, 1984.
8. S. Nagamiya, M.-C. Lemaire, S. Schnetzer, H. Steiner and I. Tanihata, Phys. Rev. Lett. 45 (1980) 602.
9. I. Tanihata, M.-C. Lemaire, S. Nagamiya and S. Schnetzer, Phys. Lett. 97B (1980) 363.
10. I. Tanihata, in : "Proceedings of Hakone Seminar on High-Energy Nuclear Interactions and Properties of Dense Nuclear Matter", K. Nakai and A.S. Goldhaber, eds., Hakone, Japan (1980), p. 382.
11. I. Tanihata, S. Nagamiya, S. Schnetzer and H. Steiner, Phys. Lett. 100B (1981) 121.
12. S. Nagamiya et al., Phys. Rev. C24 (1981) 971.
13. J. Cugnon, T. Mizutani and J. Vandermeulen, Nucl. Phys. A352 (1981) 505.
14. J. Cugnon, D. Kinet and J. Vandermeulen, Nucl. Phys. A379 (1982) 553.
15. P.J. Siemens and J.O. Rasmussen, Phys. Rev. Lett. 42 (1979) 844.
16. G. Bertsch, in : "Progress in Particle and Nuclear Physics" vol. 4, D. Wilkinson, ed., Pergamon, Oxford (1980), p. 483.
17. P. Danielewicz, Ann. Phys. 152 (1984) 239, 305.
18. R. Yvon, "Les Corrélations et l'Entropie Statistique Classique", Dunod, Paris (1965).
19. P.C. Martin and J. Schwinger, Phys. Rev. 115 (1959) 1342.
20. E.P. Wigner, Phys. Rev. 40 (1932) 749.
21. S.E. Koonin, in "Nuclear Theory 1981", ed. by G. Bertsch, WSPC, 1982.
22. L.P. Kadanoff and G. Baym, "Quantum Statistical Mechanics", Benjamin, New York (1962).
23. J. Cugnon, Nucl. Phys. A387 (1982) 191c.
24. R. Balescu, "Equilibrium and Non Equilibrium Statistical Mechanics", Wiley-Interscience, New York (1975), ch. 11.
25. A.R. Bodmer and A.D. MacKeller, Phys. Rev. C15 (1977) 1342.
26. L. Wilets, Y. Yariv and R. Chesnut, Nucl. Phys. A301 (1978) 359.

27. D.J.E. Callaway, L. Wilets and Y. Yariv, Nucl. Phys. A327 (1979) 250.
28. A.R. Bodmer, C.N. Panos and A.D. MacKeller, Phys. Rev. C22 (1980) 1025.
29. A.R. Bodmer and C.N. Panos, Nucl. Phys. A356 (1981) 517.
30. J.J. Molitoris et al., MSU preprint (1984).
31. J. Randrup, Nucl. Phys. A316 (1979) 509.
32. H.J. Pirner and B. Schürmann, Nucl. Phys. A316 (1979) 461.
33. B. Schürmann, Phys. Rev. C20 (1979) 1607.
34. B. Schürmann and N. Mancoc-Borstnik, Phys. Rev. C26 (1982) 519.
35. R. Malfliet, Phys. Rev. Lett. 44 (1980) 864.
36. R. Malfliet, Nucl. Phys. A363 (1981) 429 ; 456.
37. J. Cugnon, Phys. Rev. C23 (1981) 2094.
38. R.L. Hatch and S.E. Koonin, Phys. Lett. 81B (1978) 1.
39. B. Schürmann and J. Randrup, Phys. Rev. C23 (1981) 2766.
40. J. Hüfner and J. Knoll, Nucl. Phys. A290 (1977) 460.
41. J. Knoll and J. Randrup, Nucl. Phys. A324 (1979) 445.
42. J. Knoll and J. Randrup, Phys. Lett. 103B (1981) 264.
43. J. Cugnon, J. Knoll and J. Randrup, Nucl. Phys. A360 (1981) 444.
44. J. Knoll, Phys. Rev. C20 (1979) 773.
45. S. Bohrmann and J. Knoll, Nucl. Phys. A356 (1981) 498.
46. A.H. Blin, S. Bohrmann and J. Knoll, Z. Physik A306 (1982) 177.
47. K.K. Gudima and V.D. Toneev, Yad. Fiz. 27 (1978) 658 [Sov. J. Nucl. Phys. 27 (1978) 351].
48. E.C. Halbert, Phys. Rev. C23 (1981) 295.
49. H. Stöcker, J. Hofmann, J.A. Maruhn and W. Greiner, Prog. Part. Nucl. Phys. 4 (1980) 133.
50. J. Cugnon and J. Vandermeulen, in : "Winter College on Nuclear Physics", Trieste, March 1984.
51. G. Bertsch and G. Baym, to be published.
52. A.A. Amsden, G.F. Bertsch, F.H. Harlow and J.R. Nix, Phys. Rev. Lett. 35 (1975) 905.
53. A.A. Amsden, F.H. Harlow and J.R. Nix, Phys. Rev. C15 (1977) 1059.
54. C.F. Chapline, H.H. Johnson, E. Teller and M.S. Weiss, Phys. Rev. D8 (1973) 4302.
55. L.D. Landau and E.M. Lifshitz, "Fluid Mechanics", Pergamon, Oxford (1959).
56. H. Stöcker, M. Gyulassy and J. Boguta, Phys. Lett. 103B (1981) 269.
57. J.R. Nix, Progr. Part. Nucl. Phys. 2 (1979) 237.
58. P. Danielewicz, unpublished.
59. H. Stöcker et al., Phys. Rev. Lett. 47 (1981) 1807.
60. G. Buchwald, L.P. Csernai, J.A. Maruhn and W. Greiner, Phys. Rev. C24 (1981) 135.
61. A. Sandoval et al., Phys. Rev. C21 (1980) 1321.
62. J.R. Nix and D. Strottman, Phys. Rev. C23 (1981) 2548.
63. G. Buchwald et al., Phys. Rev. Lett. 52 (1984) 1594.
64. H.H.K. Tang and C.Y. Wong, Phys. Rev. C21 (1980) 1846.
65. R. Stock et al., Phys. Rev. Lett. 44 (1980) 1243.
66. H.A. Gustafsson et al., Phys. Rev. Lett. 52 (1984) 1590.

67. H.H. Gutbrod et al., Phys. Rev. Lett. 37 (1976) 667.
68. M.-C. Lemaire et al., Phys. Lett. 85B (1979) 38.
69. M.-C. Lemaire, in "Proceedings of the 2nd French-Japanese Collo-
    quium on Nuclear Physics with Heavy Ions", IN2P3 Publication,
    Gif-Sur-Yvette, France (October, 1979) p. 139 [Preprint :
    Lawrence Berkeley Laboratory Report LBL-10555 (1979)].
70. G. Bertsch and J. Cugnon, Phys. Rev. C24 (1981) 2514.
71. E.A. Remler, Ann. Phys. (N.Y.) 136 (1981) 293.
72. E.A. Remler, Phys. Rev. C25 (1982) 2974.
73. S.F. Butler and C.A. Pearson, Phys. Rev. 129 (1963) 836.
74. M. Gyulassy, K. Frankel and E.A. Remler, Nucl. Phys. A402 (1983)
    596.
75. S. Nagamiya, in : "Proceedings of the Symposium on Heavy Ion
    Physics from 10 to 200 MeV/A", Brookhaven 1979. See also
    Ref. 61.
76. P.J. Siemens and J.I. Kapusta, Phys. Rev. Lett. 43 (1979) 1486.
77. H. Stöcker, Lawrence Berkeley Laboratory Report LBL-12302
    (1981).
78. J. Knoll, L. Münchow, G. Röpke and H. Schulz, Phys. Lett. 112B
    (1982) 13.
79. H. Sato and K. Yazaki, Phys. Lett. 98B (1981) 153.
80. H.H. Gutbrod et al., Nucl. Phys. A397 (1982) 177c.
81. S. Nagamiya, in : "Proceedings of the International Conference
    on Nuclear Physics", P. Blasi and R.A. Ricci, eds., Tipografia
    Compositori, Bologna (1983), p. 431.
82. G. Bertsch and J. Cugnon, unpublished.
83. A.Z. Mekjian, Phys. Rev. C17 (1978) 1051.
84. J.I. Kapusta, Phys. Rev. C16 (1977) 1493.
85. J. Gosset, J.I. Kapusta and G.D. Westfall, Phys. Rev. C18
    (1978) 844.
86. J. Randrup and S.E. Koonin, Nucl. Phys. A356 (1981) 223.
87. G. Fái and J. Randrup, Nucl. Phys. A381 (1982) 557.
88. A. Sandoval et al., Phys. Rev. Lett. 45 (1980) 874.
89. J.J. Lu et al., Phys. Rev. Lett. 46 (1981) 898.
90. M. Gyulassy and S.K. Kauffmann, Phys. Rev. Lett. 40 (1978) 298.
91. R. Stock et al., Phys. Rev. Lett. 49 (1982) 1236.
92. J.W. Harris and R. Stock, in : "7th Oaxtepec Meeting on Nuclear
    Physics" (January 1984).
93. M. Cahay, J. Cugnon and J. Vandermeulen, Nucl. Phys. A411
    (1983) 524.
94. G. Bertsch, H. Kruse and S. Das Gupta, Phys. Rev. C29 (1984)
    673.
95. R. Malfliet, KVI preprint (1984).
96. R. Malfliet, Nucl. Phys. A420 (1984) 621.
97. D. Enskog, Kugl. Svenska Vet. Akad. Handl. 63 (1921) n° 4.
98. H. Pirner, Phys. Rev. C22 (1980) 1962.
99. G. Bertsch and A.A. Amsden, Phys. Rev. C18 (1978) 1293.
100. J. Kapusta and D. Strottman, Phys. Lett. 106B (1981) 33.
101. J. Cugnon, J. Knoll, C. Riedel and Y. Yariv, Phys. Lett. 109B
    (1982) 167.

102. J. Cugnon and D. L'Hôte, Nucl. Phys. A397 (1983) 519.
103. M. Gyulassy, K.A. Frankel and H. Stöcker, Phys. Lett. 110B (1982) 185.
104. H. Stöcker et al., Phys. Rev. C25 (1982) 1873.
105. P. Danielewicz and M. Gyulassy, Phys. Lett. 129B (1983) 283.
106. J. Cugnon and D. L'Hôte, Phys. Lett. (in press).
107. T.J.M. Symons et al., Phys. Rev. Lett. 42 (1979) 40.
108. G.D. Westfall et al., Phys. Rev. Lett. 43 (1979) 1859.
109. P.B. Price and J. Stevenson, Phys. Rev. C24 (1981) 2101.
110. S. Frankel and R.M. Woloshyn, Phys. Rev. C16 (1977) 1680.
111. T. Yukawa and S. Furui, Phys. Rev. C20 (1979) 2316.
112. O. Bohigas and S. Stringari, Phys. Lett. 95B (1980) 9.
113. I.A. Schmidt and R. Blankenbecler, Phys. Rev. D15 (1977) 3321 ; Phys. Rev. D16 (1977) 1318.
114. L.S. Schroeder et al., Phys. Rev. Lett. 43 (1979) 1787.
115. R.H. Landau and M. Gyulassy, Phys. Rev. C19 (1979) 149.
116. J. Knoll and S. Shyam, G.S.I. Preprint (1984).
117. D.E. Greiner et al., Phys. Rev. Lett. 35 (1974) 152.
118. A.S. Goldhaber, Phys. Lett. 53B (1974) 306.
119. H. Feshbach and K. Huang, Phys. Lett. 47B (1973) 300.
120. H. Feshbach and M. Zabek, Ann. Phys. 107 (1977) 110.
121. C.A. Whitten, Jr., Nucl. Phys. A335 (1980) 419.
122. T. Fujita and J. Hüfner, Nucl. Phys. A343 (1980) 493.
123. J. Hüfner and M.C. Nemes, Phys. Rev. C23 (1981) 2538.
124. J. Hüfner, K. Schafer and B. Schürmann, Phys. Rev. C12 (1975) 1888.
125. A.I. Warwick et al., Phys. Rev. C27 (1983) 1083.
126. N. Porile et al., Phys. Rev. C19 (1979) 2288.
127. M.E. Fisher, Physics 3 (1967) 225.
128. A.D. Panagiotou et al., Phys. Rev. Lett. 52 (1984) 496.
129. Y. Cassagnou et al., preprint Saclay 2151, April 1984.
130. X. Campi, J. Desbois and E. Lipparini, Phys. Lett. 142B (1984) 8.
131. X. Campi, J. Desbois and E. Lipparini, in : "Heavy Ion Conference", Paris, May 1984.
132. G. Bertsch and D. Mundinger, Phys. Rev. C17 (1977) 1646.
133. J. Aichelin and J. Hüfner, Phys. Lett. 136B (1984) 15.
134. X. Campi and J. Hüfner, Phys. Rev. C24 (1981) 2199.
135. M. Brack and P. Quentin, Phys. Lett. B52 (1974) 159.
136. U. Mosel, P.G. Zint and K.H. Passler, Nucl. Phys. A236 (1974) 252.
137. G. Röpke, L. Münchow and H. Schulz, Phys. Lett. 110B (1982) 21.
138. M. Barranco and J.R. Büchler, Phys. Rev. C22 (1980) 1729.
139. B. Friedman and V.R. Pandharipande, Nucl. Phys. A361 (1981) 502.
140. A. Bijl, Thesis, Leiden (1938).
141. M.E. Fisher and M.F. Sykes, Phys. Rev. 114 (1959) 45.
142. M.I. Sobel, P.J. Siemens, J.P. Bondorf and H.A. Bethe, Nucl. Phys. A251 (1975) 502.
143. G. Bertsch and P.J. Siemens, Phys. Lett. 126B (1983) 9.
144. J. Cugnon, Phys. Lett. 135B (1984) 374.
145. J.P. Bondorf, S.I.A. Garpmann and J. Zimanyi, Nucl. Phys. A296

(1978) 320.

146. M.W. Curtin, H. Toki and D.K. Scott, Phys. Lett. 123B (1983) 289.
147. V.L. Ginzburg and L.D. Landau, Zh. Eksp. Teor. Fiz. 20 (1950) 1064.
148. P. Bonche, S. Levit and D. Vautherin, Nucl. Phys. (in press).
149. B. Strack and J. Knoll, Z. Physik A315 (1984) 249.
150. P. Bonche, S. Levit and D. Vautherin, Saclay preprint (1984).
151. W.A. Friedman and W.G. Lynch, Phys. Rev. C28 (1983) 950.
152. D.H. Boal, preprint MSUCL-451 (1984).
153. S. Schnetzer et al., Phys. Rev. Lett. 49 (1982) 989.
154. J.W. Harris et al., Phys. Rev. Lett. 47 (1981) 229.
155. A. Shor et al., Phys. Rev. Lett. 48 (1982) 1597.
156. J. Randrup and C.M. Ko, Nucl. Phys. A343 (1980) 519.
157. W. Zwermann et al., Phys. Lett. 134B (1984) 392.
158. J. Cugnon and R. Lombard, Nucl. Phys. A422 (1984) 635.
159. J. Randrup, Phys. Lett. 99B (1981) 9.
160. S. Nagamiya, Phys. Rev. Lett. 49 (1982) 1383.
161. M. Gyulassy, Nucl. Phys. A354 (1981) 395.
162. S. Nagamiya et al., Phys. Rev. Lett. 48 (1982) 1780.
163. C. Sagan, "The Dragons of Eden", Ballantine Books, New York (1977).

SKYRMIONS, DENSE MATTER AND NUCLEAR FORCES

C.J. Pethick

NORDITA, Blegdamsvej 17, DK-2100 Copenhagen Ø, Denmark

and

Department of Physics, University of Illinois at
Urbana-Champaign, 1110 West Green Street, Urbana
Illinois 61801

ABSTRACT

A simple introduction to a number of properties of Skyrme's chiral soliton model of baryons is given. Some implications of the model for dense matter and for nuclear interactions are discussed.

1. INTRODUCTION

One of the challenging problems in the study of dense matter is determining when matter makes a transition to a plasma of quarks and gluons. This is relevant both in the context of collisions between ultra-relativistic heavy ions, and also in astrophysics, where it may influence the properties of neutron stars[1]. An important ingredient in the study of this phase transition is the equation of state of nuclear matter at densities up to ten or more times nuclear density. In the traditional approaches, the energy of dense matter is calculated from a two-body nucleon-nucleon interaction which is fitted to two-nucleon scattering data. Sometimes the interaction includes, in addition, many-body forces deduced partly from theoretical considerations and partly empirically. The energy density is obtained by calculating correlations in the matter using either Brueckner-Bethe-Goldstone theory or variational methods. Despite the enormous amount of work devoted to this program it is still not possible to give a quantitative account of the binding energy and equilibrium density of nuclear matter[2]. One may therefore question how reliable these approaches are at densities much greater than nuclear densities.

In this lecture I shall explore properties of dense matter from a somewhat different point of view, using a model of the nucleon introduced almost 25 years ago by Skyrme[3-6]. Already at that time, Skyrme showed that the model gave a good account of a number of properties of nucleons and nuclear forces. The model, in which nucleons are regarded as solitons in the pion field, has received increased attention in recent years following 't Hooft's[7] demonstration that at low energies QCD in the limit of a large number of colours goes over to a theory involving only meson degrees of freedom, and Witten's[8] work putting on a firmer foundation the identification of the topologically conserved quantity in the theory with baryon number.

To set the Skyrme model in context, it is helpful to briefly recall the development of bag models of nucleons[9]. The original MIT bag model had a radius of ~1F, which would imply that at a baryon density of ~$0.25 \, F^{-3}$, all space should be occupied by bags. Thus, at nuclear densities ($\sim 0.17 \, F^{-3}$), one would expect to observe very substantial quark effects. The problem with the MIT bag model is that it violates chiral symmetry, or put in other terms, it does not give a good account of the pion. This problem was removed in hybrid chiral bags in which, crudely speaking, the nucleon consisted of a bag of quarks surrounded by a cloud of pions. In such models the radius of the bag is much less than in the original MIT model. For example, in the calculations of Brown et al.[10], a radius of 0.44 F is suggested. This corresponds to a density of almost $3 \, F^{-3}$ if all space were occupied by bags. Therefore, effects of bag overlap would not become important until densities of this order of magnitude, while at nuclear densities the interaction between nucleons would be mainly via the pion clouds. This suggests that at nuclear densities the bag in the core of a baryon may not play a very important role, and that it may be a reasonable approximation to neglect the finite size of the bag. This is essentially what a baryon in the Skyrme model looks like - a pion cloud with no bag. These objects are generally referred to as skyrmions.

In this lecture we shall explore a number of properties of the Skyrme model, and especially the nature of the short-range interaction between skyrmions. This is but one of the many subjects that has been studied in the Skyrme model[11]. The emphasis will be on *qualitative* features rather than detailed quantitative calculations. As we shall see, a number of predictions of the Skyrme model are quite different from those of more conventional models, and resolving these differences may yield new insights. Even though the Skyrme model may not turn out to be correct in detail, it may be very helpful for obtaining qualitative understanding, and for suggesting fruitful lines of investigation.

The plan of the lecture is as follows. Section 2 is a simple introduction to the Skyrme model and the topological excitations in it. Section 3 describes a simple variational calculation of the single skyrmion, and sect.4 describes analogous results for multi-skyrmion states with spherical symmetry. Dense matter is considered in sect.5, and implications for nuclear forces are explored in sect.6.

2. THE SKYRME MODEL

The Skyrme model is essentially a version of the nonlinear sigma model studied extensively in connection with chiral symmetry[12]. Its basic ingredients are the pion field $\vec{\pi}$, a vector in isospin space, and an isoscalar field σ. In the nonlinear sigma model the strengths of the fields satisfy the condition

$$\sigma^2 + \pi^2 = f_\pi^2 \ , \tag{1}$$

where f_π is a constant, which can be identified with the pion decay constant. We shall often work in terms of the SU(2) matrix

$$U = \frac{1}{f_\pi}(\sigma + i\vec{\pi}\cdot\vec{\tau}) \ , \tag{2}$$

where τ_i are the Pauli isospin matrices. That U is unitary, $UU^\dagger = U^\dagger U = 1$, follows from the normalization condition (1). An alternative representation for U is in terms of a unit vector $\hat{n}$ in isospin space and an angle θ:

$$U = \exp i \, \hat{n}\cdot\vec{\tau} \, \theta \ . \tag{3}$$

This corresponds to the rotation matrix for rotation through an angle 2θ about the direction $\hat{n}$ in isospin space. The fields are related to $\hat{n}$ and θ by

$$\sigma = f_\pi \cos\theta \ , \qquad \text{and} \qquad \vec{\pi} = \hat{n} \, f_\pi \sin\theta \ . \tag{4}$$

Yet another representation for the fields is as a unit vector ϕ_i in a four-dimensional space, with $\phi_i = \pi_i/f_\pi$ for $i = 1-3$ and $\phi_4 = \sigma/f_\pi$.

Since the meson fields are described in terms of angular variables, one might expect that the model could have soliton-like solutions. Whether or not such structures are stable depends, of course, on the lagrangian, but the possibility of classifying configurations by their topology depends on the character of the quantity specifying the fields, and for a moment we concentrate on the topological features.

The simplest example to consider is a theory in one dimension (one space dimension and one time), where the configuration is specified by an angular variable θ at every point in space[13]. As an ex-

ample one might take the continuum limit of a chain of spins specified by two-dimensional unit vectors. (A special case of this is the
well-known sine-Gordon problem.) This is a U(1) version of the SU(2)
theory described above, since the quantity $U = e^{i\theta}$ is unitary. In the
simplest soliton in this theory the spin is at 6 o'clock at $x \to -\infty$,
it moves in a clockwise direction with increasing x until it is at
12 o'clock at the origin. With further increase in x, the spin continues in a clockwise direction and approaches 6 o'clock again for
large values of x. The angle θ as a function of x thus has the form
of a simple kink.

Now let us consider problems with a higher number of dimensions.
If one considers a 3-dimensional unit vector and two space dimensions,
one can imagine situations analogous to the one-dimensional one,
where the spin points up at the origin and then rotates in the plane
containing the radial vector and the vertical as one moves away from
the origin. Any one-dimensional slice through the origin will reveal
a structure similar to the one-dimensional soliton described above.
To imagine what the simplest soliton in an SU(2) theory in three dimensions looks like, it is simplest to recall that the fields can be
represented in terms of a four-dimensional unit vector. One therefore
simply generalizes the discussion of a 3-d unit vector in a 2-d space
by adding one more component to the unit vector, and one more space
dimension. In terms of the $(\hat{n}, \theta)$ representation, $\hat{n}$ could be radially
outwards everywhere, and θ varying from π at the origin to zero at
large distances. Such a structure is often referred to as a hedgehog,
since the $\hat{n}$ vectors directed radially outwards resemble the spines
of a hedgehog in the defensive position.

One important difference between the spin problem and the isospin one is that in the former case the unit vector is in ordinary
space, while in the latter case it is in isospin space.

The pion field vanishes for large distances and at the origin
but is finite at intermediate distances, while the σ field is f_π at
large distances and $- f_\pi$ at the origin.

One of the important features of Skyrme's model is that it has
a topologically conserved charge. To illustrate what this means we
shall first consider the analogous U(1) problem. It is easy to see
that the quantity

$$\int_{-\infty}^{\infty} \frac{\partial \theta}{\partial x} \, dx = \theta(\infty) - \theta(-\infty) \tag{5}$$

is conserved if θ at large distances is fixed. In addition, if one
writes

$$\rho = \frac{\partial \theta}{\partial x} , \qquad j = - \frac{\partial \theta}{\partial t} , \tag{6}$$

it is obvious that ρ and j satisfy a conservation law

$$\frac{\partial \rho}{\partial t} + \frac{\partial j}{\partial x} = 0 \ , \tag{7}$$

and therefore it is natural to interpret ρ as a density and j as a current. The quantity ρ is referred to as a topologically conserved density, since the conservation law (7) follows from the topology of the field used in the theory, in this case the angle θ, and is independent of the dynamics.

In the Skyrme theory there is an analogous topological density which is most easily expressed in terms of the currents

$$L^\mu = U^\dagger \frac{\partial U}{\partial x_\mu} \ . \tag{8}$$

In an SU(2) theory L_μ is an SU(2) matrix, but in a U(1) theory it would reduce to $i(\partial\theta/\partial x^\mu)$. The phase of the condensate wave function of an s-wave superconductor may be represented as $U = e^{i\theta}$, and in that case the spatial components of the current L_μ correspond to the momentum per particle of the condensate. In an SU(2) theory the topological current j^λ may be written as

$$j^\lambda = \frac{\varepsilon^{\lambda\mu\nu\rho}}{24\pi^2} \ \mathrm{tr}(L_\mu L_\nu L_\rho) \ , \tag{9}$$

where $\varepsilon_{\lambda\mu\nu\rho}$ is the completely antisymmetric tensor, normalized so that $\varepsilon_{1234} = 1$. The topological density $\rho = j^4$ is therefore given by

$$\rho = -\frac{1}{24\pi^2} \ \varepsilon^{ijk} \mathrm{tr}(L_i L_j L_k) \ , \tag{10}$$

where i, j and k are spatial indices. Notice that for the U(1) case in one dimension the equation analogous to (9) is $j^\lambda \sim \varepsilon^{\lambda\mu}L_\mu$, which is equivalent to eq.(6).

If one integrates the topological density over all space, one finds the winding number for the configuration. In the case of the simplest soliton described above, it is unity. Long ago, Skyrme[4] suggested that the topological density should be identified with the baryon density, and Witten[8] has placed this identification on a firmer footing.

To return to chiral bags for a minute, one finds that part of the baryon number is associated with the pion cloud, with a density (10), while the rest of it is associated with the quarks. Even though there may be three valence quarks in the bag, the total baryon number associated with the quarks in the interior is not 1 but something different, because the chiral boundary condition where the pion field is matched to the bag of quarks violates particle-antiparticle symmetry, and therefore leads to different shifts for positive and negative energy states. This results in a polarization of the vacuum. Goldstone and Jaffe[14] have studied in detail the flow of the baryon number from quarks to the pion field as conditions at the boundary of the bag change, and show that the total baryon number is conserved if one employs the usual expression for the baryon density associated with quarks, and eq.(10) for that associated with the pion cloud. This further strengthened the arguments for identifying the topological density (10) with the baryon density.

3. THE SINGLE SKYRMION

Now let us turn to a consideration of the energy of a soliton, which in the Skyrme model is usually referred to as a skyrmion. The lagrangian density is

$$\mathcal{L} = \frac{f_\pi^2}{4} \, \text{tr}(L_\mu L^\mu) + \frac{\varepsilon^2}{4} \, \text{tr}([L_\mu, L_\nu][L^\mu, L^\nu]) \, , \tag{11}$$

where ε^2 is a dimensionless coupling constant. (Note that in general we shall use units in which $\hbar = c = 1$.) The first term gives the usual lagrangian density for free massless pions plus some additional non-linear terms. If this term were the only one present, solitons would not be stable, as may easily be seen by dimensional arguments. Consider U having the form $U(\vec{r}/\ell)$. The L_μ scale as $1/\ell$, and therefore the energy density varies as $1/\ell^2$. Consequently, the total energy is given by the energy density times the volume (of order ℓ^3 in three dimensions) and therefore varies as ℓ. Consequently, it is energetically favourable for the soliton to shrink to zero size. To prevent this happening, Skyrme introduced the fourth-order term, which scales as ℓ^{-4} and therefore gives a contribution to the total energy varying as ℓ^{-1}. There are other fourth-order Lorentz invariant combinations that can be constructed, but the form Skyrme chose is the only one which has no more than two time derivatives, a desirable property for quantizing the theory.

Let us now perform a variational calculation of the energy of a single skyrmion. First we remark that we are looking for classical static solutions to the field equations and therefore all time derivatives are neglected. In the simple hedgehog, U has the form

$$U = \exp i \, \hat{r} \cdot \vec{\tau} \, \theta(r) \, , \tag{12}$$

where $\hat{r}$ is a unit vector in the radial direction (in isospin space), and the energy density, which is obtained from eqs.(12), (8) and (11) is

$$\varepsilon(r) = \frac{f_\pi^2}{2} \left(\left(\frac{d\theta}{dr}\right)^2 + \frac{2\sin^2\theta}{r^2} \right) + 8 \, \frac{\varepsilon^2 \sin^2\theta}{r^4} \left(r^2 \left(\frac{d\theta}{dr}\right)^2 + \tfrac{1}{2}\sin^2\theta \right) . \tag{13}$$

As $r \to 0$, the angle θ must tend to $N\pi$, where N is an integer, otherwise the total energy would be infinite. If θ did not do that, U would be discontinuous at $\vec{r} = 0$, since $\hat{r}$ behaves discontinuously at the origin. Likewise, θ must tend to a multiple of π as $r \to \infty$, and we shall generally adopt the convention that θ tend to zero there.

The behaviour of θ as a function of r is found by minimizing with respect to $\theta(r)$ the total energy, obtained by integrating the energy density (13) over all of space. The Euler-Lagrange equation is

$$\frac{d^2\theta}{dr^2} \left(1 + 16 \frac{\varepsilon^2}{f_\pi^2 r^2} \sin^2\theta\right) + \frac{2}{r} \frac{d\theta}{dr} + \frac{16}{r^2} \frac{\varepsilon^2}{f_\pi^2} \sin\theta \cos\theta \left(\frac{d\theta}{dr}\right)^2 - \frac{2}{r^2} \sin\theta \cos\theta$$

$$- \frac{16}{r^4} \frac{\varepsilon^2}{f_\pi^2} \sin^3\theta\cos\theta = 0 \ . \tag{14}$$

The simplest skyrmion is the case with $N = 1$. Rather than solving equation (14), we shall estimate its energy variationally, by making a linear ansatz for θ:

$$\theta = \pi\left(1 - \frac{r}{a}\right) \tag{15}$$

for $r \le a$, and $\theta = 0$ for $r > a$. We shall treat the radius of the skyrmion, a, as a variational parameter. The energy density is given by

$$E = \int_0^a 4\pi r^2 dr \left[\frac{f_\pi^2}{2} \left(\left(\frac{\pi}{a}\right)^2 + \frac{2}{r^2} \sin^2\theta\right) + 8 \frac{\varepsilon^2}{r^4} \sin^2\theta\left(\frac{r^2\pi^2}{a^2} + \frac{1}{2} \sin^2\theta\right)\right]$$

$$= \alpha_2 a + \alpha_4 a^{-1} \ , \tag{16}$$

where

$$\alpha_2 = 2\pi f_\pi^2 \left(1 + \frac{\pi^2}{3}\right) \ , \tag{17}$$

$$\alpha_4 = 16\pi^3 \varepsilon^2 (1 + c_1) \ , \tag{18}$$

and

$$c_1 = \frac{1}{\pi} \int_0^\pi dx \frac{\sin^4 x}{x^2} \simeq 0.21393 \ . \tag{19}$$

The energy has a minimum for

$$\frac{\partial E}{\partial a} = \alpha_2 - \alpha_4 a^{-2} = 0 \ , \tag{20}$$

or in other words $\alpha_2 a = \alpha_4 a^{-1}$, the second- and fourth-order terms contribute equally to the energy. The minimum value of the energy, the skyrmion mass, is therefore

$$E_{min} = 2\sqrt{\alpha_2 \alpha_4} \tag{21}$$

$$= 8\pi^2 \sqrt{2\left(1 + \frac{\pi^2}{3}\right)(1 + c_1)f_\pi \varepsilon} \tag{22}$$

$$\approx 255 \, f_\pi \varepsilon \, . \tag{23}$$

The radius of the skyrmion is given by

$$a = \sqrt{\frac{\alpha_4}{\alpha_2}} \approx 4.73 \, \frac{\varepsilon}{f_\pi} \quad . \tag{24}$$

If one solves the variational problem for θ numerically, one finds a mass $E \approx 206 \, f_\pi \varepsilon$, only 20% below our crude variational estimate.

To obtain a value for the mass we need values for ε and f_π. Jackson and Rho[15] took f_π from experiment, and determined ε from the condition that at large distances the strength of the pion field ($\pi \sim 1/r^2$) should agree with the experimental value. They found $\varepsilon^2 = 0.00552$. With the exact integration of the variational equation for θ, this gives a mass of 1425 MeV, which is not too bad given the simplicity of the model.

One important point is that the skyrmion does not have good spin and isospin, since it is invariant under simultaneous spatial and isospin rotations, but not under rotations of them separately. This is clear from the fact that at every point the isospin direction is correlated with the spatial coordinate. To construct states corresponding to physical nucleons and other baryons (n, p, Δ) one must construct superpositions of skyrmions obtained by performing isospin (or spatial) rotations on the basic skyrmion we have considered. This procedure, which is described in detail by Adkins et al.[16] and by Jackson et al.[17] is analogous to projecting out states of good angular momentum of a deformed nucleus by superposing nuclei with different orientations for the distortion, or to constructing states of a superconductor with a definite number of particles by superposing states with a definite phase for the pair wave function. However, we shall not describe these manipulations in detail here since our concerns are with rather gross qualitative features that are affected very little by the particular character of the baryon under consideration.

4. MULTI-SKYRMION STATES

We now turn to more complicated states. Skyrme[6] suggested that as a trial function for a two-skyrmion state one take for U a product of the U matrices, U_1 and U_2, for two individual free skyrmions:

$$U = U_1 U_2 \ . \tag{25}$$

One particularly useful feature of this ansatz is that the baryon number associated with U is the sum of the baryon numbers associated with U_1 and U_2, and therefore the topology of U is automatically taken care of. With this ansatz one can calculate the two-nucleon interaction, as was done by Skyrme[6], and in more detail by Jackson et al.[17] For large separations the interaction is simply the one-pion exchange potential, while at closer distances the interaction is repulsive and shows many of the same qualitive features as the actual nucleon-nucleon interaction.

To explore the skyrmion-skyrmion interaction at short distances, let us evaluate the energy of a number of identical skyrmions centred at the same point. In the product ansatz the angle θ for one skyrmion is multiplied by N, the number of skyrmions superposed. For our present purposes it is sufficient to explore the behaviour for large N, and we shall repeat our variational calculation. We therefore put

$$\theta = N\pi\left(1 - \frac{r}{a}\right) \ . \tag{26}$$

As one can easily see, the $(d\theta/dr)^2$ terms in the energy density dominate, and one finds

$$E_N \approx \frac{2}{3}\pi^3 f_\pi^2 a N^2 + 16\pi^3 \frac{\varepsilon^2}{a} N^2 \ . \tag{27}$$

In the spirit of the product ansatz one should replace a in eq.(27) by its equilibrium value for the $N = 1$ skyrmion, given by eq.(24). However, one can see by minimizing eq.(27) with respect to a that

$$\frac{a_{N \to \infty}}{a_{N = 1}} = \sqrt{\frac{1 + 3/\pi^2}{1 + c_1}} \simeq 1.036 \ , \tag{28}$$

and therefore the energy is reduced by less than one per mille by allowing a to relax. To obtain a simple expression for the energy, we evaluate it at the value of a that gives a minimum, and find

$$E_{N \to \infty} \approx N^2 8\pi^3 \sqrt{\frac{2}{3}} f_\pi \varepsilon$$

$$\approx \frac{N^2}{2} 405 \ f_\pi \varepsilon \ . \tag{29}$$

This result has been interpreted as implying that at short distances the interaction between skyrmions is a two-body interaction, since the number of ways of pairing N skyrmions is $\frac{1}{2}N(N-1)$. The rest mass term for the individual skyrmions is proportional to N and therefore it is less important in the limit of large N. The strength of this two-body interaction would be of the same order of magnitude as the skyrmion rest mass ($\sim 254\, f_\pi \varepsilon$ in our variational calculation). This behaviour for large N has been studied in more detail by Nisichenko[18] who did it numerically, and by Bogomolnyi and Fateev[19], who obtained an asymptotic expansion. The latter's results are

$$E_N \approx \frac{208.9}{2}\, N(N + 0.8726)\; f_\pi \varepsilon + \mathcal{O}(N^0)\;,\qquad (30)$$

which shows that our rough estimates are qualitatively correct.

The interpretation of the N^2 behaviour as implying that the skyrmion-skyrmion interaction has a two-body character is inconsistent with calculations of properties of dense matter, which we consider in the next section. We shall resolve this apparent contradiction in sect.7.

5.   DENSE MATTER

We now study the energy of dense matter in the Skyrme model following the treatment of Kutschera et al.[20] As in our previous discussion of skyrmions, we shall for the moment neglect time-dependent terms and shall consider only static configurations. This means that we do not take into account effects due to the motion of the skyrmions, and therefore the energy we obtain may be regarded as the potential energy of the skyrmions. If one were to try to construct U for many skyrmions in dense matter by using the product ansatz, one would quickly run into difficulties. First of all, at any point, U would have contributions from many different individual skyrmions if the size of a free skyrmion were large compared with a typical distance between skyrmions. Second, since the U's are matrices, one must specify the order in which the product is computed. This suggests that one should abandon the product ansatz and work with U itself, just as one would in many condensed matter problems, such as liquid crystals or ferromagnets. Let us try to envision what the vector $\hat{n}$ and the angle θ would look like if we placed a number of identical skyrmions on the lattice sites of a simple cubic lattice. Close to each lattice site, $\hat{n}$ has the usual hedgehog-like structure, and in a section through a lattice plane one would expect it to behave as shown in fig.1. To avoid discontinuities, θ must be a multiple of π at the lattice sites, and we shall take $\theta = \pi$ there, since we wish to consider matter that resembles a number of N = 1 skyrmions squashed together. By symmetry it is clear that θ must be zero at the points marked by crosses in fig.1. In fact, on the surface of the unit cell there are 26 points at which $\theta = 0$.

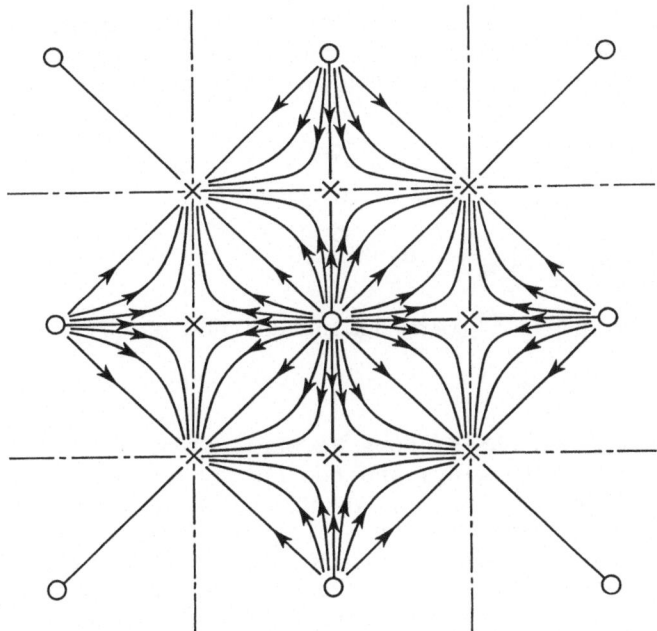

Fig.1. Schematic representation of the isospin
vector $\hat{n}$ for skyrmions on a simple cubic lattice.
The figure represents a section through one of
the lattice planes containing the centres of the
skyrmions, which are indicated by circles. At
the points marked by crosses, θ vanishes.

To solve for U is clearly a difficult task, so we make some
physically motivated assumptions. First of all, since θ is zero at
so many points on the faces of the cube, it may not be a bad approximation to put it equal to zero everywhere on the surface. If θ varied
much from zero between the points where it is forced to be zero by
symmetry, the currents L_μ would be large and this would lead to large
energy densities. We shall further assume that the cube may be replaced by a sphere of the same volume, a procedure known as the
Wigner-Seitz approximation which is often used in solid-state physics.
Had we started with a more closely packed lattice, which is probably
more realistic than the simple cubic one we chose for simplicity, the
Wigner-Seitz approximation would have been even more accurate. The
calculation of the energy per cell is now essentially the same as the
calculation of the energy of a single skyrmion, except that the
boundary condition is now that $\theta \to 0$ not for $r \to \infty$ but for $r \to a$, where
a is the cell radius, given by

$$\frac{4}{3}\pi a^3 = \frac{1}{n} \ ,$$
(31)

where n is the mean baryon density. We can easily obtain an estimate of the energy from our earlier variational calculation, provided we substitute for a in terms of n. The energy density is given by [cf. eqs.(16)-(18)]

$$E = n(\alpha_2 a + \alpha_4 a^{-1})$$ (32)

$$= \left(\frac{4\pi}{3}\right)^{1/3} \alpha_4 n^{4/3} + \left(\frac{3}{4\pi}\right)^{1/3} \alpha_2 n^{2/3} .$$ (33)

At high densities the $n^{4/3}$ term dominates, and one finds

$$E \approx 971 \ \varepsilon^2 n^{4/3} \hbar c ,$$ (34)

where we have included the factors $\hbar c$ explicitly. At lower densities the $n^{2/3}$ term comes into play, and the energy per particle has a minimum at $n = (3/4\pi)(\alpha_2/\alpha_4)^{3/2}$, which corresponds to the volume per skyrmion being given by the equilibrium radius of a single skyrmion in our variational calculation [cf. eq.(24)]. At lower densities the energy per skyrmion increases, but this is an unrealistic feature of our simple trial function. At lower densities it is clearly favourable to build up matter from single skyrmions of radius given by the equilibrium radius (24) with $\theta = 0$ at all other points. The energy per skyrmion would then remain constant at the value (24) at lower densities.

An integration of the Euler-Lagrange equation for θ leads to an energy density at high densities

$$E \approx 870 \ \varepsilon^2 n^{4/3} \hbar c ,$$ (35)

which is not far below our simple variational estimate.

Up to now we have assumed that all the skyrmions are identical. If we had put skyrmions with different isospin structures at the various lattice sites, the number of points at which $\theta = 0$ could have been reduced, and this could lower the energy. However, Skyrme[6] showed by using algebraic inequalities that the energy density is bounded below,

$$E \geq 640 \ \varepsilon^2 n^{4/3} \hbar c ,$$ (36)

without any restriction on the form of U. This shows that rather little energy can be gained by isospin 'combing' of skyrmions.

The energy density at high densities increases much less rapidly with density for skyrmions than for more conventional equations of state. In the latter case the interaction energy density typically varies as n^2, essentially because interactions of pairs of nucleons dominate, and the effective interaction is a weak function of density. In the case of interactions with significant velocity dependence, the energy may increase even more rapidly with density. Figure 2(a) shows a plot of the interaction energy for skyrmions and for two conventional equations of state, those of Bethe and Johnson[21] (B-J) and of Friedman and Pandharipande[22] (F-P). Beside the less rapid increase of the energy density of skyrmions at high density, the other noticeable feature is that at low densities the energy density of skyrmions lies well above that for the conventional equations of state. This is partly due to our ansatz for U which is least satisfactory at low densities, and, more importantly, to the fact that the Skyrme model lacks the intermediate range attraction responsible for binding nuclear matter, a point stressed by Jackson.[23] The absence of this intermediate range attraction is also partly responsible for the skyrmion mass being higher than that of the nucleon, since such attraction will play a role in determining the energy of a single nucleon as well as the energy due to interactions between nucleons.

The equation of state of cold matter is of great importance for determining the masses of neutron stars. As we have seen, the skyrmion equation of state is not reliable at densities near nuclear matter and below, so Kutschera et al.[20] constructed a composite equation of state that has the same properties as the skyrmion one at high densities, but agrees with a more conventional one, that of Friedman and Pandharipande[22], at lower densities. They fitted the interaction energy to an expression which has as its highest power of the density n^2. The interaction energy in the composite model was obtained by multiplying the fit to Friedman and Pandharipande's result by $(1 + n/n_0)^{-2/3}$, with $n_0 = 2.815$ F^{-3}, which ensured that at high densities it had the same coefficient as the skyrmion equation of state. To obtain the total energy density, the kinetic energy of a free neutron gas was added to the interaction energy. In the Skyrme model the kinetic energy term must be regarded as an approximation for the time derivatives, and it is not clear how good this is. However, at high densities the kinetic energy is dominated by the interaction energy, so the kinetic energy term does not play a crucial role. Plots of the interaction energy density and pressure as functions of density for various equations of state are shown in fig.2.

The maximum mass of a neutron star calculated with Friedman and Pandharipande's equation of state is 1.96 $M_\odot$, where $M_\odot$ is the solar mass, while for the composite equation of state it is 1.51 $M_\odot$. The smaller maximum mass is a reflection of the softness of skyrmion matter compared with conventional nuclear matter. Observationally, the best determined neutron star mass is that of the binary pulsar PSR 1913 + 16, whose mass is 1.42 ± 0.06 $M_\odot$.[24] The masses of neutron stars in accretion-driven x-ray sources are less well determined, but are not inconsistent with a maximum neutron star mass of 1.51 $M_\odot$. The lower maximum neutron star mass will possibly have implica-

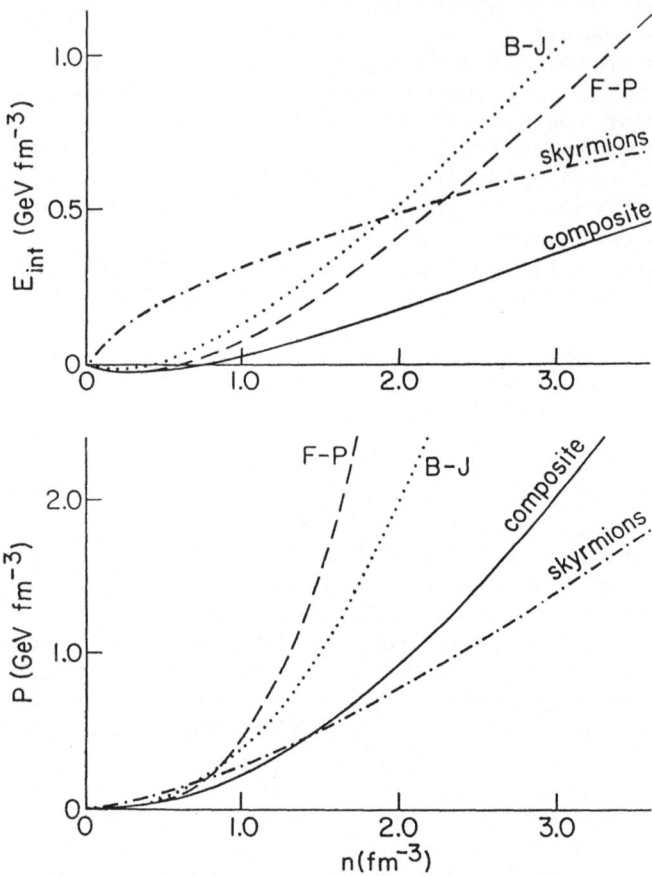

Fig.2. Interaction energy density and total pressure as functions of baryon density for pure skyrmions, for the calculations of Bethe and Johnson (ref. 21) and of Friedman and Pandharipande (ref.22) and for Kutschera et al.'s composite equation of state (ref.20).

tions for the relative numbers of neutron stars and black holes formed as a consequence of stellar collapse, but at present the statistics are so uncertain that no definite statements can be made regarding the maximum mass.

6. QUARK MATTER

At high densities it is expected that matter will consist of essentially free quarks. If we neglect the quark masses, the energy density is given by

$$E = \frac{3}{4} p_F c n_q , \tag{37}$$

where n_q, the number of quarks, is the number of colours (3) times the baryon density:

$$n_q = 3n . \tag{38}$$

The quark Fermi momentum p_F is given by

$$n_q = 3\nu_f \left(\frac{p_F^3}{3\pi^2 \hbar^3} \right) = \frac{\nu_f p_F^3}{\pi^2 \hbar^3} , \tag{39}$$

where ν_f is the number of flavours and the quantity in parentheses is the number density for a Fermi gas with a spin-degeneracy factor 2. Thus the kinetic energy density is given by

$$E = \frac{9}{4} \left(\frac{3\pi^2}{\nu_f} \right)^{1/3} \hbar c \, n^{4/3} . \tag{40}$$

If one includes the bag energy and the energy due to gluon exchange to first order in the QCD coupling constant α, one finds[26]

$$E = B + \frac{9}{4} \left(\frac{3\pi^2}{\nu_f} \right)^{1/3} \hbar c \, n^{4/3} \left(1 + \frac{2\alpha}{3\pi} \right) , \tag{41}$$

where B is the bag constant. The coupling constant is almost certainly less than the MIT value 2.2 and therefore at high densities E is between $4.83 \, n^{4/3}$ and $7.08 \, n^{4/3}$ for 3 flavours of quarks. In the Skyrme model, with the kinetic energy term included and the interaction energy (35), one finds $E \sim (2.32 + 870 \, \varepsilon^2) n^{4/3} \hbar c$, which is $7.15 \, n^{4/3} \hbar c$ for $\varepsilon^2 = 0.00552$ [15] or $5.98 \, n^{4/3} \hbar c$ for the value $\varepsilon^2 = 0.00421$ adopted by Adkins et al. [16] The skyrmion matter results are thus remarkably close to those of quarks. The skyrmion model therefore gives a reasonable model for nucleons at low densities, and has a behaviour close to that of quarks at high densities. It may therefore be a useful starting point for investigating qualitative features of nucleon interactions at high densities.

The similarity of the skyrmion and quark equation of state has interesting implications for the phase transition between nuclear matter and quark matter. For nuclear matter described by a skyrmion equation of state, the density discontinuity at the transition will tend to be smaller than for nuclear matter described by a more conventional equation of state.

Here a word of caution should be interjected. At a fundamental level there is no justification for naively extrapolating the simple Skyrme model, since it is really the first two terms in an expansion in powers of the currents L_μ. One may therefore ask whether higher-order terms should be included. Formally, of course, one clearly should, and this has in fact been done recently by Jackson et al.[27] This alters the behaviour of the energy density at very high densities, but it is not clear, since no detailed calculations have been carried out yet, what difference it will make at densities of, say, ten times nuclear density, which is of the greatest importance for the transition to quark matter. One should also bear in mind that the Skyrme model is only an approximation to a better model of the nucleon, for example a chiral bag, and the fourth-order term alone may well describe the behaviour of the chiral bag better than a formal expansion in powers of the currents would. There is some indication that this is the case, as Jackson[23] has pointed out. The energy of the N skyrmion spherical states is proportional to N^2 for large N, which is the same as for a bag model with a fixed bag volume if one filled only s-states. On the other hand, if one filled all angular momentum states, the energy would vary as $N^{4/3}$, which is what one obtains if one packs roughly spherical skyrmions together in a fixed volume. Clearly these problems need to be explored in greater detail.

7. NUCLEAR FORCES

In sect.5 we showed that the energy density in the Skyrme model behaves as $n^{4/3}$ at high densities. On the other hand, it has been argued on the basis of the behaviour of the energy of the spherical N-skyrmion state considered in sect.4 that the skyrmion-skyrmion interaction at short distances has a two-body character. If the latter were true, one would expect the energy density to behave as n^2, not $n^{4/3}$, at densities so high that the separation between skyrmions is much less than the range of the interaction. In this section we resolve this apparent contradiction, following ref.28.

First of all let us explore the baryon density in the spherically symmetric N-baryon state. This is given by

$$n(r) = -\frac{1}{2\pi^2 r^2} \sin^2\theta \, \frac{d\theta}{dr} , \qquad (42)$$

and it therefore vanishes when $\theta = \nu\pi$, where ν is an integer, provided r is non-zero. As a consequence, the baryon density has an onion-like structure, with each layer of the onion separated from the next by a shell on which the baryon density vanishes. To estimate the energy of one of these shells, it is convenient to imagine that one peel it off and flatten it out, thereby producing a skyrmion having the shape of a rectangular parallelepiped. In this peeling process we imagine that the isospin vectors are not altered, and therefore the topology is preserved. Let the parallelepiped have characteristic dimensions ℓ_1, ℓ_2 and ℓ_3 in three mutually perpendicular directions. To make a correspondence with the dimensions of the spherical N-skyrmion state, we note that the characteristic thickness of an onion layer is $\sim a/N$ in the radial direction, and a in the two angular directions, and we shall later replace the ℓ_i by these lengths. The components of the currents L_i vary as ℓ_i^{-1}, and therefore the fourth-order term in the lagrangian density contributes to the energy density an amount

$$E = \varepsilon^2 \left(\frac{\alpha_1}{\ell_2^2 \ell_3^2} + \frac{\alpha_2}{\ell_3^2 \ell_1^2} + \frac{\alpha_3}{\ell_1^2 \ell_2^2} \right) \tag{43}$$

$$= \varepsilon^2 n^{4/3} \left(\alpha_1 \left(\frac{\ell_1}{\ell_0} \right)^2 + \alpha_2 \left(\frac{\ell_2}{\ell_0} \right)^2 + \alpha_3 \left(\frac{\ell_3}{\ell_0} \right)^2 \right) , \tag{44}$$

where the α_i are constants of order unity and $\ell_0^3 = \ell_1 \ell_2 \ell_3 = n^{-1}$. This expression has a minimum, for fixed n, if $\alpha_i^{1/2} \ell_i$ is the same for all i, or, in other words, characteristic lengths in all directions are roughly the same. For the case of an onion-skin skyrmion, it is easy to see that the energy density is $\sim N^{2/3}$ times greater than for a skyrmion of the same average density having the same dimensions in all directions. Thus, it is energetically expensive to make skyrmions with different characteristic dimensions in different directions. Since the energy per unit volume of an onion-skin skyrmion is of order $\varepsilon^2 n^{4/3} N^{2/3}$, it is easy to see that the energy of the N-skyrmion state is of order $\varepsilon^2 N^2/a$, in agreement with (27). Note that the fact that we ignored the second-order term in making the estimate is unimportant, since in the N-skyrmion spherical state the second-order and fourth-order terms contribute equally to the energy.

This discussion shows that the energy per skyrmion varies as $\varepsilon^2 n^{1/3}$ times a factor depending on the shape of the volume available to a skyrmion. In terms of inter-skyrmion forces, this means that the interaction between skyrmions acts only between close neighbours, since a skyrmion's energy depends on the shape and size of the volume enclosing it, and this is determined by the position of the neighbouring skyrmions. The energy of a skyrmion is insensitive to the positions of skyrmions beyond the 'cage' of neighbours surrounding it, which implies that the interaction between skyrmions is small for distances exceeding a typical inter-skyrmion distance. From this observation one can argue that there must be significant many-body forces, which cancel the two-skyrmion interaction at distances greater than a typical inter-skyrmion distance. The reason for nuclear forces being very different in the skyrmion model at high densities is that all baryons are made up from a single U field, and therefore they can not overlap. Interactions are produced by neighbouring skyrmions distorting each other. In the Skyrme model nonlinear effects play a crucial role, whereas in most conventional models the interaction is taken to be dominated by a superposition of two-body interactions, and therefore nonlinearities are not very important.

These conclusions about the nature of the skyrmion-skyrmion interaction are at variance with the results of calculations that employ the product ansatz (25).[17,29] As we mentioned above, the latter suggest that the skyrmion-skyrmion interaction is a two-body interaction. Explicit calculations of the three-body interaction made with the product ansatz are in agreement with this conclusion. However, as we have seen, there are serious reasons to question the product ansatz for small separations of the skyrmions, since distortions of the skyrmions are likely to be important. We also showed that the baryon density in the product ansatz for small separations had the onion-like form, and did not resemble the baryon distribution one might expect for two skyrmions close together.

As a final topic in this discussion of nuclear forces, we return to the behaviour of the two-skyrmion interaction at short distances. One of the difficulties is that it is not clear what should be used as the position of the skyrmion. In the product ansatz calculation this is usually taken to be the distance between the centres of the two individual skyrmions from which the product is formed. However, such points have no particular significance in the total U, and Skyrme[6] long ago suggested choosing as the position of the skyrmions the points where $U = -1$, which correspond to points where the $\hat{n}$ vector related to U has a hedgehog-like form. In the product ansatz one finds that for two identical skyrmions such points never approach closer than a distance of order of the skyrmion size. If one chooses this definition of the position, the product ansatz does not allow one to explore the interaction energy for small separations.

To explore the interaction at smaller separations one must go beyond the product ansatz with undistorted skyrmions. Let us consider the case of two identical skyrmions separated by a distance r (the distance between the $U = -1$ points). Midway between the two skyrmions θ vanishes, just as in the dense matter case drawn in fig. 1. Thus the currents L_i are of order r^{-1}, and therefore for small r the energy density is of order r^{-4}, since the fourth-order term in the lagrangian density (11) is the more important. Consequently, the total energy is of order r^{-4} times the volume ($\sim r^3$) over which the energy density is present, that is of order ϵ^2/r. This is quite different from what the product ansatz gives, in large part because of the different definition of the skyrmion position. Numerical calculations of the skyrmion-skyrmion interaction[30] which do not employ the product ansatz confirm the $1/r$ dependence. The $1/r$ behaviour of the interaction is, however, quite consistent with our dense matter calculations, since $1/r$ for nearest neighbours varies as $n^{1/3}$, and therefore a $1/r$ interaction between nearest-neighbour skyrmions would lead to an energy per unit volume varying as $\varepsilon^2 n^{4/3}$, in agreement with (35). The energy of dense matter is therefore quite consistent with the view that, at high density, interaction between skyrmions occurs only by direct contact.

Finally we remark that for a pair of skyrmions which are not identical, the interaction energy does not generally vary as $1/r$ at small separations. This is discussed in more detail in ref.28.

8.   CONCLUSIONS

In this lecture we have considered a number of simple properties of baryons and baryon-baryon interactions in Skyrme's model. Among the interesting results of the model is that it gives an equation of state for dense matter which is much softer than most conventional equations of state, but rather similar to that of a gas of massless quarks. A second interesting result is that there are very large many-body forces which are a consequence of skyrmions distorting each other significantly when they are close together. A third conclusion is that for certain channels, the interaction between skyrmions can have a $1/r$ dependence at small separations.

Our considerations here have been largely qualitative, and more work is required to make quantitative statements. Also for realistic estimates to be made, it is necessary to incorporate effects which lead to the intermediate-range attraction between nucleons.

The Skyrme model has a number of important virtues: it is consistent with QCD, it accounts for many properties of nucleons and nucleon-nucleon interactions, and it is simple. Further study of its properties may lead to valuable insights which can be checked against the properties of more detailed models, and possibly experiment.

The work reported in this lecture was supported by NSF Grant No. NSF PHY 80-25605. I am grateful to Marek Kutschera for many helpful discussions.

REFERENCES

1.   For a general review of dense matter, see G. Baym, in: Nuclear Physics with Heavy Ions and Mesons, R. Balian, M. Rho and G. Ripka, eds., North-Holland, Amsterdam (1978) p.745.

2. For a review of work on nuclear matter, see V.R. Pandharipande and R.B. Wiringa, Rev. Mod. Phys. 51:821 (1979).

3. T.H.R. Skyrme, Proc. Roy. Soc. A247:260 (1958) and 252:236 (1959).

4. T.H.R. Skyrme, Proc. Roy. Soc. A260:127 (1961).

5. T.H.R. Skyrme, Proc. Roy. Soc. A262:237 (1961).

6. T.H.R. Skyrme, Nucl. Phys. 31:556 (1962).

7. G. 't Hooft, Nucl. Phys. B72:461 (1974).

8. E. Witten, Nucl. Phys. B223:423,433 (1983).

9. A general review of bag models has been given by A.W. Thomas, in: Advances in Nuclear Physics, J.W. Negele and E. Vogt, eds., Plenum, New York (1984) Vol. 13, p. 1. See also C.E. DeTar and J.F. Donoghue, Ann. Rev. Nucl. Part. Sci. 33:235 (1983).

10. G.E. Brown, A.D. Jackson, M. Rho and V. Vento, Phys. Lett. 140B:285 (1984).

11. We refer to the lectures by M. Rho, Chiral bags, skyrmions and quarks in nuclei, to appear in the proceedings of the Varenna Summer School, June 18-23, 1984, for an account of many other aspects of skyrmions.

12. An introduction to chiral symmetry may be found in G. Baym and D.K. Campbell in: Mesons in Nuclei, M. Rho and D.H. Wilkinson, eds., North-Holland, Amsterdam (1979) Vol. III, p. 1031.

13. This case is considered in detail by, e.g., B. Felsager, Geometry, Particles and Fields, Odense University Press, Odense (1981) Ch. 4.

14. J. Goldstone and R.L. Jaffe, Phys. Rev. Lett. 51:1518 (1983).

15. A.D. Jackson and M. Rho, Phys. Rev. Lett. 51:751 (1983).

16. G.S. Adkins, C.R. Nappi and E. Witten, Nucl. Phys. B228:552 (1983).

17. A. Jackson, A.D. Jackson and V. Pasquier, Nucl. Phys. A432:567 (1985).

18. V.P. Nisichenko, Dissertation, Patrice Lumumba University, Moscow (1981).

19. E.B. Bogomolnyi and V.A. Fateev, Yad. Fiz. 37:228 (1983) [Sov. J. Nucl. Phys. 37:134 (1983)].

20. M. Kutschera, C.J. Pethick and D.G. Ravenhall, Phys. Rev. Lett. 53:1041 (1984).

21. H.A. Bethe and M. Johnson, Nucl. Phys. A230:1 (1974).

22. B. Friedman and V.R. Pandharipande, Nucl. Phys. A361:502 (1981).

23. A.D. Jackson, private communication.

24. J.H. Taylor and J.M. Weisberg, Ap. J. 253:908 (1982).

25. P.C. Joss and S.A. Rappaport, in: Accretion Driven Stellar X-ray Sources, W.H.G. Lewin and E.P.J. van den Heuvel, eds., Cambridge Univ. Press, Cambridge (1984) p. 1.

26. See, e.g. ref.1, sect. 8.

27. A. Jackson, A.D. Jackson, G.E. Brown, A.S. Goldhaber and L.C. Castillejo, to be published.

28. M. Kutschera and C.J. Pethick, unpublished.

29. The three-body interaction has been calculated using the product ansatz by U.-G. Meissner and U.B. Kaulfuss, Phys. Rev. C, December, 1984.
30. H.M. Sommermann, H.W. Wyld and C.J. Pethick, unpublished.

NUCLEUS-NUCLEUS COLLISIONS AT HIGH ENERGIES

Martin A. Faessler

CERN
Geneva, Switzerland

INTRODUCTION

Nucleus-nucleus interactions at high energies (say, above 100 GeV incoming energy) are very complicated. Even the simpler nucleon-nucleon interaction has resisted a quantitative explanation, so far. Why do we investigate very complex reactions if the simpler reactions of the same kind are not understood?

We know that hadrons (=strongly interacting particles) consist of: a) quarks[1] which are pointlike ($r < 10^{-3}$ fm)[2], are spin 1/2 fermions[3], have fractional electric charges ($\pm 2/3$ e or $\pm 1/3$ e)[4] and a weak charge (couple to $W^{\pm}$ and Z), and have flavour (up, down, strange, charm, bottom, top); and of b) another kind of constituents which do not couple to electroweak fields but carry about 50% of the hadron momentum[5]. We believe those other constituents are gluons, i.e., the fields of the strong interaction which couple to the "colour charge" of quarks and are thus responsible for their strong interaction.

The older "naive" quark models and the modern theory of Quantum Chromodynamics (QCD)[1] describe a large range of different phenomena; some of them fairly quantitatively (the "hard" interactions, i.e. those involving large momentum transfers and thus probing the point-like nature of the constituents, e.g., deep-inelastic lepton interactions, $e^{+}e^{-}$ annihilation into hadrons, high-p_T hadron production in pp interactions); others more qualitatively (the spectrum of known resonances and particles, and "soft" interactions of hadrons, i.e. those with small momentum transfers, e.g. the normal hadron-hadron interactions).

But the trouble - or challenge - is that isolated quarks and gluons have probably not been observed yet. We may believe quarks are constituents of a conceptually new kind which are confined in hadrons and cannot exist as isolated particles. In the language of QCD, quarks come in three colours and are confined because only colour singlets can exist in nature. Colour singlets can be formed combining 3 quarks with different colours, or a quark and an antiquark with the corresponding anti-colour, or by a superposition of these combinations. All known hadrons (stable or resonances) are indeed thought to be constructed from either 3 valence quarks [$(q_1q_2q_3)$ = baryons] or a valence quark and a valence anti-quark [$(q_1 \bar{q}_2)$ = mesons], plus some amount of sea quarks [$\Sigma_i q_i \bar{q}_i$] and of gluons.

So far no hadrons or resonances have been observed which can be unambiguously assigned to theoretically possible, "exotic" colour-singlet combinations of more than 3 valence quarks or of only gluons (Examples: the search for baryonium $(q_1q_2\bar{q}_3\bar{q}_4)$[6]; the search for glueballs[7]; the K^+p reaction ($\bar{s}uuud$) or pp reaction (uuudd) show no resonances . If we cannot observe single isolated quarks because they are confined it is not only natural but compulsive to search for more complex states, of four, five or more valence quarks. However, the fact that such states have not been observed yet does not prove they do not exist, they may simply be too short-lived to be seen as resonances.

Of course, nuclei are states containing 3A valence quarks but do we need the quarks to understand the structure of nuclei? This question has been discussed actively in the past years[8]. New incentives were provided when the European Muon Collaboration (EMC) at CERN discovered that the structure functions of nuclei are not simple superpositions of structure functions of nucleons (the "EMC effect"[9]). Since this result was obtained at high Q^2 (invariant squared momentum transfer) it reflects presumably more than the trivial fact that nuclei are not superpositions of free nucleons, but are complicated structures of bound nucleons. It revived speculations that nuclei contain confinement volumes (bags) with 6, 9 or more quarks. The occurrence of such large bags within nuclei has also been hypothesized[10] to explain other odd phenomena, like the cumulative production (production of particles from a nucleon target which are outside the kinematic limit for nucleon-nucleon interactions). However, the EMC effect and these phenomena cannot be considered as a proof that large bags exist within nuclei in their ground state. Therefore, the basic question remains: Can we put more than 3 valence quarks into a single confinement volume?

A recent theoretical response to this old question came from lattice QCD (LQCD) at finite temperature[11]. LQCD predicts that at an energy density about 10 times the one of nuclei in their ground state ($\sim 10 \cdot 0.17$ GeV/fm^3), nuclear matter composed of almost separate nucleons will fuse into a quark-gluon plasma (QGP) - quarks will be deconfined; before they are essentially confined in their host nucleons, after the deconfinement phase transition they will move freely in a larger confinement volume.

Many people believe that the experimental proof of such a phase transition is almost as important a test for QCD as finding the W or Z was for the theory of electroweak interactions. According to recent theoretical estimates, the necessary temperature and pressure or baryon density can indeed be reached in nucleus-nucleus collisions at c.m. energies perhaps as low as a few GeV per nucleon-nucleon collision. This explains the excitement and the remarkable activation of interest for nucleus-nucleus collisions which has brought together again nuclear physicists, astrophysics and particle physicists, in the last 3 or 4 years.

The goal of these lectures is to review some of the experimental results from very high energy nucleus-nucleus collisions, and to compare then with pp and pA data, and with theoretical predictions or expectations. Since there _may_ be hints of new phenomena but there is no convincing proof yet, the comparisons serve the purpose of showing to which level "old physics" can explain the data.

The following lectures contain:

1) A brief survey of the sources of existing data for high-energy nucleus-nucleus collisions —experiments with cosmic rays and with ions at the CERN Intersecting Storage Rings - and of potential nucleus accelerators.
2) A presentation of data on total multiplicities of produced particles in pp, pA and AA collisions with the main gold of providing useful "rules of thumb".
3) A more detailed presentation of distributions in longitudinal phase space variables (rapidity, longitudinal momentum or Feynman-x_F) of produced particles (mainly pions) and of leading nucleons, including a comparison with models.
4) A selection of data on the distribution of particles in the transverse direction - the summed energy distribution in the central rapidity region and the single-particle p_T distribution.
5) A review and interpretation of data on strangeness production in pp and pA interactions and a comparison with the expectation for strangeness in the QGP.
6) A survey of the goals and set-ups of future experiments investigating nucleus-nucleus collisions at high energy, in particular of those experiments proposed at the CERN Super Proton Synchrotron and the Brookhaven Alternating Gradient Synchrotron.

1. SOURCES OF THE DATA

Data on nucleus-nucleus collisions at energies above that of the Bevalac at Berkeley (USA) and the Synchrophasotron at Dubna (USSR) were only obtained at the CERN Interacting Storage Rings (ISR) and from cosmic-ray interactions at high altitude. The pp and pA data used for comparison in this review come from Fermilab, the CERN Super Proton Synchrotron (SPS) and the CERN ISR.

1.1. Accelerators

A list of the high-energy proton accelerators, their beam energies and accelerated particles is given in Table 1. This table also contains projects which are going to be realized in the future and two projects which are presently being discussed.

The accelerator network at CERN is shown in Fig. 1. Deuteron and alpha beams were extracted from a modified duoplasmatron source. For alpha beams a pulsed gas stripper was used after the preacceleration to 600 keV in order to enrich the number of $^{++}$He. It requires some special RF gymnastics ($2\beta\lambda$ mode in the Linac and harmonic number switching in the PS or PS booster) to accelerate ions which have half the charge-over-mass ratio of protons[12]. After acceleration to 13 GeV per nucleon the ions were transferred to and stored in the ISR, and finally accelerated to 15 GeV per nucleon.

Technically, d and α beams could be injected into the SPS already today and be accelerated to the maximum energy 225 GeV per nucleon. A new source and preaccelerator are being built by the GSI Darmstadt and LBL Berkeley; this source will allow injection of ions up to ^{16}O into the PS and SPS. The scheduled installation of the ion source will take place in 1986 and CERN will then have a physics program for heavier ions. A survey of this program will be given in section 6 of these lectures.

At Brookhaven recently a program was proposed to transfer ions from the Tandem accelerator to the Alternating Gradient Synchrotron (AGS). Thus it will be possible to accelerate ions with masses as high as 32 (sulphur) in the AGS to 15 GeV per nucleon. The addition of a booster would allow the acceleration of ions up to iron. An even more ambitious project presently being discussed at BNL is to fill the empty Isabelle tunnel with a Relativistic Nuclear Collider, two rings with maximum energies in the range of 50 to 100 GeV per nucleon. However, several laboratores in the US are interested in a similar project and no decisions have been made so far.

LBL already proposed some years ago a relativistic nuclear collider - with energies higher than ISR energies - called VENUS but did not get the money to build it. This time a more modest proposal is being discussed at LBL: The "Minicollider" with energies up to 5 GeV per nucleon for ions up to uranium.

1.2. Cosmic Rays

The primary cosmic rays contain heavy nuclei of enormous energy.

Table 1. High-energy proton accelerators and colliders

Accelerator	Laboratory and location	Max. beam energy for protons a) (GeV)	pp or p$\bar{\text{p}}$ c.m. energy a) (GeV)	Accelerated particles a)
Proton Synchrotron (PS)	CERN Geneva, Switzerland	28	7.3	p, $\bar{\text{p}}$, d, α, (^{16}O*)
Intersecting Storage Rings (ISR) †	"	31	62	p, $\bar{\text{p}}$, d, α
Super Proton Synchrotron (SPS)	"	450	29	p, $\bar{\text{p}}$, (d, α, ^{16}O)*
Collider (SP$\bar{\text{P}}$S)	"		540	p, $\bar{\text{p}}$
Alternating Gradient Synchrotron (AGS)	BNL, Brookhaven USA	30	7.6	p, (..., ^{32}S)*
Relativistic Nuclear Collider **	"	~100	~200	p,...,^{238}U
Fermilab Accelerator	FNAL, Batavia, USA	500	31	p
Tevatron *	"	1000	43	p, $\bar{\text{p}}$
Tevatron Collider *	"	1000	2000	p, $\bar{\text{p}}$
Minicollider **	LBL, Berkeley, USA	~10	20	p,...,^{238}U

† Final shutdown in 1984

* Near future approved program.

** Project, presently being discussed

a) The maximum beam energy for heavy ions is one half of the maximum proton energy; therefore the c.m. energy per nucleon-nucleon collision is one half of the pp c.m. energy at colliding beam machines and $1/\sqrt{2}$ of the pp c.m. energy for fixed target machines.

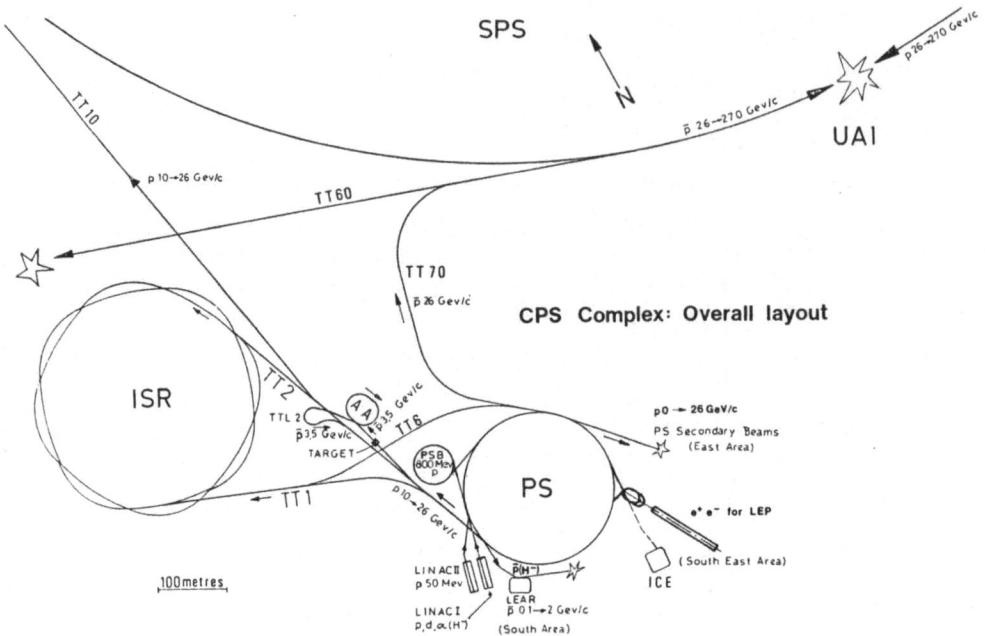

Fig.1: The CERN particle accelerator network.

Unfortunately (from the physicist's point of view) these nuclei never make it down to our towns. The atmosphere has a total thickness of 1030 g/cm^2 at sea level. The absorption length λ_{abs} (= the mean free path with respect to a strong inelastic interaction in units of [g/cm^2]) is

$$\lambda_{abs} = (\sigma_{inel} [cm^2] \ N \ [nuclei/g])^{-1} \qquad (1a)$$

for a given material with N nuclei per gram, σ_{inel} being the inelastic cross-section of the incoming particle on an average nucleus of the material. The absorption length is 90 g/cm^2 for protons with 100 GeV energy traversing air. Thus the chance that a proton penetrates down to sea level without interacting is exp(−1030/90). For nuclei the chance is correspondingly smaller.

One has to send a balloon to the stratosphere or install detectors on high mountains or satellites or (future) space platforms to measure the primary cosmic-ray spectrum and to study interactions of cosmic nuclei. Figure 2 shows the measured differential flux of primary p, e, d, He^4, C and Fe as a function of kinetic energy per nucleon[13]. Figure 3 shows the integral spectrum from 10 to 10^{11} GeV[14]. Note that the vertical axis is multiplied by E$^{1.5}$.

310

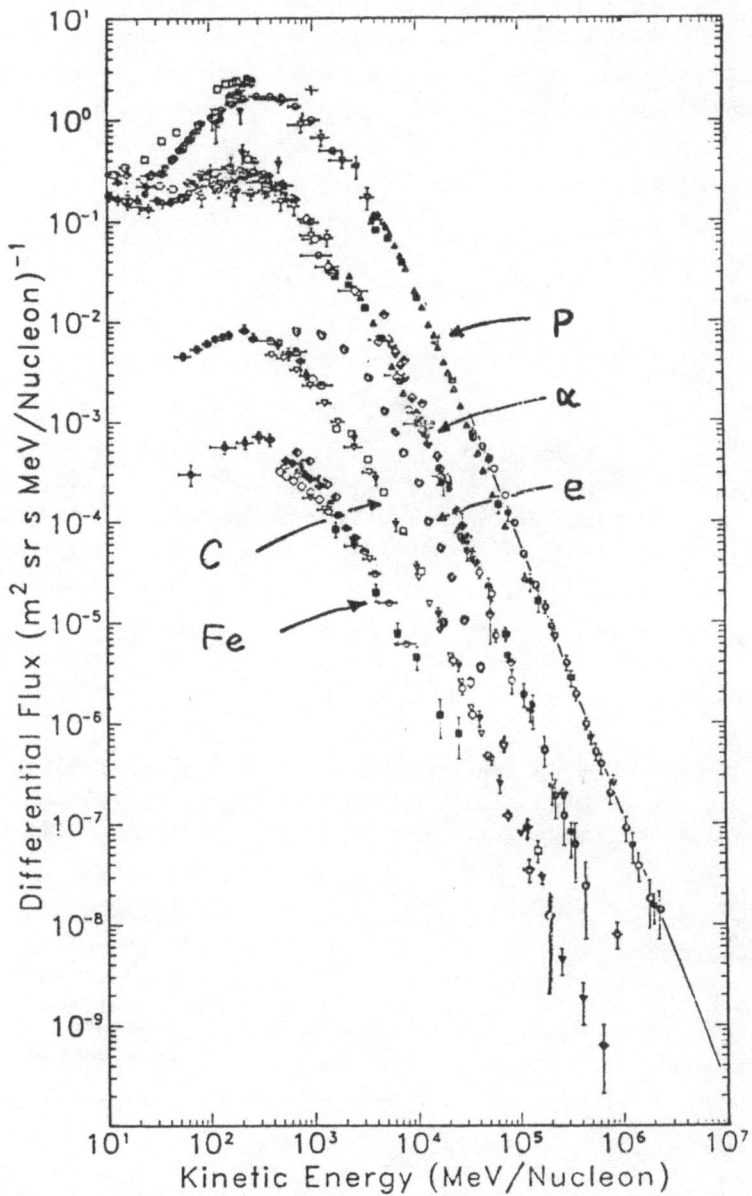

Fig.2: Differential flux of main cosmic-ray species: protons, alphas, carbon, iron and electrons as a function of kinetic energy per nucleon. This is a compilation of data from different experiments, the figure was taken from reference 13.

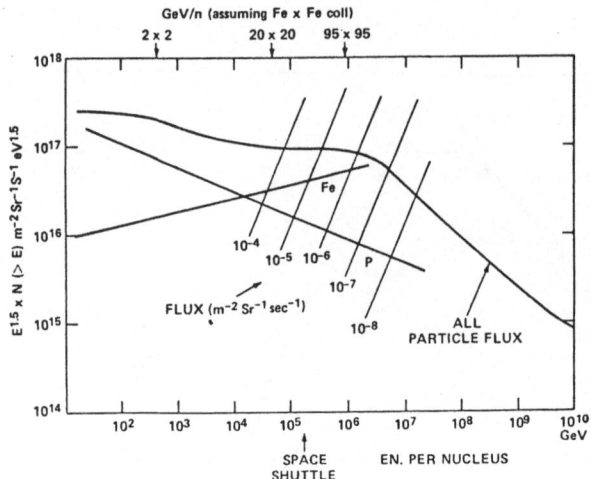

Fig.3: Integral flux of primary cosmic rays as a function of energy
(per nucleus) from 10 to 10^11 GeV. Also shown is the composition
below 10^6 GeV and lines of constant intensity for the integral
flux. Figure taken from reference 14.

The partial fluxes of protons and of iron nuclei are indicated. It
can be seen that at high energy per nucleus iron nuclei dominate.
This picture also demonstrates how rare very high-energetic cosmic
rays are.

The origin of high energy cosmic rays and the mechanisms of
their acceleration are not as well understood as man-made accelerators. The bulk is believed to originate from supernova explosions
within our galaxy. However, at least the highest energy cosmic rays
have probably an extragalactic origin; this hypothesis is supported
by their isotropy. The sun contributes only little to the total
flux observed on earth.

1.3. Typical experimental set-ups

In the following, two set-ups used for studies of nucleus-nucleus collisions will be shown, one being an example for a detector
used at an accelerator (at the ISR) and the other one being an
example for a cosmic-ray detector.

The first example is typical for detectors employed at high-energy colliders, (pp, p$\bar{\text{p}}$, e$^+$e$^-$) only with respect to its size and complexity, whereas the detailed specifications are quite unique (this example was only chosen because of the author's personal experience with the detector). The main magnet is a split-field magnet, two dipoles with opposite fields joined by a common flux return yoke, hence the name Split-Field-Magnet detector (SFM)[15] (see Fig. 4). The maximum field strength is 1 T and the airgap is 1 m high with a volume of 30 m^3. The gap is equipped with planar multiwire proportional chambers (MWPCs), arranged in such a way as to cover optimally the full solid angle. Two compensator magnets downstream of the interaction are primarily needed to correct the trajectories of the non-interacting beams but, equipped with MWPCs they also serve as part of small-angle spectrometers, in particular for elastically scattered tracks. The total number of sense wires is around 70,000. Momenta and charges of all outgoing charged particles are measured, except for angles smaller than 7 mrad with respect to the beams owing to the unavoidable presence of the beam pipes.

The second detector example is an emulsion chamber[16] carried by a baloon to the stratosphere (with atmospheric thickness of only 3-5 gm/cm^2 above) (Fig. 5). This chamber uses emulsion plates, X-ray films and plastic detectors in order to define the charge of the incident nucleus, to observe the interaction in the target layer and to measure electromagnetic energy in a calorimeter section. The calorimeter is a sandwich consisting of Pb and emulsion plates (see Fig. 5b). The whole stack of layers is 18 cm deep and 40 by 50 cm^2 wide. How does such a detector see a typical event? Assume a high-energy iron nucleus comes in from the top. It will typically interact after 3 cm. (The absorption length of protons in emulsion is $\lambda_{abs} \approx 134$ g/cm^2 = 35 cm; for an iron projectile it is approximately $\lambda_{abs}/A^{2/3} \approx 3$ cm). Hundreds of particles will be produced, the charged particles, mostly pions, ionize molecules in the emulsion plates and plastic plates and their tracks are later made visible by developing the emulsion and etching the plastic plates. The neutral pions which are produced, decay very promptly into two gammas and these initiate two electromagnetic showers. The scale for the longitudinal development of an electromagnetic shower is given by the radiation length X$_0$ which is defined similarly as λ_{abs} but with respect to the interaction of a high-energy electron:

$$\langle E \rangle = \langle E_0 \rangle \exp (X/X_0) \tag{1b}$$

where $\langle E \rangle$ is the average energy of the incoming electron after traversing a distance X. The radiation length in emulsion is X$_0$ = 11 g/cm^2 = 3 cm and in lead it is only 0.5 cm. The dominant processes in an electromagnetic shower (or cascade) are bremsstrahlung for the

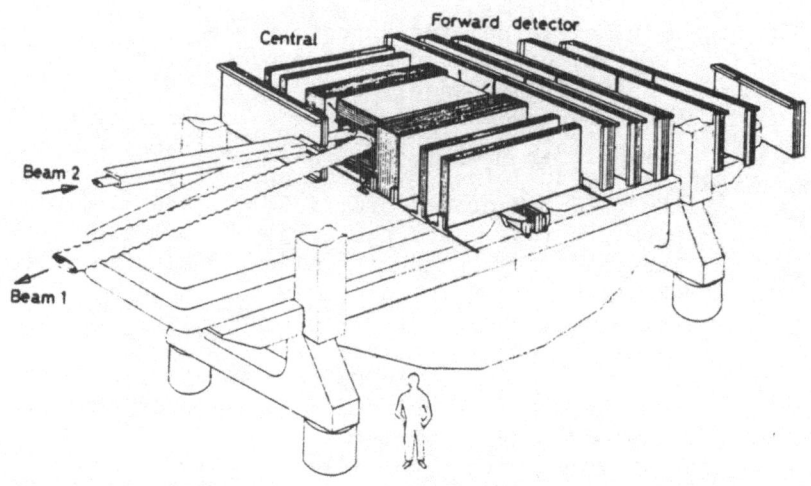

Fig.4: Split–Field–Magnet (SFM) detector at the CERN Intersecting Storage Rings (ISR)[15]. An artist's view with the upper iron yoke removed and only part of the Multiwire Proportional Chambers shown.

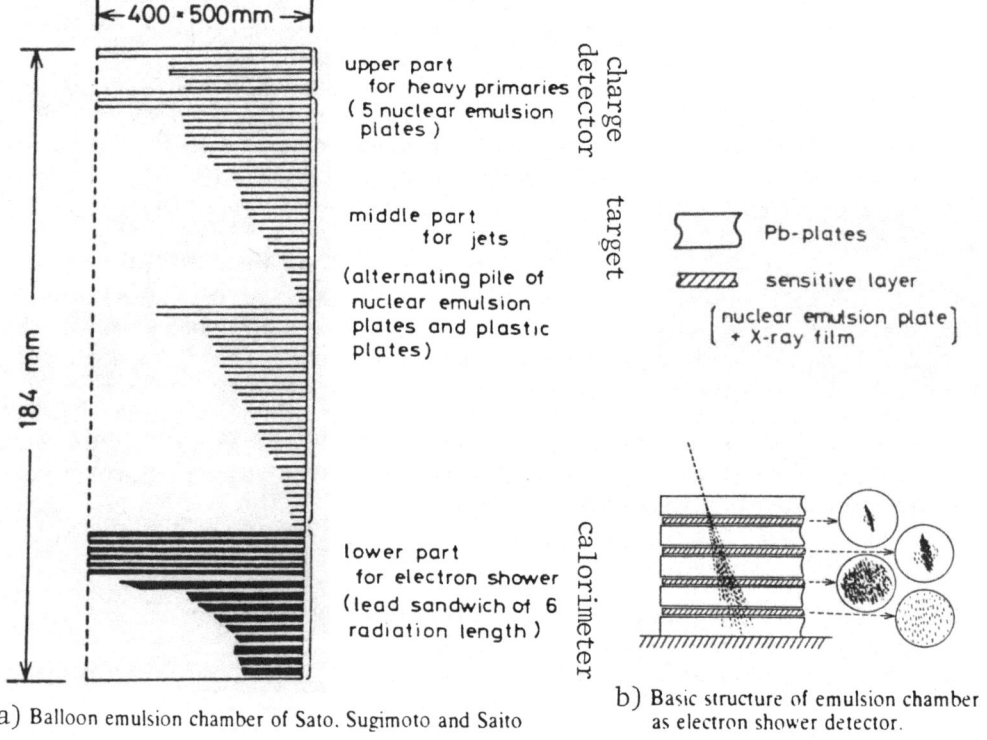

a) Balloon emulsion chamber of Sato. Sugimoto and Saito

b) Basic structure of emulsion chamber as electron shower detector.

Fig.5:   a) Emulsion chamber used in balloon experiments.
         b) Pb/emulsion–plate sandwich[16].

314

electrons and e^+e^- pair production for the photons. The calorimeter
section of the emulsion chamber, Fig. 5, is 6 X_0 deep. This is not
enough to contain the whole shower. Nevertheless the energy of the
γ can be estimated by studying the transverse developments of the
shower as a function of X. The interlaid emulsion plates record the
tracks of the generated electron shower. The combination of X-ray
film and nuclear emulsion greatly reduces the work of emulsion scann
ing with the microscope, because the X-ray films are first scanned
with the naked eye to find shower spots down to shower energies of
about 1 TeV.

In summary, what this emulsion chamber can measure, is the number and pseudo-rapidity (angular) distribution of produced particles
or of jets, the angles and energies of produced photons, and the Z of
the incoming nucleus. Whereas electronic detectors dominate today
largely at accelerators and colliding-beam machines, and are increas-
ingly employed for cosmic ray studies, the traditional visual detectors
using nuclear emulsion are still very important tools in the latter
case.

2. MULTIPLICITY OF PRODUCED PARTICLES

2.1. Multiplicity distributions

It is natural to ask as one of the first questions how many
particles are produced in an inelastic collision of nucleus A_1 with
nucleus A_2 at a given energy. This number fluctuates from event to
event so the answer will be a probability distribution: P_n is defined
as the probability of having n particles in the final state. Sometimes the topological cross-section $\sigma_n = P_n \sigma_{inel}$ is used instead of
P_n where σ_{inel} is the total inelastic cross-section.

In Fig. 6 the distribution P_{n_-} is shown for $\alpha\alpha$ and αp collisions
and is compared with the P_{n_-} distribution in pp collisions at corresponding c.m. energies per nucleon-nucleon interaction. The measurements were done with the SFM detector at the ISR[17]. Here n_- is the
number of negative particles; the number of produced charges n_{ch} is
related to n_- by charge conservation

$$n_{ch} = 2n_- + Z_1 + Z_2 \quad , \tag{2}$$

where Z_1 and Z_2 are the charges of the incoming particles or nuclei.
So the P_{n_-} distribution is also representative for the charged particles.

Firstly one notices that, going from pp to αp and $\alpha\alpha$, the
distributions get wider and the average value shifts towards higher
multiplicity, whereas the form of the distribution does not change
very much. This latter observation is confirmed by looking at a
plot (Fig. 7) of the second versus the first moment of the

315

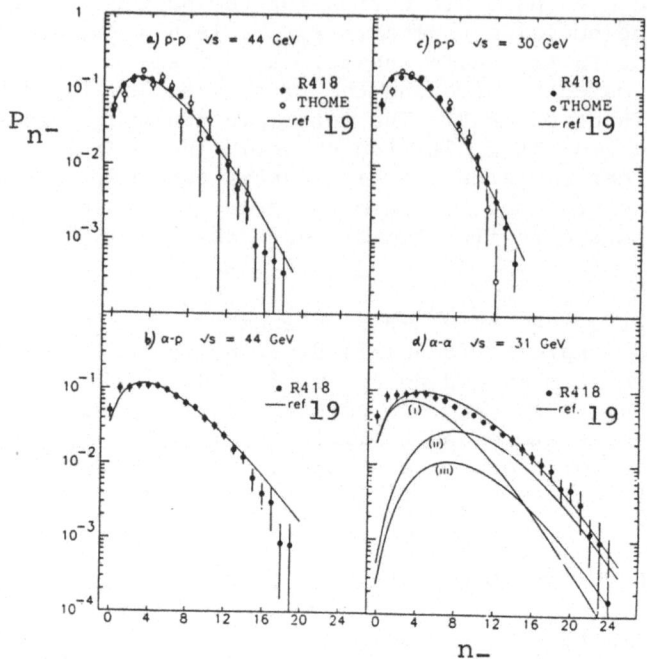

Fig.6: Multiplicity distribution of negative particles for pp, α-p and α-α interactions. Data from the SFM detector (R418)[17], and the streamer chamber[63]. Lines are calculations by Chao and Pirner[19].

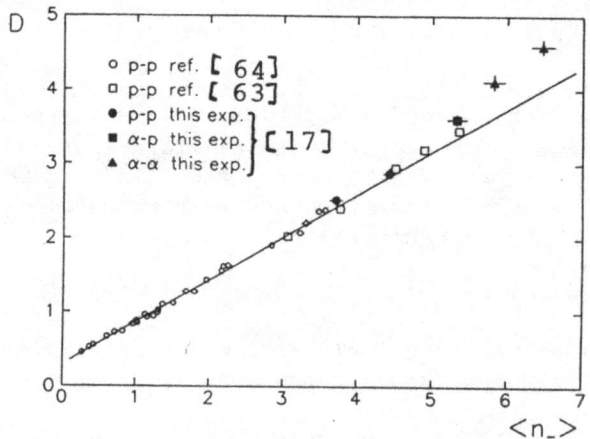

Fig.7: Dispersion $D = \sqrt{(\langle n_-^2 \rangle - \langle n_- \rangle^2)}$ of the multiplicity distributions versus average multiplicity $\langle n_- \rangle$ for pp interactions below ISR energy[64], at ISR energies[63,20] and for πS-p and α-α interactions[17]. The line indicates the original Wroblewski fit[18].

distribution, i.e. of D versus $\langle n_- \rangle$, where

$$D^2 = \langle n_-^2 \rangle - \langle n_- \rangle^2 = \Sigma \; n_-^2 \cdot P_{n_-} - \langle n_- \rangle^2$$

$$\langle n_- \rangle = \Sigma \; n_- \cdot P_{n_-} \tag{3}$$

In this plot also lower energy pp data are shown with an empirical straight line fit by Wroblewski[18]. One sees that the αp and $\alpha\alpha$ points are close to the line although not exactly on top of it.

2.2. Dependence of average multiplicity on energy, target and projectile mass.

The question is now, if we understand the rise of the average multiplicity and of the width of the distribution going from pp to αp and $\alpha\alpha$? The answer is yes; up to the precision of the data quite a quantitative explanation is provided by assuming multiple interactions inside the α particle. This is demonstrated by the curves in Fig. 6 which are calculations by Chao and Pirner[19]. At this point I will not go into the details of their model and of other models (this will be done later) but rather outline the essential experimental facts which have shaped the models. The "rules of the game", as derived from pp and pA data, are as follows:

i) The multiplicity of charged particles produced in pp interactions rises slowly with energy. The rise is proportional to a sum of logarithms of s ($\sqrt{s}$ = c.m. energy) :

$$\langle n(pp) \rangle = a + b \; \ln(s) + c \; \ln^2(s),$$

see Fig. 8[20].

ii) The increase of the multiplicity in pA interactions relative to pp interactions at the same energy depends only very weakly on the incoming energy, i.e.

$$R_A = \frac{\langle n(pA) \rangle}{\langle n(pp) \rangle} \approx \text{constant} \tag{4}$$

for incoming energies above ~20 GeV. In other words, the multiplicity for a pA interaction increases like the multiplicity for pp interactions as a function of energy with a nucleus-dependent proportionality constant R_A.

iii) There is an approximately linear dependence of R_A on $\langle \nu \rangle$, where $\langle \nu \rangle$ is defined as

$$\langle \nu \rangle = A \cdot \sigma_{pp}/\sigma_{pA} \qquad . \tag{5}$$

If one assumes that the incoming projectile propagates through the nucleus without being deflected and without changing its cross-

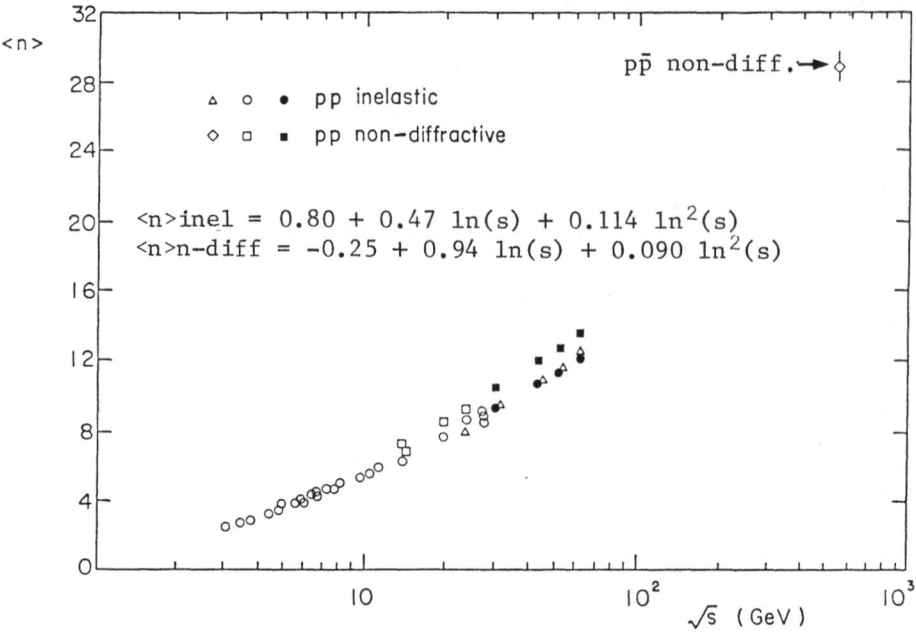

Fig.8: **Charged particle multiplicity for all inelastic and for non-diffractive pp interactions as a function of c.m. energy** $\sqrt{s}$[20].

section one can calculate the average number of collisions with the nucleons in A for every given impact parameter. Averaging over the impact parameter leads to the expression (5). In this sense $\langle\nu\rangle$ can be interpreted as the average number of nucleon-nucleon (NN) collisions for pA interactions. A good rule of thumb for the linear relation between R_A and $\langle\nu\rangle$ is

$$R_A \approx \frac{1}{2}(\langle\nu\rangle + 1).\qquad(6)$$

The properties (ii) and (iii) are illustrated in Fig. 9 (a measurement done at Fermilab[21].

According to these "rules of the game" the models factorize the multiplicity from proton-nucleus interactions into an energy-dependent part due to the yield per NN collision and a purely geometrical part due to multiple NN collisions. One should not be misled, however, to conclude that pA and AA collisions are trivial superpositions of NN collisions. Since in pA interactions the same proton interacts several times, the result can by no means be a simple superposition of several independent NN collisions. The fact that the proton interacts within very short space-time intervals renders it an interesting problem.

Just to get a feeling for the numbers, notice that the average multiplicities in pp are rather small compared to what they could be if all the incoming energy were thermalized. For a typical ISR energy, say $\sqrt{s}$ = 50 GeV, $\langle n_{ch} \rangle \approx 10$. If all energy were thermalized into pions with a Hagedorn temperature of 180 MeV, their typical energy would be 0.5 GeV in the c.m.s. and 50/0.5 = 100 of them would be created. However we know that although the transverse momentum distributions look thermal, the longitudinal momentum distributions do not - they look rather like a bremsstrahlung spectrum to first approximation. For the latter the integrated multiplicity is proportional to log(s); indeed this is the case for the pp multiplicity but only in first approximation. A little more about theoretical ideas concerning the nature of these processes will be said after having looked at the phase-space distribution of produced particles.

A straightforward extrapolation of the rule of thumb (6) to collisions of nuclei A and B is

$$ R \left(\frac{AB}{pp} \right) = \frac{\langle n(AB) \rangle}{\langle n(pp) \rangle} \approx \frac{1}{2} (\langle \nu_A \rangle + \langle \nu_B \rangle) = \frac{1}{2} P \; , \tag{7} $$

where $\langle \nu_A \rangle$ and $\langle \nu_B \rangle$ are the average numbers of interacting nucleons ("participant" nucleons as opposed to "spectator" nucleons) in nucleus A. Otterlund et al.[22] have shown that this formula describes the data very well for central nucleus-nucleus collisions. They have only considered head-on collisions of small cosmic nuclei with large target nuclei, so that all incident nucleons participate in the interactions and the number of participants in the target can be estimated fairly accurately. The five high-energy cosmic ray events are very close to the line corresponding to (7) (see Fig. 10).

Notice that according to (7) one does not gain in thermalized energy if one collides nuclei instead of protons. If all nucleons participate, there are as many pions created per incoming nucleon (or per incoming energy) as for a pp collision. For less central collisions one would even lose compared to pp. Luckily (7) is only a crude approximation and it has been shown that one can gain in thermalization using nuclei, in particular for central collisions.

3. RAPIDITY DISTRIBUTIONS OF FINAL STATE PARTICLES

3.1. Definition of variables

Turning to the phase-space distribution of the particles in the final state after a nucleus-nucleus interaction I will first introduce the kinematic variables.

Consider pp collisions in the c.m. system; let the incoming momenta be $\pm p_{in}$ along the z-axis (=beam axis). The outgoing particles have momentum components parallel to the z-axis (parallel or

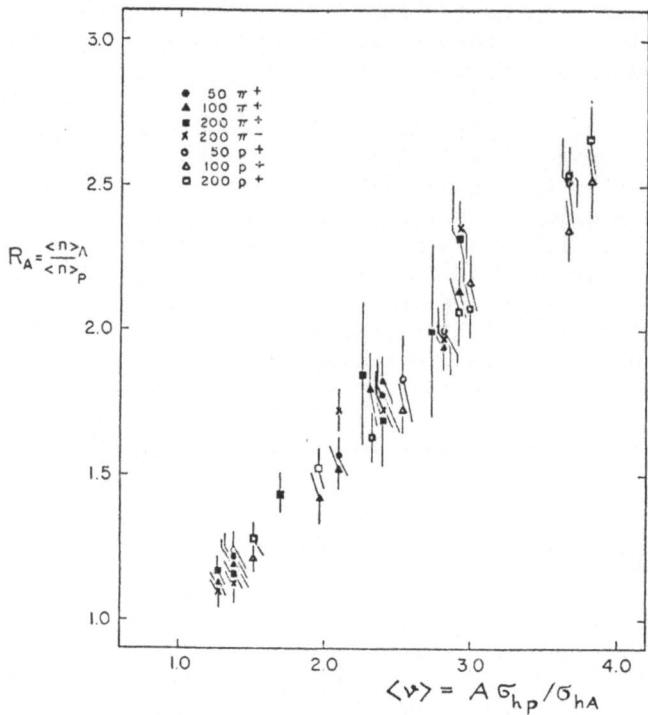

Fig.9: Ratio of average charged multiplicity in hadron-nucleus (hA) to that in hadron-proton (hp) interactions for various incoming energies, as a function of ⟨ν⟩, the average number of collisions[21].

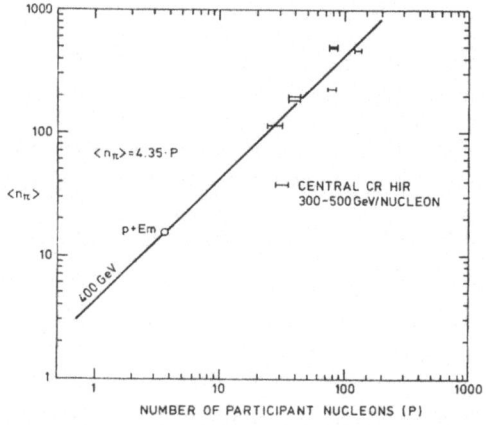

Fig.10: Multiplicities of particles produced in central cosmic nucleus interactions as a function of the number of participant nucleons[22]. The line corresponds to 0.5⟨n(pp)⟩P.

"longitudinal" momenta p_L) which range from $-p_{in}$ to $+p_{in}$. Their momenta orthogonal to the beam (transverse momenta p_T) are in general very small; the average $<p_T>$ is around 300 MeV/c, only weakly dependent on p_{in}, and the p_T distribution falls rapidly for increasing p_T - we will see this in the next section.

If one plots the distribution of particles as a function of p_L or of the Feynman variable $x_F = p_L/p_{in}$, then the shape of the distribution is not Lorentz-invariant, i.e. if we transform it for instance to a reference system where one proton is at rest, the distribution would look quite different.

Therefore, one often prefers another longitudinal phase-space variable, the rapidity, which is a function of the longitudinal velocity β_L:

$$y = \frac{1}{2} \ln \left(\frac{E + p_L}{E - p_L}\right) = \ln \left(\frac{E + p_L}{m_T}\right) = \mathrm{tgh}^{-1} (\beta_L) \qquad (8)$$

with $m_T = \sqrt{(p_T^2 + m_0^2)}$ = "transverse mass". One can easily see that under a special Lorentz transformation along the z axis the rapidity is only shifted by a constant:

$$\begin{pmatrix} p_I \\ E \end{pmatrix} = \begin{pmatrix} \gamma_0 & \eta_0 \\ \gamma_0 & \eta_0 \end{pmatrix} \begin{pmatrix} p_L \\ E \end{pmatrix} \qquad (9)$$

where $\gamma_0 = 1/\sqrt{(1-\beta_0^2)}$, $\eta_0 = \gamma_0 \cdot \beta_0$ and β_0 is the boost velocity. Hence

$$y' = \frac{1}{2} \ln \frac{(E+p_L)(\gamma_0 + \eta_0)}{(E-p_L)(\gamma_0 - \eta_0)} = y + \frac{1}{2} \ln \frac{\gamma_0 + \eta_0}{\gamma_0 - \eta_0} = y + y_0. \qquad (10)$$

The rapidity variable requires a measurement of particle mass and momentum (or of the velocity). There are only a few experiments at high energy which determine the momentum and mass or velocity over a large range in momentum. Lacking the capability to measure mass and velocity one often uses the pseudorapidity, which is defined as

$$\eta = -\ln \mathrm{tg} (\theta/2), \qquad (11)$$

where θ is the angle between the incoming beam and the produced particle. Since

$$\mathrm{tg} (\theta/2) = \sqrt{\frac{p+p_L}{p-p_L}} = \frac{p+p_L}{p_T}, \quad p = \sqrt{(p_T^2 + p_L^2)} . \qquad (12)$$

one sees that rapidity and pseudorapidity are equivalent if the rest mass is negligible, i.e., if

$$m_T \approx p_T \text{ and } E \approx p.$$

Since I mentioned that one reason for using the rapidity was to obtain Lorentz-invariant differential distributions, I should add for completeness: Lorentz-invariant infinitesimal phase-space volumes are

$$\frac{dp^3}{E} = \frac{dp_L \cdot p_T \cdot dp_T \cdot d\phi}{E} = dy \cdot p_T \cdot dp_T \cdot d\phi , \qquad (13)$$

where ϕ is the azimuthal angle ranging from 0 to 2π; here we have first used cylindrical coordinates and then the equality $dy = dp_L/E$, which can be easily derived from the definition of y. In high energy physics one usually shows Lorentz-invariant differential cross-sections, i.e. $E \, d^3 \sigma/dp^3$.

3.2. Rapidity distributions in p-nucleus interactions

Let us start this time with a survey of p-nucleus data so that we can make use of the conclusions from these data to understand better the nucleus-nucleus data. From the comparison of pp and pA data we can learn about the nature of multiple collisions of one proton with several nucleons within the nucleus; these collisions occur within very short space-time intervals Therefore it is often said that they tell us something about the space-time evolution of the strong interaction.

Figure 11 illustrates the change of the pseudorapidity distribution $(\sigma)^{-1} d\sigma/d = dN/d\eta$ of charged particles as a function of the target mass and incoming energy[23]. While for increasing A the particle density increases at lower η, for increasing energy it increases at larger η.

It is instructive to consider rapidity distributions which are normalized by rapidity distributions for pp interactions. Figure 12 shows such ratios

$$R_-(y) = \frac{(dN_-/dy)_{pXe}}{(dN_-/dy)_{pp}} \qquad \bar{R}_-(y) = \frac{2(dN_-/dy)_{pXe}}{(dN/dy)_{pp}} . \qquad (14)$$

This ratio is smaller than 1 only at the highest y (most likely a consequence of energy conservation). Going to lower y it increases, reaches a kind of plateau and then rises again. One would like to see how this ratio changes as a function of the incoming energy. Unfortunately this experiment[24] was done only at one energy (200 GeV). But the energy and projectile dependence has been measured by a different experiment. Figure 13 shows ratios of pseudorapidity distributions for $\langle \nu \rangle = 3$, normalized by those for $\langle \nu \rangle = 1$. These rapidity distributions were obtained by interpolations and extrapolations based

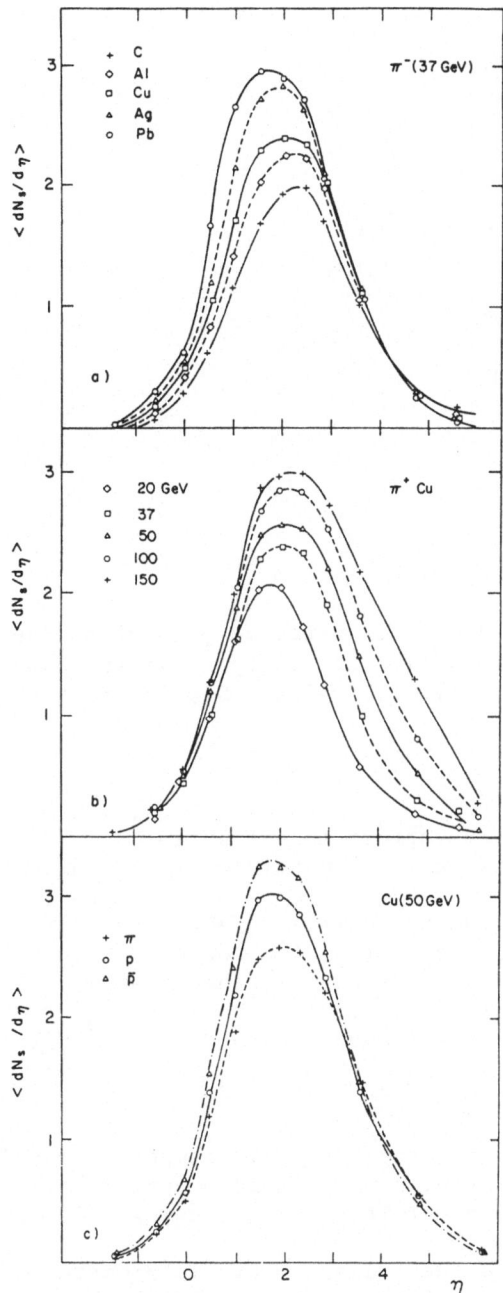

Fig.11: Pseudorapidity distributions of charged particles with
v/c > 0.7 for: a) different target nuclei, π^- at 40 GeV/c
b) different energies, π^- Cu   c) different projectiles (50 GeV/c)
on Cu[23,25].   Lines are drawn to guide the eye.

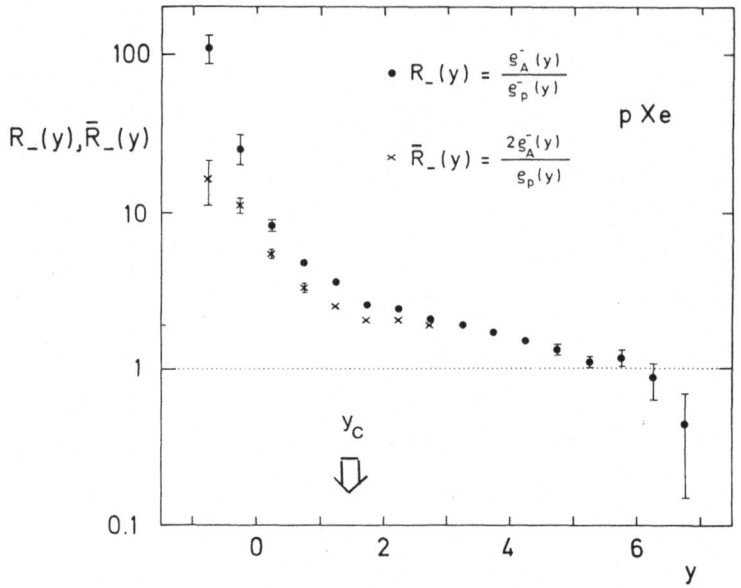

Fig.12: The ratio R_(y) for negative particles, see Eq.(14) (dots).
Also shown (crosses) is R̄_(y) with the denominator
symmetrized for y < 3, i.e. using $0.5(\rho_- + \rho_+) = 0.5\rho$[24].

on measurements with various target nuclei[23,25]. One notices a very
weak energy and projectile dependence of the ratio.

In Figure 14 the ratio R(y) is shown for a fixed rapidity, name-
ly $y_{cm} = 0$, versus $\langle \nu \rangle$[26] ($y_{cm} = 0$ corresponds to half the rapidity
of the beam in a fixed target reference system). The ratio rises as
a function of $\langle \nu \rangle$ in a way consistent with some model predictions and
inconsistent with others shown in the figure[27-32].

3.3. <u>Models and comparison</u>

In the following, the presently competing models to explain particle production in pp, pA and AA interactions are introduced on a
very qualitative and intuitive level, using "QCD jargon".

First, imagine the following simple process: Try to remove a
quark from a nucleon, i.e. a bag containing three valence quarks
(Fig. 15a). If you pull strongly and slowly enough the quark will
come out but it will pick up a sea quark from the nucleon which mat-
ches its colour, so at the end there will be two colour singlets in
the final state, a meson and a baryon. Next, do the same but kick the
quark very hard or pull fast, like it is done in deep inelastic

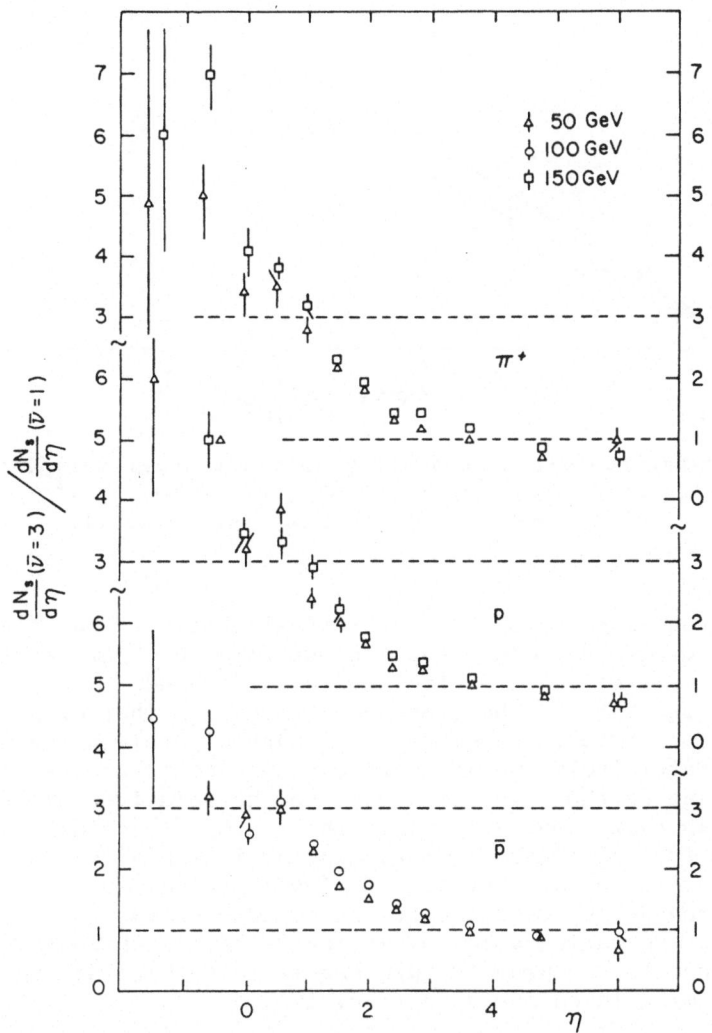

Fig.13: Normalized pseudorapidity distributions (for ⟨ ν ⟩ = 3
divided by those for ⟨ ν ⟩ = 1) for   a) incoming pions,
b) protons,   c) antiprotons[23,25].

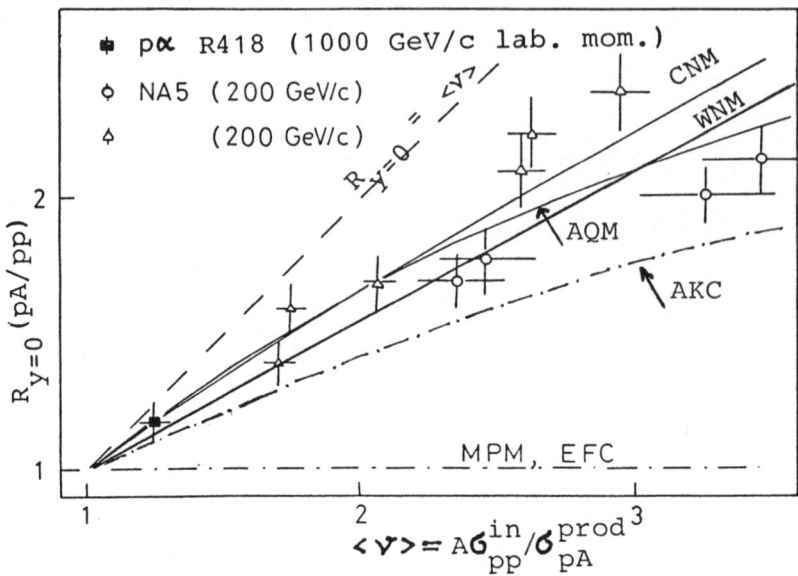

Fig.14: Normalized particle density in the central rapidity region as a function of $\langle \nu \rangle$[26] Theoretical predictions: MPM[27], AQM[28], CNM[30], EFC[31], WNM[32], AKC [Kerman, this school].

lepton-nucleon interactions by the virtual photon. Then the colour forces will create a field - the "colour flux tube" or "string" which has properties similar to a stretched elastic band. When the field energy is large enough, the tube decays into a number of particles which are distributed, on the average, approximately homogeneous in rapidity. Intuitively one may come to this result because the forces at every point of the band are equal, so the relative momentum between two particles (neighboured in y) in their c.m.s. is independent of y, and this is equivalent to equal spacing in rapidity. Next, we come to the soft pp interaction. A simple process would be that a gluon is exchanged between the two incoming p or two quarks out of them, and in separating a colour string is formed which then decays. The difference to the previous process is that there are two additional quarks coming in on one side (Fig. 16).

What will happen if an incoming nucleon encounters subsequently several target nucleons - as is the case in a pA interaction? Will the field between the projectile nucleon and the target nucleus i) be essentially the same as for a pp interaction, i.e. one colour string be created or ii) will several quarks of the incoming nucleon exchange gluons with different target nucleons thus creating several colour strings or iii) will colour pile up on both sides - the target and projectile - in a random way such that after ν interactions the colour charge is $\sqrt{\nu}$? Prediction iii) was made by A. Kerman at this summer institute. Prediction i) is essentially equivalent to the one of the Energy Flux Cascade (EFC) model by Gottfried[31] or of the Multiperipheral Parton model (MPM) by Kancheli and Bertocchi[27]. There are several predictions of class ii) which differ with respect to the possible number of colour strings: In the Additive Quark model (AQM)[28] several colour strings are created, but their maximum number is limited by the number of valence quarks of the incoming hadron (3 for a nucleon, 2 for a meson), whereas in the Dual Parton Model (DPM)[29] and in the Colour Neutralization model (CNM)[30] the effective number of colour strings is only limited by the available energy, since also sea quarks of the incoming hadron can exchange colour.

How can one test these model predictions, i.e. measure the number of colour strings? The best approach is to study the particle density in the central region, at $y_{cm} = 0$, normalized to the one-string particle density (as obtained in a pp collision). This was done in Fig. 14. The reason is that at the rapidity edges there are complications (which at the given highest energies of present accelerator data might still extend to the central region!) Firstly, energy conservation shapes the edges - this is one of the reasons for the depletion of particle density (ratio below 1) for large positive y or η in Fig. 12 and 13; how this depletion occurs quantitatively is model dependent and, in fact, most models neglect energy conservation. Secondly, cascading of produced secondaries within the nucleus - as well neglected in several models, is responsible for the steep rise at low η. This kind of cascading is limited by the "formation time"[33]. Secondaries have first to be formed before they can interact and this takes some time. This time is Lorentz-dilated in the laboratory

$$\tau = \tau_o \cdot E/m_T = \tau_o \cosh (y) \qquad , \qquad (15)$$

where τ_o is a characteristic formation time in the rest system of the secondary (one assumes that τ_o is of the order of 1 fm/c). If τ is larger than the linear dimension of the nucleus, the secondary cannot interact (it does not exist yet). Thus above a certain critical energy E_c, which is given by the requirement

$$\tau_c = \tau_o \cdot E_c/m_T = r_o A^{1/3} \qquad (16)$$

there is no cascade of secondaries. The value one obtains for the corresponding critical rapidity $y_c = \cosh^{-1} (E_c/m_T)$ is around 1 to 2 (using $m_T = 400$ MeV) in the fixed target reference system; note that y_c is independent of the incoming energy.

The formation time constraint has been deduced and explained in the Multiperipheral Parton Model (MPM) by Kancheli and Bertocchi[27]. However, in this model, above the critical rapidity there is no cascading and the ratio R(y) should be equal to 1. The fact that the ratio $R(y_{cm} = 0)$ is larger than 1 (see Fig. 14), even at the highest energy (αp interactions) is probably an indication that particle multiplication is possible even above y_c or E_c. However, it has to be said that the value of y_c is not well determined and thus it cannot be fully guaranteed that $y_{cm} = 0$ is above y_c at the available incoming energies (1000 GeV/c for αp collisions at the ISR corresponding to a rapidity span of ln (2000) = 7.6). If we could compare pA with pp interactions for a rapidity span of, say 10 units, corresponding to lab energies of about 10 TeV, and for a very heavy nucleus A (at least uranium!) the situation would presumably be clarified immediately. (One would perhaps expect to see a ratio similar in shape to the one in Fig. 12 with a steep increase at low y and an extended plateau-like central region). The increase of particle densities in pA compared to pp collisions at y above the critical rapidity y_c has to be of a different nature than the one below y_c; it leads to the conclusion that more than one colour string is created by constituent interactions. These constituents (the quarks) exist already before the interaction so their interaction is not limited by a formation time constraint. The additivity of independent constituent interactions is indeed incorporated in all particle production models presently "on the market".

By means of present data one cannot decide between the potential number of interacting constituents (see Fig. 14) but the data suggest that more colour strings between the projectile-proton and the target are formed in pA than in pp interactions and the weak energy dependence of the ratio R(y) at small y clearly favors a model which incorporates a formation time.

I have tried to describe very briefly and qualitatively the level of understanding of the basic particle production mechanism in pp and pA interactions. A more thorough discussion of the models and their comparison with existing data is clearly outside the scope of this lecture.

3.4. Rapidity distributions after nucleus-nucleus interactions

In nucleus-nucleus collisions two new kinds of multiple interactions occur in addition to the pp and pA type interactions (see Fig. 17).

i) Parallel interactions of nucleon-pairs, which may or may not communicate with each other - in the second case they are trivial superpositions of NN interactions, in the first case, of course not (Fig. 17b).
ii) "High density" interactions where two rows of nucleons collide (Fig. 17c).

In terms of the quoted constituent models these new multiple interactions are no new physics - they lead to a predictable increase of the number of colour strings. In fact, the models do quite well in predicting the average rapidity distributions in $\alpha\alpha$ and αp

328

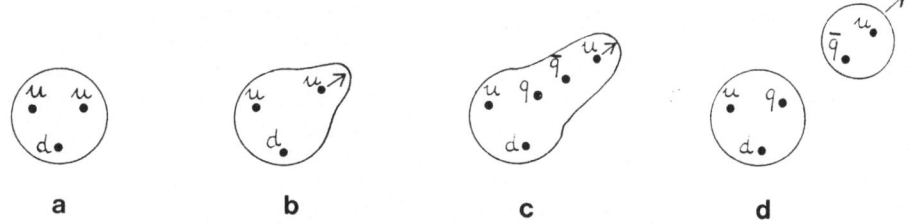

Fig.15:   a) A bag containing 3 valence quarks   [u = up, d = down quark; the bag shown represents a proton p = (u,u,d)].
b) Trying to remove a quark from its bag.
c) A sea-quark pair (q,q) appears; (q,q) can be (u,u), (d,d), (s,s) [s = strange quark] or a pair of any other existing quark flavour.
d) A new quark bag has been created; the final state consists of a baryon and a meson (q1, q2).

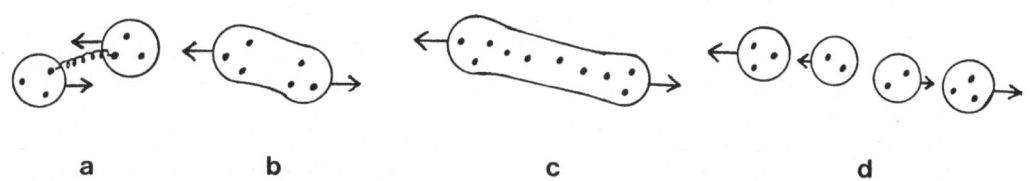

Fig.16: Sequence of processes leading to multiparticle production in hadron—hadron collisions.
a) Two incoming hadrons (colour—singlet bags) exchange colour.
b) and c) As the quarks move on, the new bag is stretched.
d) The colour flux tube decays into colour-singlet hadrons.

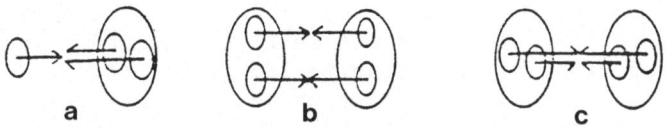

Fig.17:   Different types of multiple NN collisions in nucleus—nucleus collisions.   a) hadron—nucleus type,
b) parallel interaction, c) row-on-row collision.

collisions (see Figs. 18) which show αα, αp and pp data together with a calculation by Capella et al.[35] in the framework of the DPM. These are rapidity distributions for negative particles produced in the interaction; since negative particles are known from other experiments to be over 80% π^- one can assume a pion mass for all negative tracks and calculate the rapidity from the measured momenta.

Next we make a big jump in energy, projectile mass and target mass, and consider the pseudorapidity distributions of the two most spectacular cosmic nucleus interactions (found by the JACEE Collaboration[36]): a Si-Ar and a Ca-C interaction at estimated incoming energies of a 4 TeV/nucleon and 100 TeV/nucleon (see Fig. 19). The charged multiplicities are 1020 for the first and 760 for the second event. These rapidity distributions can be understood in the framework of the multichain model (MCM)[37] (similar to the DPM or CNM) as very head-on collisions, but Otterlund et al.[22] point out that one needs to assume an almost unbelievable amount of target participant nucleons in the first case: all nucleons in Ag must interact. Gyulassy[38] finds that the energy density in the central rapidity region and in the target fragmentation region exceeds the energy density needed for the deconfinement phase transition and he raises the question about the apparent fluctuations of the rapidity distribution. Are these events "plasmons"[5]; are we already seeing the transition to a quark-gluon phase in the large fluctuations of dN/dy? No one has performed a simple statistical (e.g. χ^2) analysis of the rapidity distributions to test whether the fluctuations of the bin contents exhibit non-statistical deviations from a smooth distribution[39] (notice the bin size is 0.1 units, so the content is 0.1 dn/dη) so one should doubt the perhaps premature conclusions about the fluctuations. A somewhat complicated power spectrum analysis of the η distributions was performed by Takagi[40]; he concludes that in the Si+AgBr event there are likely some non-statistical fluctuations which cannot be explained by the known short-range correlations whereas in the Ca+C event the fluctuations are probably more statistical. But something else seems unusual in these events, namely their p_T distribution. We will come back to this in the section about p_T distributions.

3.5. Leading protons and the stopping power of nuclear matter

The rapidity distribution in Fig. 18 was for negative tracks only and we assumed they were all π^-. It would be interesting to know more about the distribution of other particles such as kaons, protons or antiprotons. There is not much one can learn about K's and p's if an experiment is not able to identify particles at least for a number of rapidity points. But by means of a simple trick the proton rapidity distribution in αα collisions can be obtained. One can safely assume that the yields of positive and negative π must be equal in αα collisions at every rapidity because of isospin symmetry: α's contain the same number of neutrons as protons and

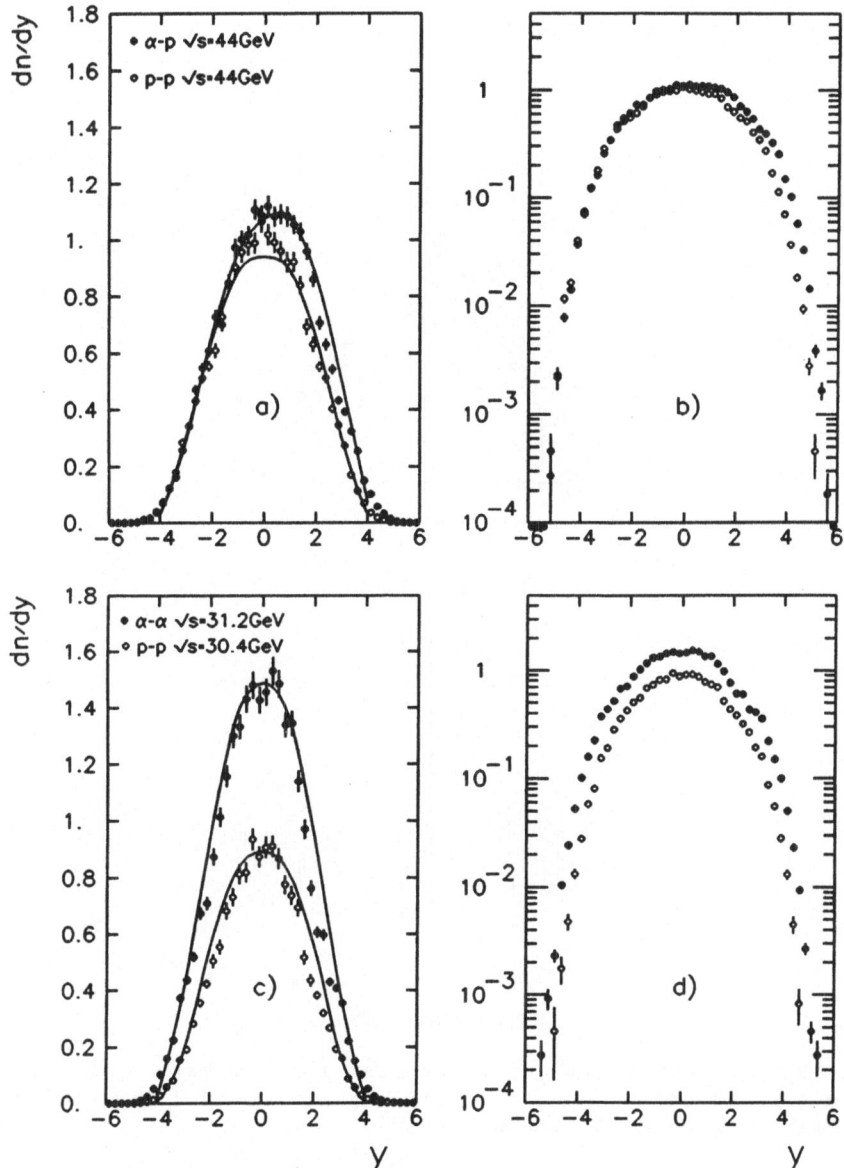

Fig.18: Rapidity distributions of negative hadrons produced in
α–p, α–α and pp interactions[34] The solid lines are
calculations by Capella et al.[35].

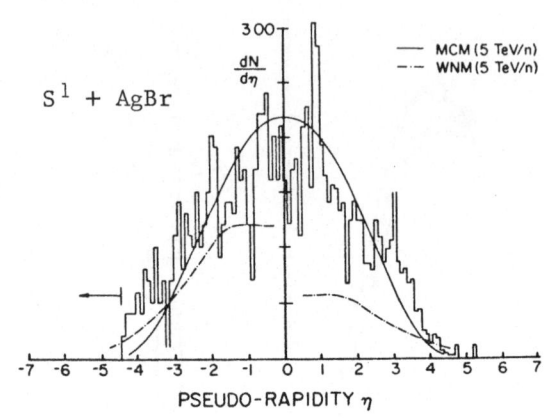

Fig. 6

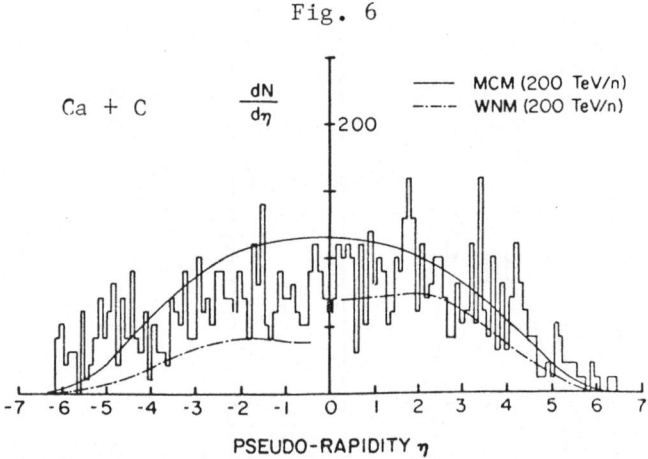

Fig.19: Pseudorapidity distributions for two very high energy cosmic-ray events: a)Si+Ag at around 4 TeV/nucleon; b) Ca+C at around 100 TeV/nucleon[36].

$$\frac{d^3\sigma}{dp^3} (p \to \pi^{\pm}) = \frac{d^3}{dp^3} (n \to \pi^{\mp}) \quad . \tag{17}$$

For K^- and K^+ there could be a difference but since both yields are small the difference is assumed to be negligibly small . So by forming the difference of positive and negative tracks in $\alpha\alpha$ collisions one can obtain the distribution of leading protons (protons created in pairs with antiprotons will cancel as well in the difference):

$$dN_{1p}/dy = dN_+/dy - dN_-/dy \quad . \tag{18}$$

The result is shown in Fig. 20[34]. One recognizes two peaks of "spectator" or "quasi-diffractive" protons around the beam rapidities (± 3.5). (Spectator protons are those which did not directly interact; quasi-diffractive protons are those bound protons which scattered elastically or inelastic-diffractively on nucleons of the other α particle). Towards lower absolute rapidities there are shoulders, about 2 rapidity units wide, which have to be attributed to interacting protons. Their distribution in rapidity ($y = \mathrm{tgh}^{-1}(\beta_L)$, remember) measures directly how much the protons have been slowed down; it measures the inelasticity, or the "stopping power" of an α particle for nucleons out of the other α particle. Of course, the same information can be obtained as well if one sums for each event all the energy carried away by the produced particles, since this is where the lost energy goes.

The question of inelasticity or stopping power is an interesting one and is presently hotly debated because it is of relevance to the search for the quark-gluon plasma. Whether nuclei are transparent to each other at high energy (as it was believed for many years because the ratio R(y) is close to one at high y), or whether they are more opaque (i.e. have larger stopping power), and how much the opaqueness changes with incoming energy is important to know before the design of the planned relativistic nuclear collider. A large stopping power means that one can more easily reach high energy and baryon densities in the nucleus-nucleus collisions; thus it is easier to reach the condition favourable for the deconfinement than if the nuclei were transparent.

The next three figures show data from pp, pA and $\alpha\alpha$ collisions which are interesting in this context. In the first, Fig. 21, the inclusive cross section pp $\to$ p (antiprotons subtracted) is shown as a function of $x_F = 2p_L/\sqrt{s}$ for two values of p_T, at ISR energies[41]. Note that this is not an invariant cross-section. Part of the diffraction peak at $x_F = 1$ is seen and the distribution is rather flat down to the lowest x_F. The density of protons which are stopped in the c.m. system ($x_F = 0$) is about the same as the density of protons which lost half their momentum ($x_F = 0.5$).

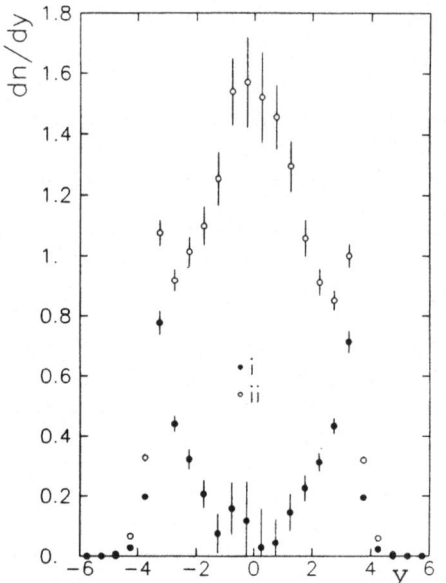

Fig.20: Rapidity distribution
of leading protons dN(1p)/dy =
$dN_+/dy - dN_-/dy$ and of positive particles dN_+/dy in α-α
interactions[34].

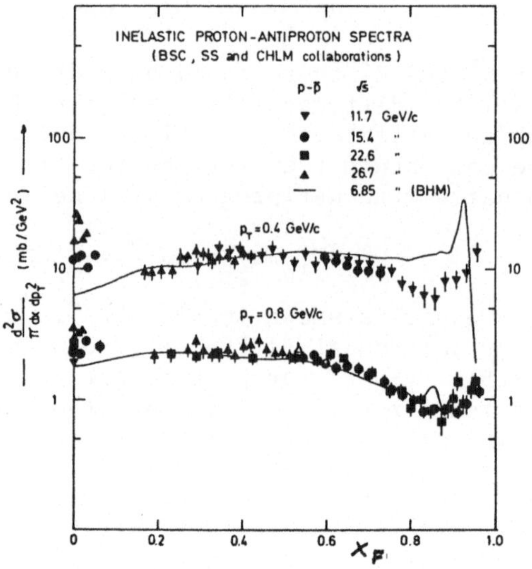

Fig.21: Inclusive differential
cross-section for pp → (p-p̄) + X
versus $x_F=2p_L/\sqrt{s}$ for 5 energies
and two values of p_T[41].

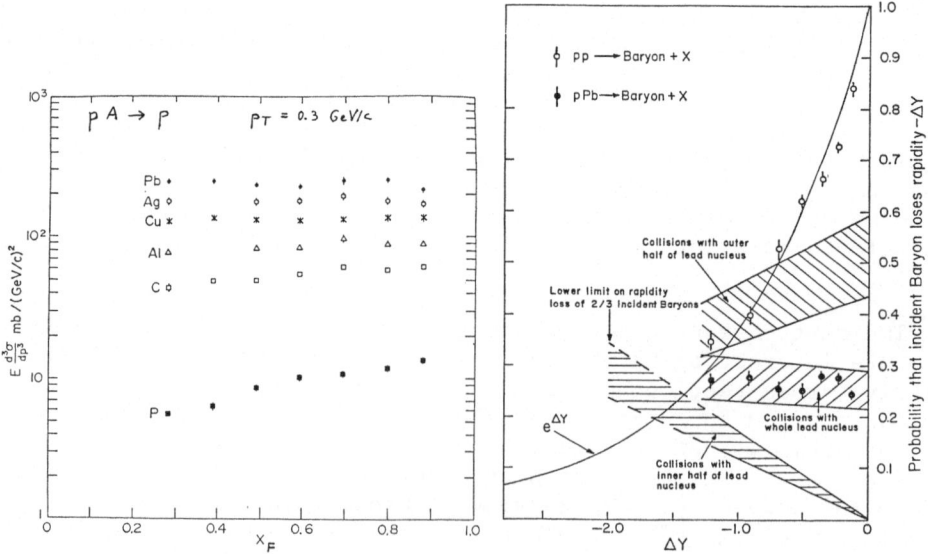

Fig.22: a) Inclusive invariant cross-section pA → pX versus x_F.
b) $dN/dy = \sigma^{-1}d\sigma/dy$ versus Δy for pp and and pPb → p.
Fixed p_T = 0.3 GeV/c[42].

Figure 22A shows the invariant cross-section as a function of x_F (this cross-section contains a factor $2E/\sqrt{s}$ compared to the one in the previous figure

$$\frac{E d^3\sigma}{dp^3} = \frac{E \, d^2\sigma}{2\pi \, d(x_F \cdot \sqrt{s}/2) \, dp_T^2/2} = \frac{2E}{\sqrt{s}} \left(\frac{d^2\sigma}{\pi \, dx_F dp_T^2}\right) \quad , \qquad (19)$$

therefore it decreases for pp interactions at the lowest x_F)[42].

One sees that the weight of the distributions moves towards lower x_F for increasing target mass. This change appears more dramatic if one plots the same cross-section but divided by the total cross-section in order to obtain the distribution normalized to one event) as a function of $\Delta y = y - y_{beam}$ (Fig. 22b). Protons are slowed down much more effectively in a pPb collision than in a pp collision. The authors[42] have estimated, based on the observed A dependence, what the distribution would look like for a central (head-on, impact parameter = 0) collision and infer that the nuclear stopping power is really very large; the average y-shift of the incoming proton is -2.4 for a central collision on lead.

It would be desirable to see the same distribution if one integrates over p_T instead of measuring at a single p_T value. If the p_T distribution at fixed x_F are different for pp and pPb collisions, then the interpretation of this result as a measurement of the stopping power is not so clear.

How does the x_F or p_L distribution of protons change if one increases the total multiplicity of particles produced in the event? A priori one expects two features:

i) A decrease in the number of protons in the spectator or quasi-diffractive peak and a corresponding increase in the number of leading protons outside the peak -- more protons undergo inelastic interactions and cause an increase in the multiplicity.
ii) A shift of the leading protons to lower p_L -- higher inelasticity of the individual proton interaction will also cause increasing multiplicity.

The second expectation is clearly satisfied in Fig. 23, which shows the proton distribution in $\alpha\alpha$ collisions (cf. Fig. 20) as a function of p_L (instead of y) and for different multiplicity bins (instead of averaged over multiplicity)[34]. However, the decrease of spectator protons starts only at the highest multiplicity. At lower multiplicity this peak rather increases. Apparently, in peripheral (low multiplicity) interactions, protons are frequently carried away by heavier fragments (d, t, ^{3}He) and thus do not show up in the p-spectator peak.

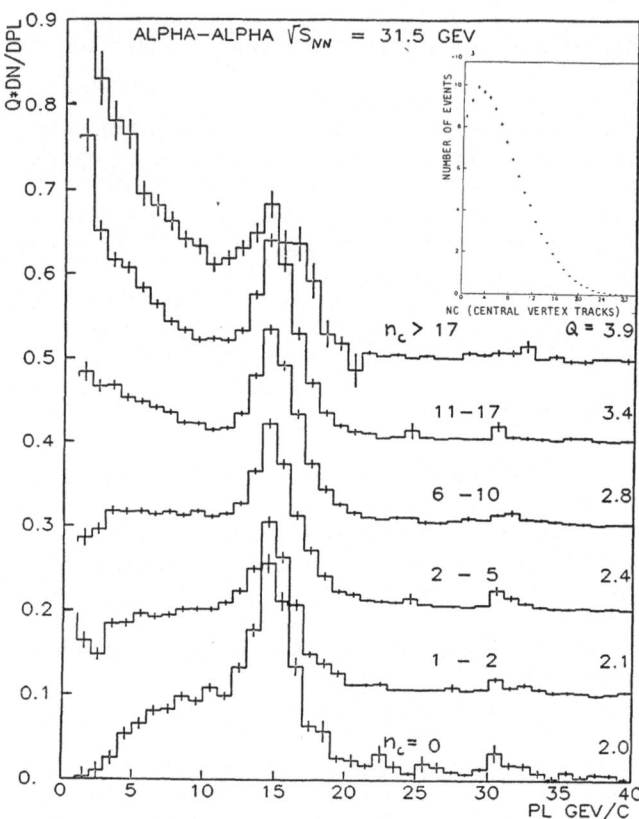

Fig.23: The p_L distribution of leading protons $dN(1p)/dp_L$ in α-α interactions for various multiplicity bins[34]. The distributions are shifted up by units of 0.1 for increasing multiplicity. The inset shows the frequency of events as a function of the multiplicity in the central rapidity region $|y| < 2$.

4. TRANSVERSE ENERGY AND MOMENTUM

4.1. Transverse Energy Distribution

Total transverse energy triggers using large solid angle calorimeters have been applied in several recent experiments at Fermilab, SPS, ISR, and the CERN Sp$\bar{\text{p}}$S Collider in order to observe unbiased jets from hard parton scattering events. Only at the highest ISR energy ($\sqrt{s}$ = 63 GeV) and at the p$\bar{\text{p}}$ Collider ($\sqrt{s}$ = 540 GeV) these triggers were a clear success. At lower energy they failed to isolate the wanted events with the properties of hard parton scattering; these events are buried under a background of events with a large number of tracks nearly isotropic in azimuthal angle. However, with respect to the search for the QGP these background events are interesting because they are candidates for high local energy densities or thermalized energy. Here one is tempted to partly reverse the well known saying "Yesterday's discovery is today's calibration and tomorrow's background" into "Yesterday's background is today's discovery".

In the last light-ion run, R110[43] has measured the total energy (E_{tot}^O) of neutral pions in the central rapidity region ($|y|<1.1$ for most of the azimuthal angle.) Figure 24 shows the distributions for pp, dd, and $\alpha\alpha$ interactions, all at $\sqrt{s_{NN}}$ = 31 GeV). The cross-section falls by 10 orders of magnitude over the measured range in E_{tot}^O for $\alpha\alpha$ interactions! The integrated luminosities which led to these remarkable plots were 5.3 x 10^{35} cm^{-2}, 4.2 x 10^{35} cm^{-2}, and 1 x 10^{35} cm^{-2} for pp, dd, and $\alpha\alpha$ respectively.

The curves shown in Fig. 24 correspond to one-, two-, 4- and 5-fold convolutions of the E_{tot}^O distribution for pp collisions. It can be seen that the one for dd collisions - only the tail of which has been measured - has the same slope as the two-fold convolution. However, the slope of the E_{tot}^O distribution for $\alpha\alpha$ collisions at E_{tot}^O larger than 17 GeV is flatter than the four-fold convolution which corresponds to the superposition of four independent pp collisions. Since four independent nucleon-pair collisions is the maximum possible number of multiplicity-producing collisions in the so-called "wounded nucleon model", this observation leads to the conclusion that the wounded nucleon model fails at very large E_{tot}^O. In the framework of "wounded quark" models (like the AQM), where more than one constituent of each nucleon can produce multiplicity in the central region, the flatter slope at high E_{tot}^O could presumably be accommodated. For instance, in the AQM the maximum number of convolutions which has to be included in a multiple-inelastic interaction expansion would be 12 (for 3 x 4 independent quark-quark interactions).

It should be mentioned that the transverse energy E_T distribution of charged particles has been measured previously (in 1980) in the AFS[44] for $\alpha\alpha$ and αp collisions. The conclusions then were that these distributions can be quantitatively reproduced by a convolution of

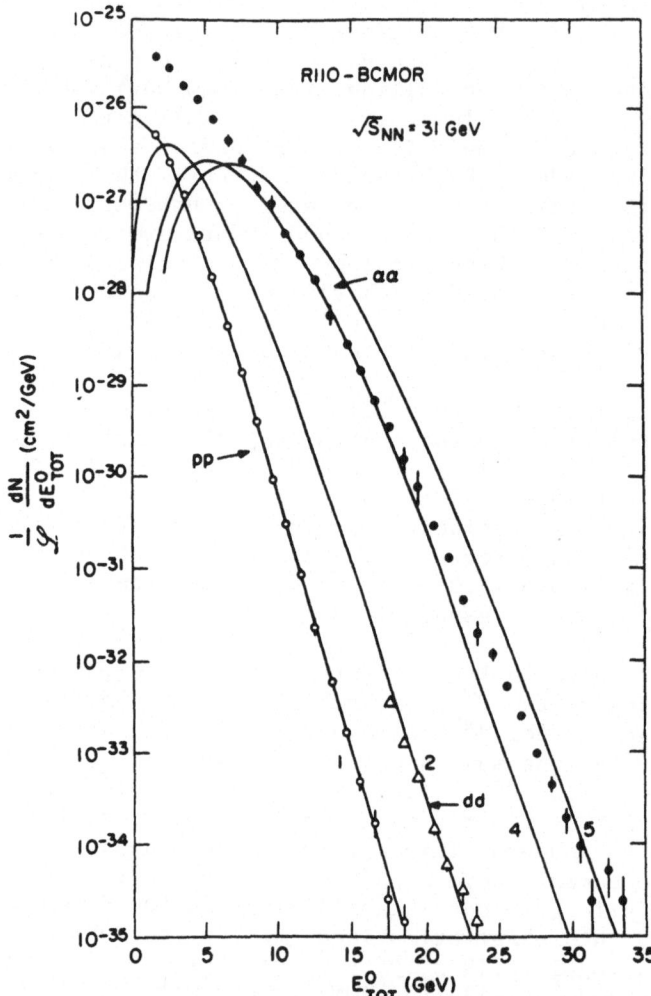

Fig.24: The total neutral energy distribution for pp, dd, and α-α interactions at √s = 31 GeV. The curves represent the distributions corresponding to 1, 2, 4, and 5 simultaneous pp collisions. All curves are normalised to the pp cross-section[43].

the experimental inclusive single-particle p_T distribution and the multiplicity distribution of charged particles. In other words, the information contained in the E_T distribution is equivalent to the combined information contained in p_T and multiplicity distributions.

As an exercise we can calculate the local energy density in the fireball created at $y_{cm} = 0$ shortly after the interaction, by extrapolating the observed (π^o) particles back to their origin. I use the word fireball because the particles measured in this experiment are emitted isotropically in azimuthal angle ϕ and also in polar angle θ as long as the window in θ is limited to $\theta = 90^o \pm 45^o$ corresponding to $|y| < 1$.

Assume an event with $E^o_{tot} = 20$ GeV; the total energy E^{ch}_{tot} of charged hadrons is unknown, but it must lie somewhere between 0 and 2 E^o_{tot} – the first value would apply for a very strong trigger bias towards neutral energy, the second value applies for no bias when π^+, π^- and π^o share the energy equally. We will assume the value to be between these two extremes, hence $E^{ch}_{tot} = 20$ GeV. mean Having estimated the total energy $E_{tot} = 40$ GeV, we next have to estimate the volume. Since the collision is most likely central, we assume a transverse interaction radius equal to one α radius. The longitudinal radius is somewhat more difficult to estimate; one has to account for the formation time, $L = 2hE/m_T^2$.[33] One obtains $L \approx 1$ fm since $E = \sqrt{(p_L^2 + m_T^2} \leq \sqrt{(3/2)}m_T$ and $m_T \approx 0.5$ GeV. Thus one obtains

$$\varepsilon = E_{tot}/V = 40/(3.14 \cdot 1.1^2 \cdot 1.6^2 \cdot 1) = 4 \text{ GeV/fm}^3.$$

This is perhaps already above the energy density where, according to LQCD, the phase transition takes place ($\varepsilon \approx 2$ GeV/fm^3). These events are therefore serious candidates for interesting physics and further studies of their properties, like $\langle p_T \rangle$, particle composition and so on, are desirable.

4.2. Single particle inclusive cross sections

At first sight, the shape of the p_T distribution of particles produced in a nucleus-nucleus collision looks very similar to the one in pp collisions, see Fig. 25.[45] This plot shows the inclusive, differential invariant cross-section for $\alpha\alpha \rightarrow h^-$ (h^- = negative hadrons π^-, K^-, $\bar{p}$) and compares it with pp $\rightarrow h^-$ as a function of p_T. (To obtain p_T distributions normalized to one event it is necessary to divide the differential cross-section by the total inelastic cross-section ($\sigma_{inel}(\alpha\alpha) \approx 275$ mb and $\sigma_{inel}(pp) \approx 32$ mb).

The characteristics of the p_T distribution of particles produced in hadronic interactions at high energies have been interpreted many years ago as thermal and a limited hadronic temperature was inferred from this[46]. However, there are deviations from a simple thermal distribution at high p_T ($p_T \geq 1$ GeV/c, at ISR energies and above).

A somewhat unorthodox modern interpretation of the p_T spectrum which has been proposed by Feinberg, Shuryak, Hagedorn and others[47] is indicated in Fig. 25. These authors assume that a QGP is created even in pp collisions. They divide the p_T spectrum into three regions: the low-p_T region of hadrons emitted after the plasma has cooled down to the Hagedorn temperature, followed by a region of quark or gluon evaporation from the hot plasma -- these quarks or gluons produce jets of hadrons -- and finally, above 6 GeV/c the regime of direct parton-parton scattering -- the perturbative QCD regime. This view of the p_T spectrum implies a reinterpretation of the Hagedorn temperature, which was originally considered the ultimate temperature of a hadron gas, as the critical temperature at which the phase transition from a hadron gas to a quark-gluon gas takes place. The difference between this view and the orthodox view is only subtle. The latter assumes QCD processes (quark-gluon dynamics) of increasing complexity to be at work, from the truly perturbative region of high p_T all the way down to about 1 GeV/c, and it resigns facing the clearly non-perturbative region below $p_T = 1$ GeV/c.

For completeness the basic features of a high-p_T interaction are illustrated in Fig. 26 by the same kind of pictures as applied previously (Figs. 15 and 16): a pair of constituents interacts and is scattered to large angles; the two scattered quarks or gluons and two groups of "spectator" quarks build up two colour strings in separating, yielding 4 jets of hadrons in the final state.

In order to make a sensitive comparison between cross-sections in $\alpha\alpha$ and pp interactions, the ratio of the cross-sections displayed in Fig. 25 is shown in Fig. 27 (data from various experiments at the ISR[45],[48-50]. A horizontal line is drawn at a value of 16 -- this is the value one expects for the ratio if four independent nucleons would collide with four independent nucleons and if there were neither shadowing nor any collective or multiple scattering effects. The ratio is below this line at low p_T, it crosses it at ~2 GeV/c and it rises to a value around 40 at the highest measured p_T. The qualitative behaviour of this ratio was not a complete surprise - except for the large value it reaches at the highest p_T ($p_T > 6$ GeV/c).

Already many years ago Cronin et al.[51] discovered the "anomalous enhancement" of high-p_T hadrons produced in pA interactions. Parametrizing the inclusive cross-section for $p + A \rightarrow h^{\pm} + x$ (where $h^{\pm}$ stands for $\pi^{\pm}$, $K^{\pm}$, p and $\bar{p}$) by a power law dependence on the mass number A,

$$d\sigma/dp_T^2 \propto A^{\alpha(p_T)} \qquad , \qquad (20)$$

they found that $\alpha(p_T)$ increases to values above 1 for p_T higher than 2 GeV/c (see Fig. 28).

340

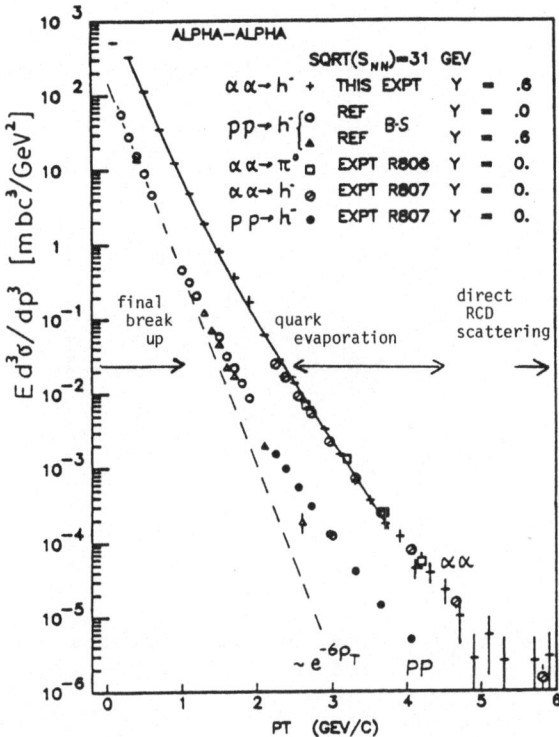

Fig.25: Inclusive invariant cross-section $\alpha-\alpha \to h^-$ and $\alpha-\alpha \to \pi^0$ compared with $pp \to h^-$ versus p_T[45,48-49].

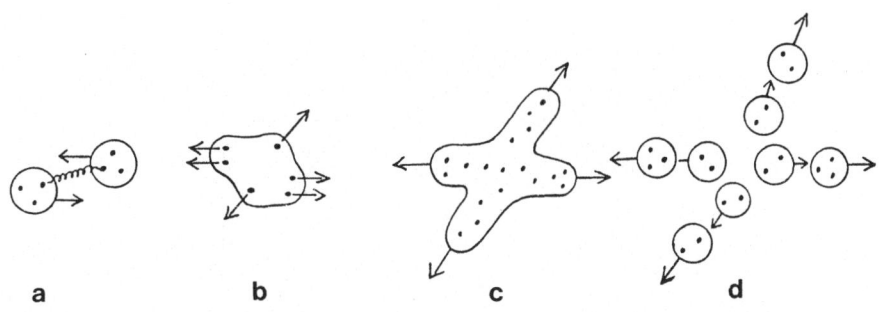

Fig.26: Sequence of processes in a typical high-p_T interaction.
a) Two incoming quarks exchange a hard (high-momentum) gluon.
b) and c) The quarks separate.
d) The colour flux tubes decay (into four jets).

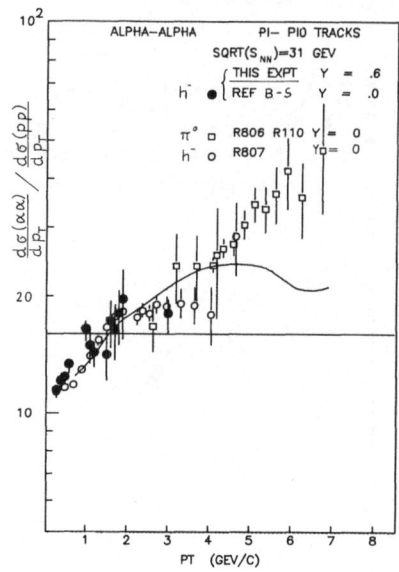

Fig.27: Ratio of inclusive invariant cross-sections
$(\alpha\alpha \to h)/(pp \to h)^{45,49}$ and $(\alpha\alpha \to \pi^0)/(pp \to \pi^0)^{48,50}$ versus p_T.

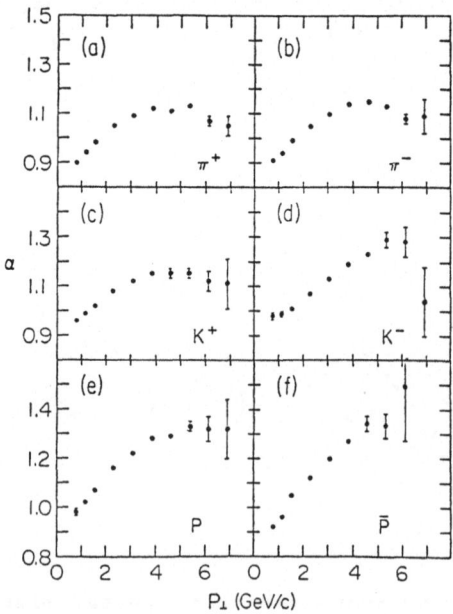

Fig.28: The power $\alpha(p_T)$ describing the A-dependence of the inclusive cross-sections $d\sigma/dp_T$ $(pA \to h)$ versus p_T for various charged hadron species[51].

This observation, later confirmed by other pA experiments, has triggered many speculations[52]. The most widely accepted hypothesis was that the effect is due to some kind of multiple parton scattering.

Now we may add another piece of experimental information which comes from the very high energy cosmic-ray interactions. Remember, these events have such high multiplicity that one can plot a whole p_T distribution from one event with a reasonable number of entries per bin! The p_T distribution of γ's (from π^0 decay) from the Ca + C event (see Fig. 19b) is shown in Fig. 29.[36] It appears that this distribution is flatter than the p_T distribution at similar energy for $p\bar{p}$ interactions (measured at the CERN $p\bar{p}$ collider at $\sqrt{s}$ = 540 GeV corresponding to E_{lab} = 150 TeV). The average p_T of π^0's $<p_T(\pi^0)>$ = $2<p_T(\gamma)>$ is 700 MeV/c compared to 400 MeV/c for $p\bar{p}$ interactions at the same energy. Similar observations have been reported for other very high energy cosmic nucleus interactions: values of $<p_T>$ up to a factor 3 higher than for pp events at the same energy have been found.

To end this section on transverse physics I am tempted to show a picture which has caused some excitement in quark-matter circles recently[53]. It shows (Fig. 30) the average p_T of π^0's (= 2 $<p_T>$ of the γ-initiated showers) as a function of the local energy density ε for some cosmic ray events in the energy range 10 - 100 TeV together with data from the $p\bar{p}$ Collider ($\sqrt{s}$ = 540 GeV). The energy density ε was calculated in a similar way as it was done in section 4.1. At a density around 2 GeV/fm^3, $<p_T>$ rises rapidly. This may indicate some new physics, perhaps QGP formation but even though the effect is dramatic it will be difficult to prove that it is actually related to deconfinement. Remember the beginning of this subsection where the intermediate p_T range was attributed to jets of hadrons resulting from a QGP. If this kind of processes gain more weight at higher ε, one expects of course, an increase of $<p_T>$. On the other hand, one also knows that jet production and ε both increase with increasing incoming energy because more and more sea quarks can undergo hard scattering and consequently $<p_T>$ rises along with ε. Is it only a matter of taste whether one calls it QGP or QCD?

5. STRANGENESS PRODUCTION

What will be the signatures of QGP formation in high-energy nucleus-nucleus collisions? Several signals have been proposed so far[54]:

a) a change in the spectrum of dileptons (Drell-Yan pairs),
b) a change in the spectrum of direct photons,
c) a dramatic increase of the yield of strange or charmed hadrons,
d) fluctuations in the rapidity spectrum,
e) an increase of $<p_T>$.

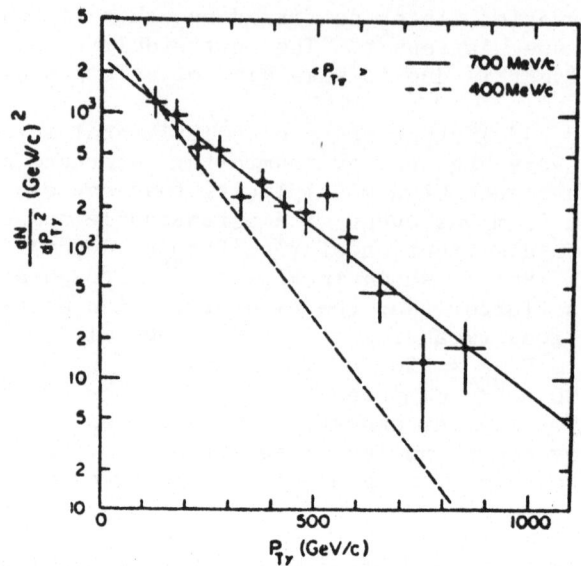

Fig.29: The p_T distribution of gammas in the cosmic-ray event Ca + C[36].

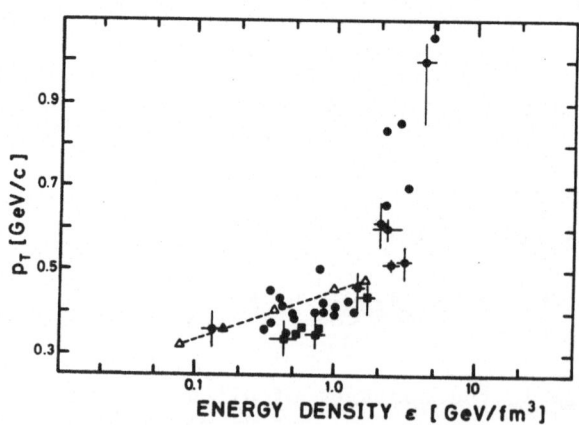

Fig.30: Average transverse momenta in cosmic ray events (solid symbols) and in $\bar{p}p$ collisions at the CERN collider[53].

Only few relevant data exist for nucleus-nucleus collisions at high energy. The K/π ratio has been measured as a function of the multiplicity in αα collisions at the ISR[55] and it does not show any interesting trends. The ratios $R^{\pm} = (K^{\pm} + \pi^{\pm})/\pi^{\pm}$ have been measured for p_T around 3 GeV/c, as well as for αα collisions at the ISR[56]: R^+ increases significantly compared to the same ratio in pp collisions but R^- barely changes. From this, and from a comparison with the dependence of K^+, K^-, π^+, π^-, p and $\bar{p}$ yields on the target mass A in pA collisions[51] one has to conclude that the increase of R^+ is due to an increase of the proton yield rather than the K^+ yield.

Three experiments have been proposed to measure strangeness production in interactions of ions (up to 160) with heavy nuclear targets at the CERN SPS[57]. In such a state it is useful to study and try to understand the data available from pp and pA interactions and to compare them with the theoretical expectations for AA interactions, or rather for the case of quark-gluon plasma formation. This is the goal of this section 5. I believe that specializing on only one of the signatures (strangeness production) is enough to learn about the general difficulties and the open questions on the way to a search for the deconfinement phase transition.

The theoretical expectation is, in short, that the ratio of the density of strange (heavy) sea quarks s, $\bar{s}$ to light sea quarks of one kind (q = u, $\bar{u}$, d, or $\bar{d}$) in the QGP is of the order of 1, more precisely:

$$R_{QGP}(s/\bar{q}) = 2/3 \exp (\mu - m_s)/T \quad , \tag{22}$$

where μ is the light quark chemical potential with typical values varying from 0 in the central rapidity region to $m_p/3 = 300$ MeV in the baryon rich fragmentation region; T is the temperature, typically T = 200 MeV, and m_s is the strange quark mass, $m_s \approx 150$ MeV; the mass of light quarks is negligibly small, $m_q \approx 10$ MeV.[58]

This ratio is compared[59] with the ratio one obtains when one counts sea quarks in the asymptotic final state after a pp collision

$$R_H(s/\bar{q}) = 3\% \tag{23}$$

for a pp collision at 14.5 GeV beam energy. The latter ratio rises up to about 7% at SPS energies, see Fig. 31.[60] The much lower measured ratio R_H compared to the expected R_{QGP} is attributed to the fact that hadronic matter consists mostly of pions. The prediction of an increased yield of strange hadrons in the case of QGP formation is based on this comparison between R_{QGP} and R_H.

Two questions have to be raised with respect to this comparison and the resulting prediction:

- What will be the ratio $R_F(s/\bar{q})$ in the final state hadrons after the QGP has gone through the confinement transition, i.e. after it has cooled down and hadronized?

- What was the initial ratio $R_I(s/\bar{q})$ in the hot hadronic phase before the asymptotic final state was reached, if no QGP was created?

During the cooling down and hadronization the production of pions is always favoured over K production owing to the small π mass. This suppression of K mesons has been estimated many years ago by Anisovich et al.[61]. They assumed that after a pp collision a quark plasma is produced (this was before the days of the gluon) and that the first generation of hadrons - low mass resonances - are formed by combining quarks of the plasma to mesons and baryons. In order to calculate the intial ratio $R_I(s/\bar{q}) = \lambda$ in the quark plasma, they used the measured branching ratios for the decay of the known low-mass resonances into long-lived final-state hadrons (π, K, p, Λ, etc.). In these decays, kaons are suppressed by a factor of about 4 compared to pions because of their higher mass. Accounting for this suppression, they were able to determine λ. Figure 32 shows λ obtained by applying their procedure to existing data as a function of $\sqrt{s}$.[60]

The ratio λ reaches a value around 0.4 at the highest energies; this "primary" suppression of strange quarks in the sea has been explained by the heavy mass m_s or more precisely by the Boltzmann factor $\exp[(m_q - m_s)/T]$.

Another piece of experimental data fits nicely together with this observation. Figure 33 shows the ratio K^+/π^+ as a function of p_T for pp and pW interactions[51]. The ratio rises with increasing p_T and reaches a kind of plateau at 0.45 for pp and at 0.55 for pW interactions. How can we interpret this behaviour? The ratio at low p_T corresponds to the one found if one integrates over all p_T (see Fig. 31) - it includes the primary suppression of strange quarks as well as the secondary suppression from the decay of resonances, hadronization of jets, etc. However, at higher p_T the K^+/π^+ ratio measures presumably the quark composition at an early stage of the interaction. A K^+ is composed of an $\bar{s}$ and a u quark ($K^+ = (u\bar{s})$), whereas $\pi^+ = (u,\bar{d})$. Higher-p_T hadrons are leading particles of a jet and as such they remember the flavour of the initial scattered quark (see Fig. 26). At high p_T it is more likely the incoming u quark which was scattered at large angle. (Since valence quarks carry most of the nucleon momentum, they are more likely to contribute at high p_T than sea quarks). Although the sea quarks s and d in the K^+ and π^+, respectively, are not, in most cases, the actively scattered quarks, they were picked up by the u quarks at an early stage of the interaction. (The evidence for this comes from the special

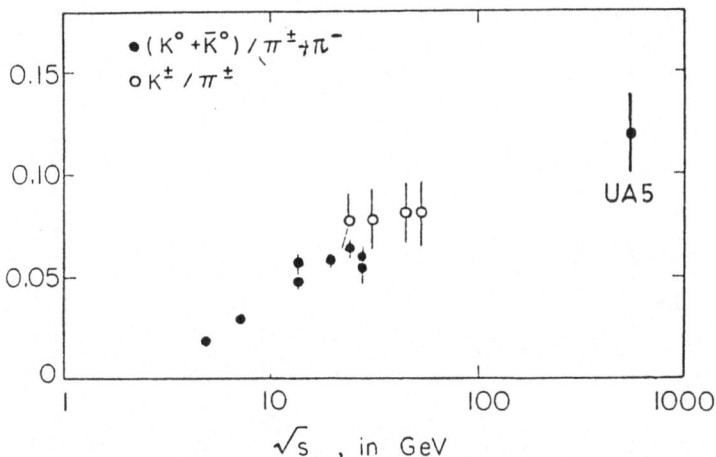

Fig.31: Ratio of strange to non-strange mesons observed in pp and pp interactions as a function of $\sqrt{s}$[60].

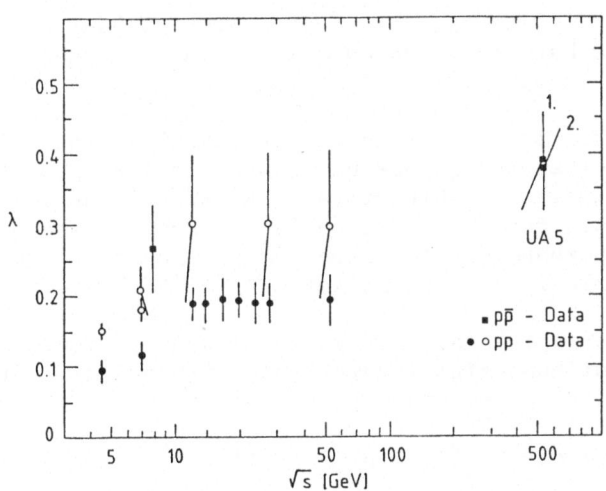

Fig.32: Ratio of strange to non-strange sea quarks at an early stage of the interaction (before the resonances decay) versus $\sqrt{s}$[60].

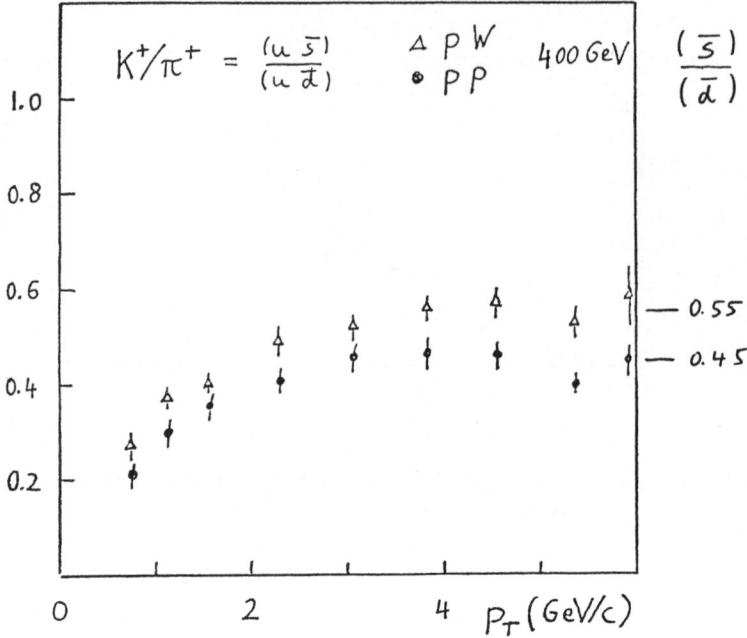

Fig.33: The ratio of K^+/π^+ yields as a function of p_T for pp and pW interactions[51].

properties of the jets associated with high-p_T hadrons: the high-p_T hadron tends to have no neighbours in momentum space, it carries most of the jet momentum).

If the outlined interpretation of the K^+/π^+ ratio is correct, one may ask whether the higher plateau value for pW compared to pp interactions signifies a higher ratio $R_I(s/\bar{q})$ in nuclei. Unfortunately, the answer to this question is not known. The difference of the ratios is perhaps due to different reabsorption probabilities of s and $\bar{d}$ quarks in nuclear matter.

The last experimental information I want to mention comes from neutrino and anti-neutrino interactions with nuclei. From the reactions leading to two opposite charged muons in the final state:

$$\nu + d \to \mu^- + c \to \mu^- + \mu^+ (+ s + \nu_\mu)$$

$$\nu + s \to \mu^- + c$$

$$\bar{\nu} + \bar{d} \to \mu^+ + \bar{c} \to \mu^- (+ \bar{s} + \bar{\nu}_\mu)$$

$$\bar{\nu} + \bar{d} \to \mu^+ + \bar{c} \tag{24}$$

the ratio of strange to non-strange sea in cold nuclear matter was
determined[62]:

$$R_\nu [2s/(\bar{u} + \bar{d})] = 0.52 \pm 0.09 \quad .$$

Again, the apparent SU(3) breaking of the quark-sea in nuclear matter
has been attributed to the higher mass of the strange quark.

All these data indicate in a consistent way that the initial
strange to non-strange quark ratio measured in pp, pA and even νA
interactions is close to the one expected for the quark-gluon plasma.
Does this mean that a QGP is always created? Of course not, because
the notion of a plasma implies thermodynamic equilibrium, a certain
critical minimum volume and other features. But the data certainly
tell us that it will be difficult to distinguish the properties of
the QGP from the conventional nonequilibrium quark-gluon dynamics by
which most processes are explained today.

6. FUTURE EXPERIMENTS

6.1. Measurement Goals and the Ideal Detector

At the beginning of section 5 most of the proposed signatures
for QGP formation were mentioned and it was then shown for one special
case (strangeness production) that one should not be too optimistic
to see a dramatic effect in any one of these signals. (Of course,
the rise of $<p_T>$ versus ε shown in Fig. 30 is a dramatic effect and it
may justify optimism, but it has to be shown first how it is related
to the deconfinement phase transition). Thus, it is generally agreed
upon that an experiment searching for a phenomenon which is difficult
to grasp, such as deconfinement, has to measure as many properties of
the final state as possible.

Therefore, let us start with outlining the ideal detector which
can measure everything. Such a detector would consist of shells 1
to 6 which serve the following functions:
1) beam tagging (identify beam particle, measure momentum vector and
   impact point in space and time on target),
2) tracking of charged particles in a magnetic field (momentum vector,
   vertex),
3) identification of charged particles (e, π, K, p separation),
4) electromagnetic calorimetry (gamma and e energy, direction,
   match with 2), 3) if it was an electron,π, η identification),
5) hadron calorimetry (energy direction, match with 2), 3) if it
   was a charged hadron),
6) muon filter to further reject hadron punch-throughs, match with
   2), 3),
7) neutrino detection would make it too ideal a detector.

The ideal detector is hermetical (covers 4π solid angle), thus only neutrino-like particles escape - they will show up as missing energy and momentum. The sequence of the shells is imposed by the amount of material needed for each function. The thin or low - g/cm^2 shells come before those which cause absorption or multiple scattering. Thus, for instance, charged particle tracking is always done first because wire chambers and the like contain little material, and the electromagnetic shower detection is done before hadron shower detection because the radiation length X_o is shorter than λ_{abs} except for A>9 nuclei.

The performance requirements of each function are dictated by the physics (properties of the events to be studied), the physics goals (i.e. the selected properties of the selected events to be studied) and by the usual space, time, money, manpower constraints (talking about real detectors, now). For instance, for charged particle tracking, the spatial and time resolution and the resulting momentum resolution, given a certain number of space point measurements; the two-track resolution and multitrack pattern recognition are relevant. For calorimetry (electromagnetic and hadron shower detection), the energy resolution shower separating capability, directional sensitivity are important criteria. In Table 2 properties of some common detector components in high energy physics are listed[65]. These lists are very incomplete; neither do they contain prices per channel nor properties which cannot be expressed by a few numbers - like the pattern recognition capability in a multitrack environment for charged particle tracking.

There are quite a few detectors today which come close to the ideal detector. Those presently installed at PEP and PETRA, the two highest-energy colliders at present and the two detectors (UA1,UA2) at the CERN p$\bar{\text{p}}$ Collider. Even closer to the ideal detector are the four detectors (OPAL, DELPHI, ALEPH and L3) being built for LEP, the future Large Electron Positron Collider at CERN, the two detectors for the SLC, the SLAC Linear Collider, and the two detectors (CDF and DO) being constructed for the Fermilab Tevatron p$\bar{\text{p}}$ collider. In order to learn state of the art technology in detector building, it is almost sufficient to study some of these latter 8 proposals - most of them are projects in the 10^8 \$ range.

The main task faced in an experiment which aims to study all particles produced in a nucleus-nucleus collision as it was supposed for the ideal experiment, would be to handle final states with an enormous number of tracks: For a central ^{16}O - Pb collision the number of charged tracks is around 300 at SPS energies. Track densities in the forward region are much higher than for typical jets at LEP - this is illustrated in Fig. 34.[66] Therefore, real experiments have to make compromises, look only into selected regions of phase space or detect only certain particle species.

In the next, last subsection a survey of proposed experiments

Table 2

A) <u>Tracking devices</u> [a]

detector type	spatial resol. (rms in mm)	time resolution	dead time
bubble chamber	0.01	O(ms)	O(50ms)
streamer chamber	0.3	2 μs	100 ms
high press. hol. streamer chamber[b]	0.07		
emulsion	0.001	exposure time	
proportional chamber (1 mm pitch)	0.3	50 ns	200 ns
drift chamber	0.05	2 ns	100 ns
silicon strip (20 μm pitch[c])	0.003	10 ns	
scintillator	0.3	150 ps	10 ns
scint. glass fibre[d]	0.003		

B) <u>Charged particle identification</u>[e]-[h]

time of flight for non-relativistic secondaries
energy loss
threshold Cherenkov counters (gas, liquid, plastic, aerogel)
ring imaging Cherenkov detectors
transition radiation

C) <u>Electromagnetic shower detectors</u>[a], [e]-[h]

detector type	energy resolution $\Delta E(FWHM)/E$, E in GeV	depth(X_0)
NaI(Tl)	$2\%/E^{-1/4}$	20
Pb glass	$11\%/\sqrt{E}$	14
Pb/liquid Argon sandwich	$16\%/\sqrt{E}$	16
Pb/scintillator sandwich	$17\%/\sqrt{E}$	13
BGO	$2\%/\sqrt{E}$	20
scintillating glass	$2\%/\sqrt{E}$	

D) <u>Hadron calorimeters</u>[e]-[i]

detector type	energy resolution $\Delta E(FWHM)/E$, E in GeV
Fe/scintillator sandwich	$70\%/\sqrt{E}$
U/scintillator compensated	$35\%/\sqrt{E}$

a) Particle Properties Data Booklet from "Review of Particle Properties" Reviews of Modern Physics <u>56</u>,No.2,Part II (1984).
b) V.Eckardt and S.Wenig, Nucl.Instr.Meth. <u>213</u>(1983)217.
c) R.Klanner, Max-Planck-Inst.Munich preprint MPI-PAE-Exp-El 135.
d) J.Kirkby,private communication; D.Binnie et al.,Notre-Dame Univ. preprint UND-HEP-84-RR01 (1984); N.Atkinson et al.,NIM <u>225</u>('84)1.
e) Proceedings 1983 DPF Workshop on Collider Detectors,Feb.28-Mar.4, 1983, LBL Berkeley, eds.S.C.Loken,P.Nemethy, LBL-15973 (April '83).
f) Proc. Int. Conf. on Experimentation at LEP,Phys.Script.<u>23</u>('81) 317.
g) Proc. ECFA-CERN Workshop on Large Hadron Collider, Lausanne and Geneva 21-27 March 1984, ECFA 84/85 CERN 84-10 (Sept.1984).
h) Proc. Int. Conf. on Instrumentation for Colliding Beam Physics SLAC-Report 250 (1982).
i) C.W.Fabjan and T.Ludlam, Ann.Rev.Nucl.Part.Sci.<u>32</u>(1982)335.

for the CERN–SPS (^{16}O) program and for the ^{32}S program at the BNL–AGS will be given, and some of them will be sketched very briefly to make clear what kind of compromises have been made in these experiments.

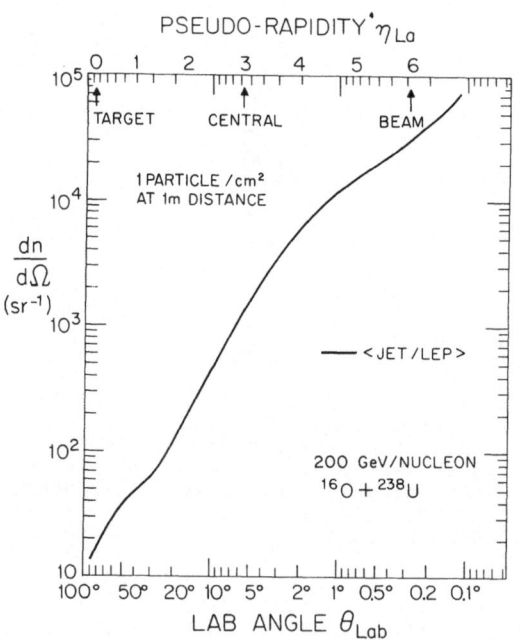

Fig.34: Track density (per solid angle) as a function of lab-angle and lab-pseudorapidity.

6.2. Proposed Experiments

About 10 experiments have been proposed so far for the future (>=1986) Oxygen runs at the CERN-SPS. The first proposal[67] was made for the Proton-Synchrotron (13 GeV/nucleon). However, for several practical reasons CERN decided to extract the beams via the SPS. All following experiments were then proposed for the SPS and requested the maximum SPS energy of 225 GeV/nucleon. The collaboration which proposed the first two (joint) experiments will provide the ion source and a Radio Frequency Quadrupole (RFQ) preacceleration stage.

One of the two first experiments uses a large volume streamer chamber inside a superconducting magnet as vertex detector for charged particles either from the primary vertex or from strange neutral hadron decays (Λ,K^0). The energy flow is measured by an azimuthally and radially segmented electromagnetic and hadron calorimeter (Fe/Pb/scintillator sandwich) which covers the polar angular region from 0^0 to 135^0 (in the c.m.s. at 200 GeV/nucleon). The calorimeters are also used to trigger on transverse energy or to veto on forward energy.

The second experiment uses the "plastic ball" as vertex detector and aims to detect direct photons as signature for the QGP by means of a special type of electromagnetic calorimeter. The forward region contains segmented electromagnetic and hadron calorimeters.

There are two proposed (and approved) experiments applying visual techniques. One will search for fractionally charged nuclei leaving tracks in plastic sheets[68]; the other one will study pseudorapidity and multiplicity distributions of produced particles and nuclear fragments in nuclear emulsion[69].

Four experiments were proposed to measure charged and neutral strange hadron production using different multiparticle spectrometers. Two of them were turned down, one (letter of intent) has not made it yet to the official level and the fourth, using a Time-Projection-Chamber (TPC) equal to TPC modules presently being constructed for one of the future LEP experiments, is still under discussion.

Two experiments propose to measure lepton production. Again, one interesting letter of intent has not made it to the official proposal level. The other one[70] has already been approved and I will outline its features in somewhat more detail than the previously mentioned proposals. Fig. 35a shows the full experimental lay-out which contains the following four main parts:

1) a muon spectrometer at forward angles; PC1 to PC6 are multiwire proportional chambers and the wall behind PC6 is an additional muon filter;

353

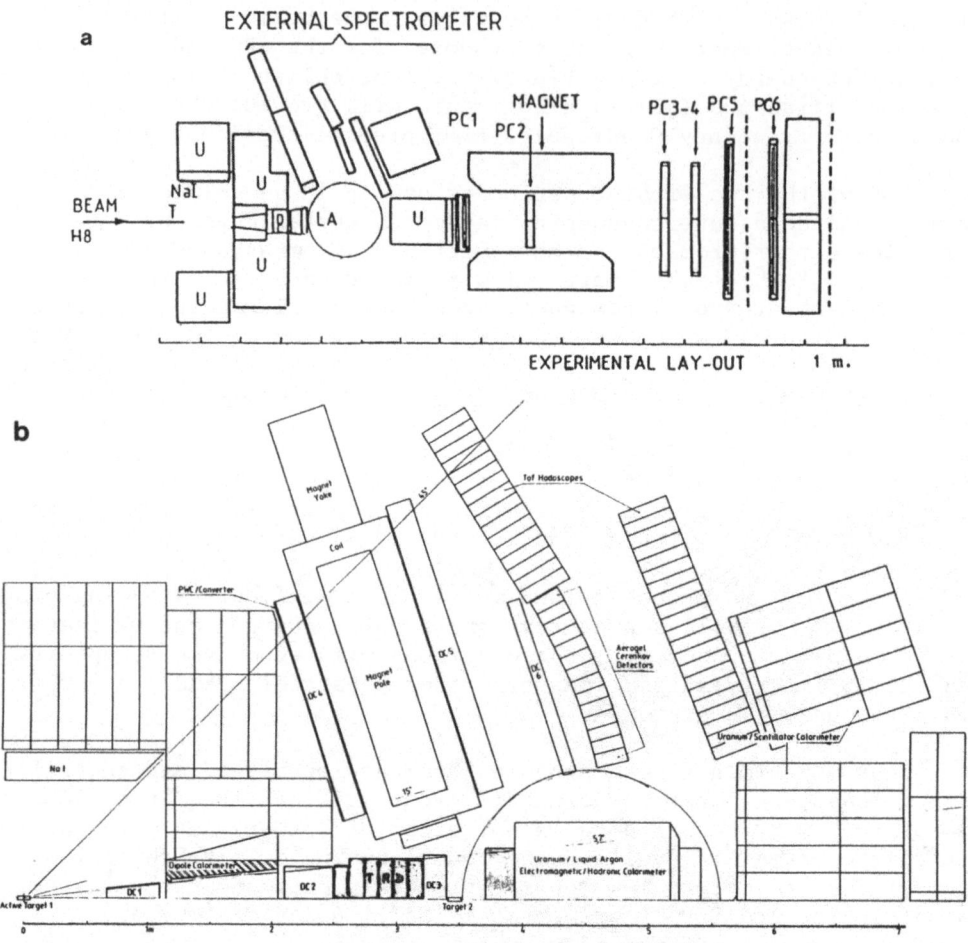

Fig.35: Set-up of experiment NA34 at the CERN–SPS (Study of lepton production); a) complete lay-out, b) target region.

354

2) segmented compensated Uranium/scintillator sandwich calorimeters (U) with electromagnetic and hadronic sections implemented by a fine-grained Uranium/liquid Argon (LA) calorimeter to measure electromagnetic and hadronic energy flow, and an NaI crystal wall mainly for photon detection;

3) an external spectrometer containing a C-type magnet, drift chambers, time-of-flight scintillator hodoscopes and aerogel Cherenkov counters;

4) an electron spectrometer at forward angles (not recognizable in Fig. 35a).

Fig. 35b is a blow-up of the detectors around the target. Target 1 is an active target; it consists of 200 target wires, each wire 0.1 mm thick, spaced by 0.5 mm. These wires act as cathode wires being part of a proportional chamber. Thus the wire(s) where an interaction took place can be recognized. Target 2 is further downstream and as close as possible to the forward calorimeters; this target minimizes the decay path of the pions and kaons (most of the muon background results from these decays). The drift chamber modules DC1 to DC3, the spherical weak-field dipole magnet between DC1 and DC2, and the 6 transition radiation detector modules are part of the electron spectrometer which is used for studies in hadron-nucleus interactions to be performed by essentially the same detector set-up[71]. The external spectrometer "sees" the target through a small crack in the U calorimeter walls.

Across the Atlantic Ocean, at the Brookhaven AGS, preparations are under way for a run (in 1986) with ^{32}S ions of 15 GeV/nucleon. An experiment has been approved recently[72] which aims to study semi-inclusive spectra of p, π, K, d, α and ϕ with a 25 msr single-arm spectrometer. The set-up is shown in Fig. 36. The large-angle spectrometer is equipped with a 1.5 Tm C-type magnet and sets of high resolution drift chambers (Ti) for tracking. Particle identification is done by time of flight, using scintillation counter hodoscopes (Si) and a segmented threshold gas Cherenkov counter (SC). The forward Cherenkov complex is needed to extend particle identification to smaller angles, i.e. higher momenta. In addition, the set-up will contain a streamer tube array (ST) around the target for counting charged particle multiplicities, a lead-glass array for photon energy flow measurements and a Cherenkov counter array consisting of plastic detectors for counting the number of forward particles with v/c = 1/n > 0.7.

Clearly the unprecedented amount of preparations for experiments using high-energy ions reflects a very strong new interest in this field of physics.

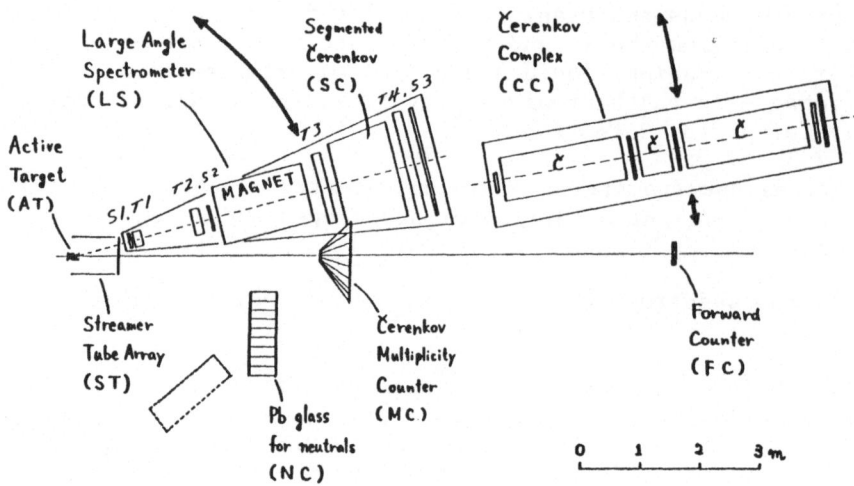

Fig.36: Set-up of an experiment at the BNL-AGS.

ACKNOWLEDGEMENTS

    I thank the organizers and participants of this summer school
for providing a very pleasant, stimulating and informative atmosphere.
The written version of these lectures was completed during a stay at
the University of Pennsylvania, Philadelphia, Pa., where I enjoyed
the hospitality of the Physics Department, in particular of S. Frankel
and his group. I am indebted to D. Hofford for her most professional
and rapid typing of the manuscript. The Deutsche Forschungsgemein-
schaft gave their support by awarding me a Heisenberg grant.

REFERENCES

1. F.E.Close, An Introduction to Quarks and Partons,Academic Press,
    London 1979.
2. B.Adeva et al.,(Mark J Collaboration), Phys.Rep.109(1984)179ff
    and references therein.
3. S.L.Wu, Phys.Rep.107(1984)120ff. Reference 1, p.185ff.
4. This is only true if quarks have three colours, see reference 1,
    p.161f, 231ff, and 253ff.
5. Reference 1, p.233.
6. D.Flamm and F.Schoeberl, An Introduction to the Quark Model of
    Elementary Particles, Gordon and Breach, New York 1982.
    L.Montanet et al., Phys.Rep.63(1980)149.
7. D.Hitlin, Proc. 1983 Int.Symposium on Lepton and Photon Inter-
    actions, August 4-9,1983, eds. D.G.Cassel,D.L.Kreinick, publ.by
    F.R.Newman Lab. of Nuclear Studies, Cornell University, Ithaca,
    New York 14853 USA, Ithaca 1983, page 746.
    See any of the latest high-energy physics conference proceedings.

8. D.Wilkinson (ed.) Progress in Particle and Nuclear Physics, vol.8, Quarks and the Nucleus, Proc. of the Int. School of Nuclear Physics, Erice 21—30 April 81, Pergamon Press Oxford 1982. K.Bleuler (ed.) Quarks and Nuclear Structure, Proc. Bad Honnef, FR Germany 1983, Lecture Notes in Physics 197,Springer Berlin Heidelberg 1984,

9. S.J.Wimpenny, Proc.10th Int.Conf. on Particles and Nuclei, Heidelberg, July 30 — August 3, 1983, to be publ. in Nucl.Phys.A.

10. Proceedings of the 7th Int. Seminar on High Energy Physics Problems, Dubna (USSR) 19—23 June 1984, ed. A.M.Baldin.

11. Proc.4th Int. Conf. on Ultrarelativistic Nucleus—Nucleus Collisions, Helsinki, Finland June 12—21,1984, ed. K.Kajantie. Proc. Third Int. Conf. on Ultrarelativistic Nucleus—Nucleus Collisions, Brookhaven National Lab.,Upton,New York, Sept.26—30, 1983, Nucl. Phys. A418 (1984).

12. H. Haseroth, Nuclear Beams from PS and SPS, Proc. of the Workshop on SPS Fixed—Target Physics in the Years 1984—1989, CERN, Geneve,CERN 83—02, Decembre 1982, Vol. III, p. 443.

13. A. Dar, Normalons, Anomalons and Anomalies in High Energy Cosmic Rays, in Proc. of the 6th High Energy Heavy Ion Study, LBL UC Berkeley 1983,LBL—16281, p. 523.

14. G.B. Yodh, in Proc. of the First Workshop on Ultrarelativistic Nuclear Collisions, LBL UC Berkeley, May 1979, LBL—8957, p. 139.

15. R. Bouclier et al., Nucl. Instr. Meth. 115 (1974) 235, R. Bouclier et al., Nucl. Instr. Meth. 125 (1975) 19, W. Bell et al., Nucl. Instr. Meth. 156 (1978) 111.

16. G.M.G. Lattes et al., Phys. Rep. 65 (1980) 151.

17. W. Bell et al., Phys. Lett. 128B (1983) 349.

18. A. Wroblewski, Acta Phys. Pol. 4B (1973) 857.

19. W.Q. Chao and H.J. Pirner, Z. Phys. C14 (1982) 165.

20. A. Breakstone et al., Charged Multiplicity Distributions in Proton—Proton Interactions at ISR Energies, preprint CERN/EP 83—165, submitted to Phys. Rev.

21. C. Halliwell et al., in Proc. Topical Meeting on Multiparticle Production on Nuclei at very High Energies, Trieste,10—15 June 1976,(eds. G.Bellini, L.Bertocchi, P.G.Rancoita), IAEA—SMR—21.

22. I. Otterlund et al., Z. Phys. C 20 (1983) 281.

23. K. Braune et al., Z. Phys. C 17 (1983) 105.

24. C. De Marzo et al., Phys. Rev. D 26 (1982) 1019.

25. M.A. Faessler, Ann. Phys. 137 (1981) 44.

26. M.A. Faessler, Proc. Workshop on Quark Matter Formation and Heavy Ion Collisions, Bielefeld, May 1982 (eds. M. Jacob and H. Satz),World Scientific, Singapore, 1982, p. 169.

27. O.V.Kancheli, Pisma v. Zh. Eksp. Teor. Fiz. 18 (1973) 465, Transl. JETP Lett. 18 (1974) 174. L. Bertocchi, in Proc. High Energy Physics Involving Nuclei, Trieste,Sept.1974 (Editrice Compositori,Bologna,1974), p.197. L. Bertocchi, in Proc. 6th Int. Conf. on High Energy Physics and Nuclear Structure, Santa Fe, 1975 (eds. D.E. Nagle et al.) AIP, New York 1975, p. 238.

28. V.V. Anisovich and V.M. Shekhter, Nucl. Phys. B55 (1973) 455.
    V.V. Anisovich et al., Nucl. Phys. B55 (1973) 474.
    A. Bialas et al., Acta. Phys. Pol. B8 (1977) 585.
    N.N. Nikolaev et al., Preprint CERN TH 2541-CERN (1978).
29. A. Capella and Tran Thanh Van, Phys. Lett. 93B (1980) 146.
30. S.J. Brodsky et al., Phys. Rev. Lett. 39 (1977) 1120.
31. K. Gottfried, Phys. Rev. Lett. 32 (1974) 957.
32. A. Bialas et al., Nucl. Phys. B111 (1976) 461.
33. L. Stodolsky in Proc. 6th Int. Colloquium on Multiparticle
    Reactions,Oxford, 1975 (eds. C. Hong-Mo, R. Phillips and
    D. Roberts),Rutherford Lab.,Chilton, Didcot, 1975, RL 75-143,
    p. 577.
34. W. Bell et al. Preprint CERN-EP/84-133 (1984),subm. to Z.Phys.C.
35. A. Capella et al., Phys. Lett. 108B (1982) 347.
36. T.H.Burnett et al.(The JACEE Colaboration) πPossible Candidates
    for Quark-Gluon Plasma in the JACEE Emulsion Chamberπ in Proc.
    of the 6th High Energy Heavy Ion Study at LBL, UC Berkeley,
    1983, LBL-16281, p. 563.
    T.H.Burnett et al., Nucl.Phys.A 418 (1984) 152c.
37. K. Kinoshita et al., Z. Phys. C8 (1981) 205.
38. M. Gyulassy, Nucl, Phys. A400 (1983) 31c.
39. I thank A.Dar for drawing my attention to this.
40. F. Takagi, Phys.Rev.Lett.53 (1984) 427.
41. F.C. Erne, Phys. Lett. 49B (1974) 356.
42. W. Busza and A.S. Goldhaber,Phys.Lett.139B (1984) 235.
43. A.L.S.Angelis et al.,Phys.Lett. B (1984)   .
44. H.Gordon et al., Phys.Rev.D28 (1983) 2736.
45. W. Bell et al., Phys. Lett 112 B (1982) 271.
46. R. Hagedorn, Suppl. Nuovo Cimento III (1965) 147.
    R. Hagedorn, Nucl. Phys. B24 (1970) 93.
47. E. Feinberg. Nuov. Cim. 34A (1976) 391.
    E. Shuryak, Phys. Lett. 78B (1978) 150.
    E.M. Friedlander and R.M. Weiner, Phys. Rev. Lett. 43 (1979) 15.
    E.V. Shuryak, Phys. Rep. 61 (1980) 71.
    R. Hagedorn, Preprint CERN TH 3684 (1983).
48. A. Karabarbounis et al., Phys. Lett. 104B (1981) 75.
49. T. Akesson et al, Nucl. Phys. B209 (1982) 309.
50. A.L.S. Angelis et al., Phys. Lett. 116B (1982) 379.
51. J.W. Cronin et al., Phys. Rev. D11 (1975) 3105.
    D. Antreasyan et al., Phys. Rev. D19 (1979) 764.
52. M.A. Faessler, Phys. Rep. 88 (1982) 401.
53. H.Satz, Closing Talk ,reference 11b.
54. Proc. Workshop on Quark Matter Formation and Heavy Ion
    Collisions,Bielefeld, May 1982 (eds. M. Jacob and H. Satz),
    World Scientific, Singapore, 1982.
55. W.Frati, Nucl.Phys.A418 (1984) 177c.
56. A. Breakstone et al.(Ames-Bologna-CERN-Dortmund-Heidelberg-LBL-
    Lund-Warsaw Collaboration) (R418), to be published.
    M.A.Faessler, preprint CERN-EP/84-112 (1984).

358

57. N. Angert et al. (GSI—LBL—Heidelberg—Marburg—Warsaw
    Collaboration),Proposal to the PSCC, CERN/PSCC/82-1, PSCC/P53.
    C.R.Gruhn et al., Prop. to the SPSC, CERN/SPSC 84-13, SPSC/P196.
    M.A.Faessler et al.,(CERN—GSI—Lund—Pennsylvania—Riverside—
    Serpukhov—Strasbourg—Vienna—Warsaw Collaboration),
    Proposal to the SPSC, CERN/SPSC 84-40, SPSC/P202.
58. J. Rafelski in Proc. Workshop on Future Relativistic Heavy Ion
    Experiments, GSI Darmstadt, FRG, 1981, (eds. R.Bock and R.Stock),
    GSI Report 81-6, p. 282.
    J. Rafelski, Phys. Rep. 88 (1982) 331.
59. B. Mueller, πThe Physics of the Quark—Gluon Plasmaπ, Lectures
    given at Univ. de Liege, April 1983, University of Frankfurt
    preprint UFTP 125/83.
60. Th. Mueller, πStrangeness Suppression at Collider energiesπ,
    preprint CERN/EP 83-141 (1983).
61. V.V. Anisovich and V.M. Shekhter, Nucl.Phys. B55 (1973) 455
                                  and Nucl.Phys. B63 (1973) 542.
    V.V. Anisovich and M.N. Kobrinski, Phys.Lett.52B (1974) 217.
62. H. Abramowicz et al., Z. Phys. C15 (1982) 19.
63. W. Thome et al., Nucl. Phys. B129 (1977) 365.
64. E. De Wolf et al., Nucl. Phys. B87 (1978) 325.
65. Particle Properties Data Booklet, M.Aguilar—Benitez et al.,from
    Rev.Modern Phys.56 No2 Part II (April 1984).
66. Figure taken from H.J.Specht.
67. N. Angert et al., CERN/PSCC/81-1, PSCC/P 53 (1982).
68. P.B. Price et al.,CERN/SPSC/84-38,SPSC/P201 (1984).
69. K.B.Bhalla et al.,CERN/SPSC/84-27,SPSC/P198 (1984).
70. H. Gordon et al., CERN/SPSC/84-34,SPSC/P203 (1984).
71. H. Gordon et al., CERN/SPSC/83-51,SPSC/P189 (1984).
72. D. Alburger et al., (BNL—Hiroshima—LBL—MIT—Tokyo Collaboration)
    Proposal submitted to the AGS PC 1984.

COMMENTS ON MULTIPLICITY IN PROTON NUCLEUS COLLISIONS*

Arthur K. Kerman, Tetsuo Matsui, and Benjamin Svetitsky

Center for Theoretical Physics
Laboratory for Nuclear Science
Massachusetts Institute of Technology
Cambridge, MA 02139

We model particle creation in the central rapidity region in ultra-relativistic nuclear collisions by assuming an intermediate state of color flux. Our starting point is that nucleon-nucleon collisions are dominated by single gluon exchange.[1] This causes the color to flip from singlet to octet so that the two receding nucleons become linked by a flux tube with the diameter of the nucleon. This flux tube then materializes to produce the observed phenomena. In the case of a proton-*nucleus* collision there will be several gluon exchanges for each quark in the projectile, which will change the details of the flux tube's formation and material- ization. In what follows, we make some comments on this problem.

Suppose that the projectile nucleon suffers ν interactions while traversing the target nucleus. In each collision, the pro- jectile exchanges one soft gluon with a target nucleon. This pro- cess can be viewed as a random walk in the intrinsic color space. Therefore, the strength of the color charge built up after the collision will be proportional to the square root of ν:

$$Q \propto \sqrt{\nu} \tag{1}$$

(Here, we consider a simple Abelian version of the color charge. The essential feature of the random walk process does not change for groups of higher rank.[2]) Gauss' law demands that a color elec- tric flux tube form between the projectile and the target. Since the color charge at the ends of this tube is greater than that pro- duced in *pp* collisions, the field strength in the tube is initially stronger.

*What follows is an outgrowth of material presented in an informal talk by one of us (AKK).

The particle production process can be modelled as the quantum creation of $q\bar{q}$ pairs in the strong color field[3,4] (the Schwinger mechanism[5] in QED) and the subsequent combination of quarks into hadrons. The energy originally stored in the color field is gradually converted into the kinetic energy of the quarks. This process continues until the field energy is exhausted.

According to Schwinger, the pair creation rate (the probability of producing a pair per unit four-volume) is determined by the strength of the external color electric field E as

$$P = aE^2, \tag{2}$$

where a is a dimensionless numerical constant. (We have taken the massless quark limit.) The total number of pairs produced out of the color field is given by the space-time integral

$$N_{pair} = \int d^4x \, P . \tag{3}$$

To evaluate this integral, we first assume that the transverse cross section of the tube is fixed at A. We define the longitudinal coordinate z and time t in the center-of-mass frame and define the light cone variables

$$\tau = \sqrt{t^2 - z^2} \tag{4}$$

$$y = \frac{1}{2} \ln\left(\frac{t + z}{t - z}\right) \tag{5}$$

On the average, the longitudinal velocity v_z of a secondary hadron would be related to the position (t,z) where the particle is created by $v_z = z/t$. The position of hadron creation is approximately equal to the position where $q\bar{q}$ pair creation occurs. Thus, y defined by (5) can be identified with the rapidity of hadrons. Since the four-volume element is given by $d^4x = dy \, d\tau \, \tau \cdot A$, Equations (2) and (3) lead to

$$\frac{dN_{pair}}{dy} = a \, A \int_0^\infty d\tau \tau \, E^2 . \tag{6}$$

In the central rapidity region we assume that the physics is approximately invariant under Lorentz boosts in the z direction. Then the field strength E becomes a function only of the proper time τ and hence takes the form

$$E(t) = E_0 f(\tau/\tau_0) , \tag{7}$$

where E_0 is the initial field strength at $\tau = 0$ and f(x) is a dimenionless function which satisfies f(0) = 1. The constant τ_0 sets the time scale for attenuation of the color field. Since the attenuation of the field due to pair creation is controlled by the local strength of the field, and since E_0 is the only parameter which has a dimension (of energy^2 or length^{-2}), τ_0 must be inversely proportional to $\sqrt{E_0}$:

$$\tau_0 = \frac{b}{\sqrt{E_0}} \quad . \tag{8}$$

Substituting (7) into (6) and using (8) we find

$$\frac{dN_{pair}}{dy} = aAE_0^2\tau_0^2 \int_0^\infty dx \; x \; f^2(x) = c_1AE_0 \tag{9}$$

where c_1 is a dimensionless constant. The initial field strength E_0 is related by Gauss' law to the color charge Q built up in the collision as

$$AE_0 = Q \quad . \tag{10}$$

Then we see that the particle density in the central rapidity region increases in proportion to Q,

$$\frac{dN_{pair}}{dy} = c_1Q \quad . \tag{11}$$

In the above discussion, we fixed the cross section of the flux tube. The tube will expand, however, because the field pressure ($= E_0^2/2$) is greater than the equilibrium pressure. The above derivation is right only if the field attenuation due to the pair creation is much faster than the expansion of the tube. Now let's consider the other extreme case, $viz.$, that the pair creation is a very slow process and that the tube first expands and attains its equilibrium shape. The equilibrium cross section of the tube is given by

$$A_{eq} = \frac{Q}{E_{eq}} \quad , \tag{12}$$

where the equilibrium field E_{eq} is defined to balance the external bag pressure via

$$\frac{1}{2} E_{eq}^2 = B \quad . \tag{13}$$

In this case the number of pairs produced after the expansion can be estimated as

$$\frac{dN_{pair}}{dy} = aA_{eq} \int_0^\infty d\tau \; \tau \; E^2(\tau)$$
$$= c_2A_{eq}E_{eq} = c_2Q \quad . \tag{14}$$

Thus, in both cases one sees that the particle density increases in proportion to $\sqrt{\nu}$,

$$\frac{dN_{pair}}{dy} = c\sqrt{\nu} \qquad (15)$$

where c is a universal constant. Assuming that the number of $q\bar{q}$ pairs is proportional to the number of hadrons, this leads to

$$\left(\frac{dn}{dy}\right)_{pA} \Big/ \left(\frac{dn}{dy}\right)_{pp} = \sqrt{\nu} \qquad (16)$$

This result seems to fit the data at the central rapidity region reasonably well[6] if we use the phenomenological relation

$$\nu = A\sigma_{pp}^{in} \Big/ \sigma_{pA}^{prod} . \qquad (17)$$

The use of Equation (17) is equivalent, in effect, to the assumption that the number of proton interactions ν in the target nucleus with atomic number A scales as $A^{1/3}$. This observation, combined with the foregoing discussion, allows us to make an interesting prediction for particle production and energy deposition in ultrarelativistic nucleus-nucleus collisions. The central collision of two identical heavy nuclei may be considered as the incoherent creation of many flux tubes whose cross section is given by that of a proton. In this case, the number of interactions (color exchange processes) which take place in each tube creation would be proportional to $A^{1/3} \times A^{1/3}$. Thus, average local color charge density per unit transverse area built up after the nucleus-nucleus collision grows as $\sqrt{A^{2/3}} = A^{1/3}$. This implies that one can expect an energy density in the central rapidity region $\sim A^{2/3}$ times that in a *pp* collision. This initial condition also leads to faster quark pair production characterized by $\tau_0 \propto A^{-1/3}$. Under such circumstances, we may suppose that the matter produced by the decay of color flux takes the form of a plasma of unconfined quarks and gluons, which later materializes into hadrons.

ACKNOWLEDGEMENTS

The authors are grateful to M. Faessler and S. Nussinov for stimulating discussions. The correspondence of M. Faessler is greatly appreciated.

REFERENCES

1. K. Johnson, *Acta. Phys. Pol.* B6 (1975) 865.

2. T. S. Biro, H. B. Nielsen, and J. Knoll, GSI preprint 84-23 (1984).

3. A. Casher, H. Neuberger, and S. Nussinov, *Phys. Rev. D* 20 (1979) 179; N. K. Glendenning and T. Matsui, *Phys. Rev. D* 28 (1983) 2890.

4. A. Białas and W. Czyż, Jagellonian University Preprint TPJU-3/84 (1984).

5. J. Schwinger, *Phys. Rev.* 82 (1951) 664.

6. This fact was pointed out by M. Faessler.

AN INTRODUCTION TO LATTICE QCD

G. Martinelli

INFN - Laboratori Nazionali di Frascati

P.O. Box 13, 00044 - Frascati, (Italy)

1. - INTRODUCTION

In the last decade an impressive collection of converging experimental results indicated Quantum Chromodynamics (QCD) as a unique candidate to describe strong interactions. Many important predictions from perturbative QCD have been already experimentally confirmed to a certain quantitative level. Among the others:

a) The value of R= σ (e$^+$e$^-$ $\longrightarrow$ hadrons)/ σ (e$^+$e$^-$ $\longrightarrow \mu^+ \mu^-$)
b) The properties of jets at PETRA energies ($\sqrt{s}$ $\simeq$ 30 GeV)
c) The dependence of the average transverse momentum < p_T > on $\sqrt{s}$ in hadronic collisions
d) The photon structure function
e) -Onium decays
f) Scaling violations

However no single test of the theory has been so far decisive and still some important predictions, like the existence of the glueball, have not been verified. Moreover many basic questions remain unanswered by perturbative computations. Mainly:

a) confinement
b) dynamical symmetry breaking
c) hadron spectrum (glueball mass)
d) deconfinement phase transition at high temperature.

To solve these basic problems a lot of theoretical effort has been devoted to develope techniques able to explore the phenomenological consequences of QCD in the non-perturbative domain. The most exciting results have been obtained by using the lattice formulation of QCD which was started in 1974 by K.G. Wilson[1].

These lectures have been conceived to serve as an elementary introduction to this fascinating field and to offer to non-experts a survey of the most relevant results.

The plan of the lectures is the following: in Sect. 2. I will introduce general ingredients and basic concepts of lattice gauge theories and I will present the results for quark confinement in the strong and weak coupling regime. In Sect. 3 I will briefly discuss the problems which arise when one puts fermions on the lattice. In Sect. 4 the strategy of the computation of the hadron mass spectrum and of related quantities relevant for hadron physics will be described and the most important results will be reported. Finally in Sect. 5 I will present the results that have been so far obtained for QCD at finite temperature, a subject which is of great interest for people working in heavy ions collisions.

2. - GENERALTIES ON LATTICE GAUGE THEORIES

2.1. - The Gauge Action on the Lattice

All strong interaction physics can be in principle derived starting from the following action:

$$S= \int d^4x \left[- \frac{1}{2} \overline{q}(x) \gamma^\mu \overleftrightarrow{D}_\mu \, q(x) - m_q \overline{q}(x)q(x) - \frac{1}{4} F_{\mu\nu}(x)^2 \right] \qquad (1)$$

where q(x) are the quark fields; $D_\mu = \partial_\mu + ig \, A^a_\mu \, \lambda^a$ is the covariant derivative and $F^a_{\mu\nu} = \partial_\mu \, A^a_\nu - \partial_\nu \, A^a_\mu - g \, f^{abc} A^b_\mu A^c_\nu$, where A^a_μ (a=1, ..., 8) are the gluon fields; g is the bare coupling constant. The action in Eq. (1) is a generalization to a non abelian group (SU(3) of colour for QCD) of the usual, abelian local gauge invariant action of electrodynamics[2].

Unfortunately the simple action of Eq. (1) gives rise to a very complicated physics because the fields in the Lagrangian do not correspond to observable particles but are permanently confined to form hadrons (pions, protons, ...). Thus, unlike for weak and electromagnetic interactions it is possible to use perturbation theory only for deep inelastic (very high energy) phenomena where asymptotic freedom is at work[3].

Actually the only formulation which is (will become) able to give quantitative predictions for hadron physics at low energy (masses, widths, decay amplitudes etc.) is lattice QCD. To put QCD on the lattice is a very natural way of introducing an ultraviolet cutoff (necessary to any renormalizable theory) in a gauge invariant way[1]. The continuum limit is obtained when the lattice spacing a→0 (i.e. the ultraviolet cutoff goes to infinity).

We proceed in two steps:

a) The quantum theory in a euclidean space-time.

In the usual Minkowsky space-time any physical amplitude

can be expressed in terms of the vacuum generating functional and its derivatives (vacuum expectation values of operators)[4]. They have the form:

$$Z(J) = \sum_{\{\phi\}} e^{i \int d^4 x \, \mathcal{L}(\phi, \partial_\mu \phi, J)}$$

$$\langle \hat{O}(\phi) \rangle_{J=0} = \frac{1}{Z(0)} \sum_{\{\phi\}} O(\phi) e^{i \, S(\phi, \partial_\mu \phi, 0)} \tag{2}$$

ϕ is any quantum field. $\sum_{\{\phi\}}$ indicates the integral over all the possible values of the fields, at any space-time point (this is formal definition to which is possible to give a precise mathematical meaning at least in perturbation theory).

We make an analytic continuation in the time coordinate which defines the model in four dimensional Euclidean space[5]:

$$t \longrightarrow i \, x_4 \tag{3}$$

Eqs. (2) now have the form:

$$Z(J) = \sum_{\{\phi\}} e^{-\int d^4 x \, \mathcal{L}(\phi, \partial_\mu \phi, J)}$$

$$\langle \hat{O}(\phi) \rangle_{J=0} = \frac{1}{Z(0)} \sum_{\{\phi\}} O(\phi) e^{-S(\phi, \partial_\mu \phi, 0)} \tag{4}$$

Z can be interpreted as the partition function of a four dimensional classical system whose Hamiltonian is given by the relation:

$$\beta H = S \tag{5}$$

We have trasformed the quantum field theory problem in the study of the physics of a statistical system of classical fields.

b) The theory on a lattice

To regularize the infinities of the theory we replace the continum space-time by a mesh of discrete lattice points. The most simple case is a hypercubic lattice with point separation a. The lattice coordinates of a point are denoted by a four integer vector $n = (n_1, n_2, n_3, n_4)$ and the fields are defined only on the points of the lattice.

A very simple example is given by a d-dimensional free field scalar theory:

$$S_{continuum} = \int d^d x \, \frac{1}{2} (\partial_\mu \phi)^2$$

$$\phi(x) \longrightarrow \phi(n); \qquad \partial_\mu \phi(x) \longrightarrow \frac{\phi(n + \hat{\mu}) - \phi(n)}{a} \tag{6}$$

$$S_{lattice} = \frac{1}{2} \sum_{n, \hat{\mu}} \left[\phi(n + \hat{\mu}) - \phi(n) \right]^2$$

where $\phi(n) = a^{(d/2)-1} \phi(x)$ at x=na and $\hat{\mu}$ is the unit vector in the μ-direction. It is trivial to see that:

$$\lim_{a \to 0} S_{lattice} = S_{continuum}$$

In the case of a gauge theory, we require that the lattice action satisfies two requisities[1]

i)   Local (lattice) gauge invariance
ii) Formal (tree level) a→ 0 limit

An example of a pure gauge (no quarks) action which satisfies i) and ii) was originally proposed by Wilson

$$S_G(U) = \frac{1}{g^2} \sum_P tr \left[U(P) + U^+(P) \right] \tag{7}$$

where U(P) is the product of link matrices belonging to an elementary plaquette as explained in Figs. 1a and 1b:

$$U(P) = U_\mu (n) U_\nu (n+\hat{\mu}) U_\mu^+ (n+\hat{\nu}) U_\nu^+ (n) \tag{8}$$

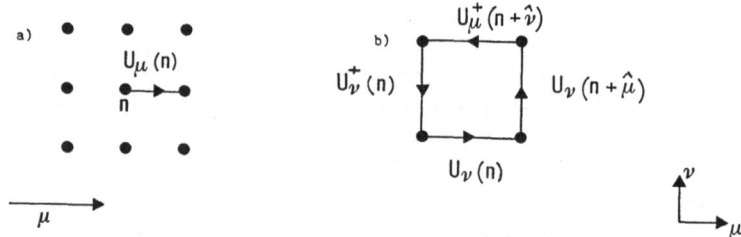

FIG. 1 - a) We associate to any point of the lattice and direction an oriented link variable $U_\mu(n)$ which connects the point n with the point $n+\hat{\mu}$. The relation between $U_\mu(n)$ and the corresponding gluon field is given in Eq. (9); b) the ordered product of four links belonging to an elementary plaquette defines U(P).

The link variable $U_\mu (n)$ is an element of the symmetry group and it is related to the gluon field A_μ^a by the equation:

$$U_\mu (n) = e^{i a g A_\mu^a (n) \lambda^a} \tag{9}$$

where λ^a are the generators of the group. $S_G(U)$ is invariant under the local gauge transformation:

$$U_\mu (n) \to g(n) U_\mu (n) g^+(n+\hat{\mu}) \tag{10}$$

where g(n) is an element of the group. It is easy to verify that for a→ 0:

370

$$S_G(U) \longrightarrow -\frac{1}{4g^2} \int F_{\mu\nu}(x)^2 \, d^4x + O(a^2) \tag{11}$$

which is the usual expression for the continuum action (up to a normalization of the gauge fields).

Two comments are necessary at this point. The first comment is that there is an arbitrariness in defining the lattice action due to the extra $O(a^2)$, $O(a^4)$ ... terms present on the lattice. We could for example define an action made by the product of 6 links which still satisfies i) and ii) as illustrated in Fig. 2. The second comment is that going to the lattice formulation we have lost Lorents invariance: only a residual symmetry under rotations of ninty degrees is left. In general, by putting the theory on the lattice (see also Sect. 3), we obtain symmetry properties which are different from the corresponding properties in the continuum. A

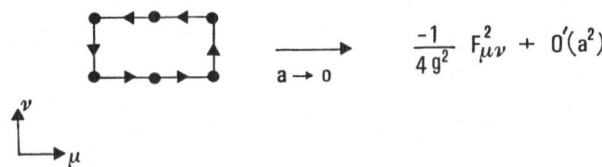

FIG. 2 - The action defined by the product of 6 links will tend to the continuum action for a $\longrightarrow$ 0. At finite a it differs from the definition shown in Fig. 1b.

proper a$\longrightarrow$ 0 limit of the lattice action should recover the symmetry properties of the original theory.

The vacuum functional for the lattice theory in absence of external currents can be written as

$$Z = \int \prod_{n,\hat{\mu}} d\,[U]_{n,\hat{\mu}} \, e^{S_G(U)} \tag{12}$$

where d $[U]$ is the invariant Haar measure with the properties:

$$\begin{aligned} d\,[U] &= 1 \\ d\,[U]\, f(U) &= \int d\,[U]\, f(gU) \end{aligned} \tag{13}$$

where g is an arbitrary element of the group.

2.2. - Removing the ultraviolet cut-off

The full quantum continuum limit of the lattice action is a much more subtle problem than the formal limit given in Eq. (11) and it involves the full restoration of the symmetry (and eventually topological) properties of the theory.

The only parameters which enter in the action of Eq. (7) are the coupling constant g and (implicitly) the lattice spacing a. All masses and lengths may be expressed in terms of these parameters:

$$m_i = \frac{1}{a} \, f_i(g) \tag{14}$$

In the limit in which the lattice spacing becomes smaller and smaller we expect that it is possible to vary g and a in such a way that the physical properties of the system remain unaltered. Cut-off independence for masses implies:

$$a \, \frac{dm_i}{da} = 0 \tag{15}$$

(This is the condition that in the limit $a \longrightarrow 0$ physics becomes independent of the ultraviolet cut-off). At small g, because of asymptotic freedom, we expect:

$$- a \, \frac{dg}{da} = - \beta_o g^3 - \beta_1 g^5 + \dots = \beta(g) \tag{16}$$

where $\beta_{o,1} > 0$ are universal coefficients that can be computed in perturbation theory.

Eqs. (15) and (16) imply that, for $a \longrightarrow 0$ all masses become proportional to a unique fundamental scale:

$$m_i = c_i \, \Lambda_{latt} \tag{17}$$
$$m_i/m_j = \text{constant}$$

where c_i are dimensionless constants and also that:

$$a = \frac{1}{\Lambda_{latt}} \, \exp\left[\frac{-1}{2 \, \beta_o g^2}\right] \, (\beta_o g^2)^{\frac{-\beta_1}{2\beta_o^2}} \left[1 + 0(g^2)\right] \tag{18}$$

At Λ_{latt} fixed (i.e. masses fixed in physical units) Eq. (18) shows that the lattice spacing a goes to zero exponentially in $1/g^2$ as $g \longrightarrow 0$. Correspondly all dimensionless correlation lengths ξ i/a= $1/m_i a$ go to infinity: in the language of statistical mechanics the system undergoes a second order phase transition. The size of the physical particles becomes bigger and bigger in units of the lattice spacing as g is decreased so that we have a better and better description of the continuum physics as much in the same way one obtains a more and more accurate result in a numerical evaluation of an integral by increasing the number of subdivisions of the interval of integration.

In the continuum limit different lattice action should produce the same continuum physics (universality principle). On one side this means that ratios of masses (e.g. the mass of the proton over the mass of the ϱ) must be independent of the action we used. On the other side to different forms of lattice actions, which have different scale parameters Λ_{latt}, should correspond the same physical quantities:

$$m_i = c_i \Lambda_{latt} = c_i' \Lambda_{latt}'$$

$$R = \frac{\Lambda_{latt}'}{\Lambda_{latt}} = \frac{c_i}{c_i'} \tag{19}$$

R can be predicted in perturbation theory and compared with the results for c_i and c_i' (obtained for example by Montecarlo techniques) as a test of universality.

Finally we should recover the symmetry properties of the continuum theory. In Fig. 3 I report as an example of the restoration of rotational invariance, the quark-antiquark equipotential lines, obtained in SU(2) gauge theory at two different values of g: $\beta = 4/g^2 = 2.0$ and 2.25. At $\beta = 2.0$ the equipotential lines clearly remember the lattice structure. The result at $\beta = 2.25$ is expected to be much more in the (quasi) continuum regime.

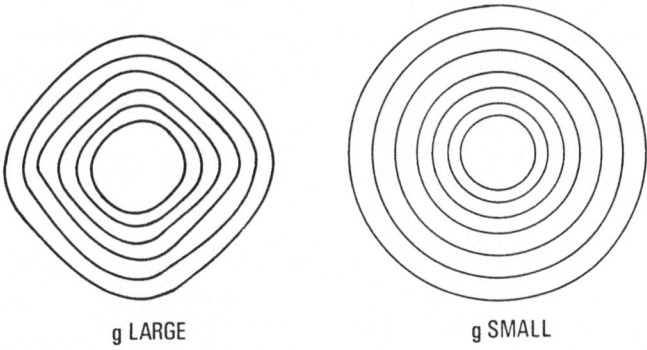

g LARGE   g SMALL

FIG. 3 - Equipotential lines of two heavy quarks in SU(2) gauge theory: $4/g^2 = 2.00$ on the left side; $4/g^2 = 2.25$ on the right side.

2.3. - Computational Techniques

We have seen that all the relevant physical informations are given by expectation values of (gauge invariant) operators:

$$\langle \hat{O}(U) \rangle = \frac{\int d\,[U]\; e^{-S_G(U)}\, O(U)}{\int d\,[U]\; e^{-S_G(U)}} \tag{20}$$

The computation of the functional integral in Eq. (20) on an euclidean lattice is not easier then the original computation in the Minkowky continuum space-time. For most of the theories (including QCD) an exact evaluation is an impossible task. Several approximation techniques are available: in the following I will briefly discuss the strong coupling expansion technique only to show that quarks are confined in the limit $g \longrightarrow \infty$. The rest of the subsection is devoted to explain Montecarlo techniques and their application to lattice QCD. Other approximations like mean field theory, 1/N expansion, 1/d expansion etc. will not be covered in these lectures.

a)   Strong coupling expansion

Let us write the exponential of the action appearing in the functional integral of Eq. (20) as:

$$e^{-S_G(U)} = e^{-\beta H} = e^{\frac{\beta}{2N} \sum_P tr\left[U(P)+U^+(P)\right]} \tag{21}$$

$$\beta = \frac{2N}{g^2} \qquad\qquad N = \text{number of colour}$$

β plays the same role of the inverse temperature in statistical physics. In the strong coupling limit one makes a Taylor expansion of the exponential of Eq. (21) around $\beta = 0$.

It is interesting to evaluate in this limit the potential of an infinitely heavy quark-antiquark pair as a function of the distance. The quark (antiquark) can be treated as a classical external source of the gluon field:

$$S(U) \longrightarrow S(U) + i\,g \int A_\mu^a \cdot J_\mu^a\, d^4x$$
$$J_\mu^a \sim \frac{\lambda^a}{2}\; \partial_{\mu o}\delta^3(x) \tag{22}$$

From Eqs. (22) the energy of the quark-antiquark system is easy evaluated using the path integral formalism:

$$\frac{Z(J)}{Z(o)} = \langle\, P\, (e^{ig \oint_C (A_\mu^a \frac{\lambda^a}{2})\, dx_\mu})\, \rangle = \tag{23}$$

$$= e^{-V(R)\,T} \qquad \text{when } T \longrightarrow \infty$$

P is a path ordered product. The contour C of the line integral of the gauge field is shown in Fig. 4 and V(R) is the quark-antiquark potential at a distance R. On the lattice the line integral becomes the ordered product of links belonging to the contour C:

$$e^{i\,g\oint_C (A_\mu^a \frac{\lambda^a}{2})\,dx_\mu} = \prod_C U_C = W(C) \tag{24}$$

$$\frac{Z(J)}{Z(o)} = \frac{\int d[U]\, e^{\frac{1}{g^2}\,tr\,[(U(P)+U^+(P)]}\,W(C)}{\dot{Z}(o)} \tag{25}$$

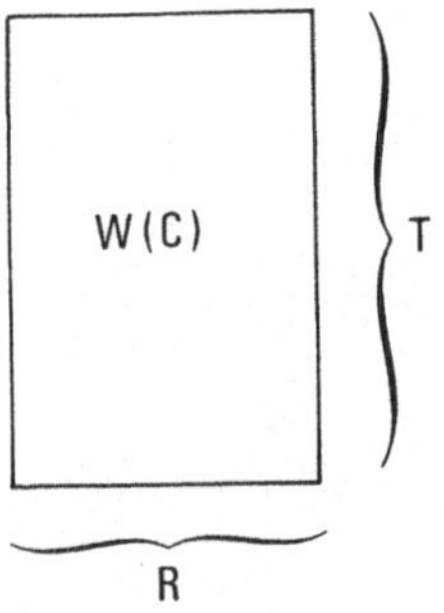

FIG. 4 - Wilson loop: the contour of the line integral of the gluon field for the computation of the quark-antiquark potential at a distance R.

At the lowest non trivial order in $1/g^2$, expanding the exponential and recalling that:

$$\int d[U]\,(U,U^+,U^2,U^{+2}) = 0$$
$$\int d[U]\,U_{ij}\,U_{kl}^+ = \frac{1}{N}\,\delta_{il}\,\delta_{jk} \tag{26}$$

One finds:

$$exp-V(R)T = (\frac{1}{g^2 N})^{TR/a^2} = exp - \frac{TR}{a^2}\,ln\,(g^2 N) \tag{27}$$

$$V(R) = \left[-\frac{1}{a^2}\,ln(g^2 N) \right]\,R = KR$$

K is the so called "string tension". In this limit, since the potential increases linearly with the distance, quarks (but also electrons!!) are confined. Many terms of the $\beta = 2N/g^2$ expansion have been calculated in order to extrapolate the series to the (physically relevant) small coupling region. Due to singularities of the serie near the real axis[7] which are probably due to an infinite order roughening phase transition[8] it is not obvious at present that informations on the interesting scaling region can be extracted by strong coupling expansion.

An alternative method to compute the quark-antiquark potential at any value of the coupling constant is given by Montecarlo techniques.

b)   Montecarlo Methods

In a Montecarlo simulation one attempts a direct integration of the gauge fields an a <u>finite</u> lattice. The accuracy of the results is dictated by limitations imposed on the size of the lattice by computer memory and speed.

The situation is illustrated in Fig. 5: at large values of g the size of the hadron is of the order the lattice spacing and strong effects due to the (too small) ultraviolet cutoff are present; at small g, with a limited number of lattice points, the hadron becomes as large as the lattice and it feels strong finite

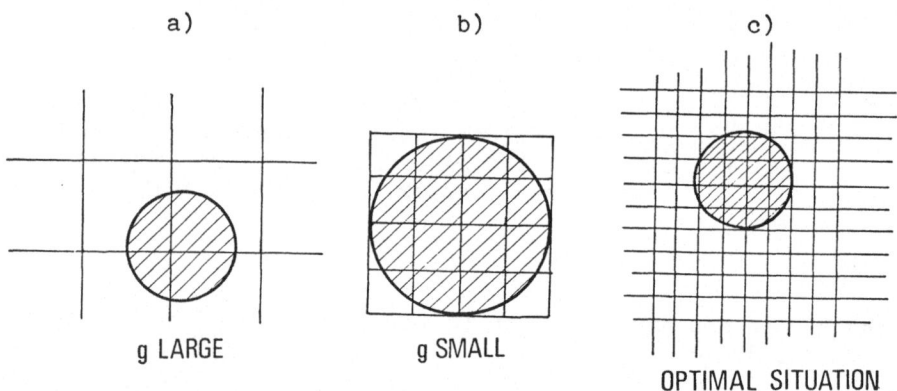

<u>FIG. 5</u> - a) Hadron on the lattice at g large: the hadron is smaller than the lattice spacing; b) when g⟶ o the hadron becomes as large as the whole lattice; c) the hadron size is much smaller of the lattice and much larger of the lattice spacing.

size (infrared) effects. The ideal situation is shown in Fig. 5c: the hadron is much larger than the lattice spacing but much smaller than the lattice size. In actual simulations this situation is really never fully achieved; we will return on this point when discussing Montecarlo results. A useful example of what occurs in practical cases is also given in Fig. 6: the data with errors are measurements of some hadron mass by Montecarlo simulation; the full curve comes from strong coupling expansion; the dashed lines correspond, up to a multiplicative constant, to the expected asymptotic renormalization group behaviour of Eq. (18); the dotted-dashed line is the spin-wave (g ⟶ o on a finite lattice) prediction. If the lattice is not large enough, finite size effects could prevent us to see the region where masses scale according to Eq. (18) (scaling region indicated by the arrow in Fig. 6). Even in the most favourable case only the results on a small range of g can be used for the extrapolation to the continuum since correlation lengths increase exponentially in $1/g^2$.

376

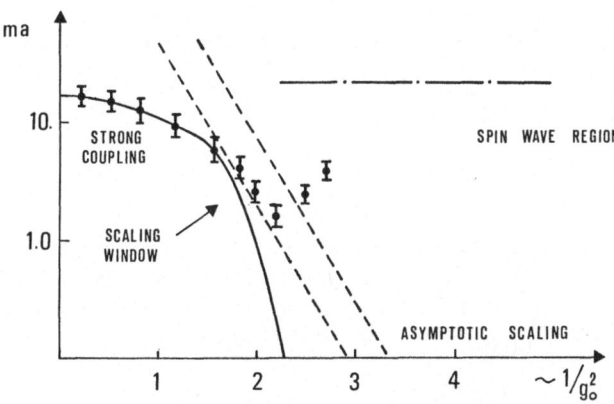

FIG. 6 - Schematic representation of the behaviour of some hadron mass as a function of $\beta = 6/g^2$. The scales of the plot are arbitrary.

To avoid finite size effects, the lattice size must be increased linearly in the correlation length, that is the number of lattice points increases as the fourth power of the correlation length. Notice that on a lattice of 10^4 points link variables possess 3×10^5 degrees of freedom on which we have to integrate. It is clear that to make a multidimensional integral on 10^5 variables one needs to use a Montecarlo technique. Several of them have been developed in the past; the most popular ones are:

i)   Metropolis[9];
ii) Heath bath[10];
iii) Langevin equation[11];
iv) Microcanonical Ensemble[12];
.
.
.

In the following I will describe as an example the Metropolis algorithm.

METROPOLIS

    -      Start with an arbitrary gauge field configuration $\{ U \}$

1 -   Extract at Random a new link $U' \rightarrow \{ U' \}$

    -      If $\exp \left[-S(\{ U' \}) + S(\{ U \}) \right] > x$
           Where x is a Random number with flat distribution in the
           interval $\left[0,1 \right]$
           Then the new configuration is $\{ U' \}$
           Otherwise $\{ U \}$

           Go to 1

This algorithm satisfies the detailed balance conditions:

$$P_{eq}(\{ U \}) W \left[\{ U \} \rightarrow \{ U' \} \right] = P_{eq}(\{ U' \}) W \left[\{ U' \} \rightarrow \{ U \} \right] \qquad (27)$$

where $P_{eq}(\{U\})$ is the equilibrium distribution probability:

$$P_{eq}(\{U\}) = e^{-S_G(U)} \qquad (29)$$

and $W\left[\{U\} \rightarrow \{U'\}\right]$ is the transition probability from one configuration to another. It can be (not rigorously) shown that the system will tend to the equilibrium with a probability distribution $P_{eq}[\{U\}]$. This means that, when the number of field configurations N_{CONF} goes to infinity:

$$\langle \hat{0}(U) \rangle = \frac{\int d[U]\, e^{-S_G(U)} 0(U)}{\int d[U]\, e^{S_G(U)}} =$$

$$= \lim_{N_{CONF} \rightarrow \infty} \frac{1}{N_{CONF}} \sum_i 0\left[\{U_i\}\right] \qquad (30)$$

Thus a further limitation to the accuracy of the results is set by the fact that the statistical error on the determination of $\langle 0(U) \rangle$ will decrese as slowly as $1/N_{CONF}^{1/2}$.

2.4. - Montecarlo Results for the Quark-antiquark potential

As we have seen in Sect. 2.3, it is possible to study the quark-antiquark potential by measuring expectation values of Wilson loops. There we considered the potential in the limit $g \rightarrow \infty$ and found that in this limit quarks are confined. By Montecarlo simulations one can measure Wilson loops in the entire range of g and in particular in the perturbative region ($g \rightarrow o$) where we can extract informations on the continuum physics. The main results from Montecarlo simulations (made on lattices with volumes up to 16^4 points) can be summarized as follows:

1) For SU(3) the asymptotic scaling predicted by Eq. (18) set up at values of $\beta = (6/g^2) \gtrsim 6-6.2$. In this region we have to focus our computational efforts[13].

2) The quark-antiquark potential is well described by the formula:

$$V(R) = \frac{a}{R} + bR \qquad (31)$$

where a and b are (logathimically varying) constants[14,15]. This form of potential was successfully used for the description of energy levels of the J/ψ and Y systems[16]. The form of the potential in Eq.(31) indicates that quarks feel a Coulomb-like potential at short distances but are permanently confined because of the term linearly increasing with R.

From the measurement of b (which is related to the string tension) one finds:

$$\Lambda_{\overline{MS}} \simeq 100 \text{ MeV}$$

$\Lambda_{\overline{MS}}$ is also measured in deep inelastic experiments. A value of around 100 MeV is in agreement the high energy results.

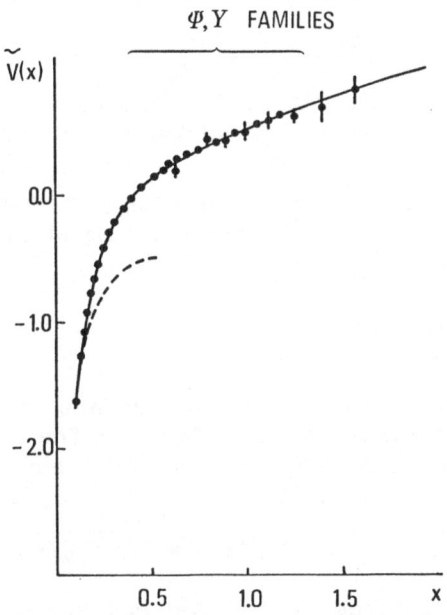

FIG. 7 - $\tilde{V}(x)$ vs x. x= bΛ_{latt}R; $\tilde{V}$= V/(bΛ_{latt}). The solid line is fit to $\tilde{V}$ of the form V=Ax-β/x. The dashed line is from two loop perturbation theory. The data (obtained for β=6 − 7.2 and V = 8^3x12) are from Ref. 14.

In Fig. 7 the potential V(R) is reported together with the fit to the expression given in Eq. (31) and with the expected quark-antiquark potential from perturbation theory. If scaling is working the values of V(β) at different values of should map on a single curve representing the potential at fixed physical distances. This appears to be the case in Fig. 7.

3. - FERMIONS ON THE LATTICE

3.1. - The Fermion Doubling Problem

A special problem arises when one tries to write on the lattice the fermionic part of the action given in Eq. (1) along the lines explained in Sect. 2.

A rather general form of the fermionic action is given by[17]:

$$S_\psi = \sum_{f,n} \int - \frac{1}{2} \sum_\mu \left[(\bar\psi^f(n)(r - \gamma_\mu)U_\mu(n) \psi^f(n+\hat\mu) \right.$$
$$\left. + \bar\psi^f(n+\hat\mu)(r + \gamma_\mu)U_\mu^+(n) \psi^f(n) \right] \qquad (32)$$
$$+ (m_f + 4r) \bar\psi^f(n) \psi^f(n) \int$$

ψ^f is the quark field with flavour f and mass m_f; γ_μ are the Dirac matrices and r is an arbitrary parameter which defines the lattice action.

Let us first consider the action with r=0. This corresponds to the naive lattice action for fermions obtained by the replacement in the continuum action:

$$\partial_\mu f(x) = \frac{f(x+\hat\mu) - f(x)}{a} \qquad (33)$$

For r=0 the action in Eq. (32) has apparently the same chiral properties of the continuum one. In fact, in the limit in which the explicit mass term m_f=0, the action is invariant under the following (global) transformations:

$$\psi(x) \rightarrow e^{i\alpha} \psi(x) \qquad\qquad \psi(x) \rightarrow e^{i\alpha\gamma_5} \psi(x) \qquad (34)$$

More generally the lattice action for r=0 and m_f=0 is invariant under the global continuum symmetry groups:

$$SU(n_f)_V \times SU(n_f)_A \times U(1)_V \times U(1)_A \qquad (35)$$

exactly as it was for the continuum QCD action. In QCD the $U(1)_V$ symmetry corresponds to the conservation of the baryon number; the $SU(n_f)_A$ symmetry is expected to be dynamically broken giving rise to massless Goldstone bosons (pions, kaons, ...) while the $U(1)_A$ is broken because of the triangle anomaly (see Fig. 8) and this should explain why the η' is not a light particle.

However, on the lattice something strange is happening: in fact the triangle anomaly is identically zero.

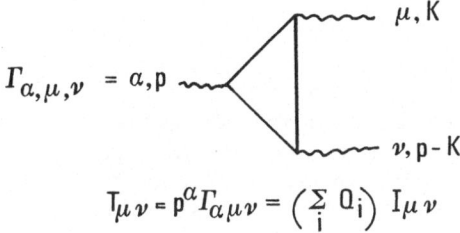

$$\Gamma_{\alpha,\mu,\nu} = \alpha,p$$

$$T_{\mu\nu} = p^{\alpha}\Gamma_{\alpha\mu\nu} = \left(\sum_{i} Q_{i}\right) I_{\mu\nu}$$

<u>FIG. 8</u> - The Feynman diagram for the triangle anomaly.

The fermion spectrum on the lattice (if the corresponding continuum theory has only one flavour) is constituted by 16 different replicas of the original Dirac field, each of the replicas giving an opposite sign contribution to the triangular anomaly. The simplest explanation of the doubling of fermions can be found by looking at the free fermion propagator on the lattice:

$$S(p) = \frac{1}{\frac{1}{a} \sum_{\mu} \gamma_{\mu} \sin p_{\mu} a} \tag{36}$$

The propagator in Eq. (36) has 16 poles corresponding to $p_{\mu}a = 0, \pi$. The triangle anomaly is proportional to the sum of the axial charges of the looping fermions (see Fig. 8) which have charges [17]:

$$Q(0, 0, 0, 0) = +1, \qquad Q(\pi, 0, 0, 0) = 1$$

The unexpected properties of the lattice action we considered here are general in character and have been stated precisely by a "No go" theorem by Nielsen and Ninomiya [18] which proves that it is not possible to write a fermion action on the lattice which is local, has the same chiral properties of the continuum and avoids the fermion doubling problem.

3.2. - Two Popular Fermionic Actions on the Lattice

Two approaches have been widely used to overcome (but never completely because of the N.N. theorem) the fermions doubling problem

a) Staggered fermions
b) Wilson fermions

a)   Consider as a simple instructive example a one space-one time non interacting fermionic system. In the continuum, the action and the Dirac equation are:

$$S = \int dt\ dx\ (\overline{\psi}\ \partial\!\!\!/\ \psi) \tag{37}$$

$$\dot{\psi} = - \gamma_5 \, \partial_z \, \psi = - \begin{pmatrix} 0 & 1 \\ 1 & 0 \end{pmatrix} \psi_z \tag{38}$$

If we discretize the spatial direction we find two massless fermions corresponding to the points pa=0, π. However, we can define a single two component Dirac field by decomposing the lattice in an even and a odd sub-lattice:

$$\psi_1(n) = \phi(n) \qquad\qquad n \text{ even}$$
$$\psi_2(n) = \phi(n) \qquad\qquad n \text{ odd} \tag{39}$$

The Equation of motion becomes:

$$\dot{\psi}_1(n) = \frac{-1}{2a} \Big[\psi_2(n+1) - \psi_2(n-1) \Big]$$

$$\dot{\psi}_2(n) = \frac{-1}{2a} \Big[\psi_1(n+1) - \psi_1(n-1) \Big] \tag{40}$$

which perfectly corresponds to the continuum case (Eq. 38). However while the action had originally a continuum γ_5 symmetry we are left with a discrete γ_5 symmetry which corresponds to translations of an odd number of lattice spacings:

$$n \to n+1 \qquad \begin{pmatrix} \psi_1 \\ \psi_2 \end{pmatrix} \to \begin{pmatrix} \psi_2 \\ \psi_1 \end{pmatrix} = \gamma_5 \begin{pmatrix} \psi_1 \\ \psi_2 \end{pmatrix} \tag{41}$$

In four dimensions, starting from a single flavour in the continuum, we end up with a four flavour theory whose fields are defined by suitable conbinations of the fermionic fields at different lattice points. Of the original continuum symmetries of the action only two symmetries survive; one vector symmetry corresponding to baryon number conservation and an axial flavour singlet symmetry which is a remnant of the full chiral structure of the theory. This symmetry, together with other discrete chiral symmetries, prevent fermions to acquire a mass if the bare explicit mass term is put equal to zero in the lattice action. Staggered fermions are particulary usefull to study the chiral properties of the theory: when the survival continuum chiral symmetry is spontaneously broken we should find the corresponding Goldstone boson in the spectrum.

b)   Wilson fermions

For r≠0 the propagator has no pole around p_μ a= π (even when m_f=0):

$$S(p) = \cfrac{1}{\cfrac{1}{a} \sum_\mu \sin p_\mu a + \cfrac{2r}{a} \sum_\mu \sin^2(\cfrac{p_\mu a}{2})} \qquad (42)$$

The unwanted fermion copies acquire an effective "mass" of the order of 2r/a and disappear from the spectrum as a $\rightarrow$ 0. The price to pay, according to the N-N theorem, is that we explicitly broke the chiral symmetry of the action so that (even if we put the explicit mass term to zero) mass terms will originate in perturbation theory (see Fig. 9); no symmetry protects the fermion from acquiring a mass. Also in this case one may expect that the correct chiral properties will be recovered in the continuum limit. The Wilson formulation happened to be particularly suitable for the computation of the hadron mass spectrum. Also naive fermions have been used to this purpose.

FIG. 9 - The quark self energy diagram which gives a mass to a Wilson quark also if in the bare action m_f=0.

3.3. - Montecarlo Simulations with Fermions and the Quenched Approximation

The fermion action in Eq. (32) has the form:

$$S_\psi = \sum_f \bar{\psi}^f \Delta_f (U) \psi^f \qquad (43)$$

Since S_ψ is quadratic in the fermion fields, we can formally integrate out the fermionic degrees of freedom in the functional integral. Let us take the vacuum expectation value of a generic gauge invariant operator depending on the fermion fields ψ, $\bar{\psi}$ and on the link variables:

$$\langle \hat{O}(\psi, \bar{\psi}, U) \rangle =$$

$$= \cfrac{\int d[\bar{\psi}] d[\psi] d[U] \, e^{-S_G(U)-S_\psi} O(\bar{\psi}, \psi, U)}{\int d[\bar{\psi}] d[\psi] d[U] \, e^{-S_G(u)-S_\psi}} \qquad (44)$$

The gaussian integral over ψ, $\bar{\psi}$ is easily done giving:

$$\langle \hat{0} \rangle = \frac{\int d\,[U]\ e^{-S_G(U)}\ \underset{f}{\Pi}\,\det\left[\Delta_f(u)\right]\ \tilde{0}\left[U, \Delta_f^{-1}(U)\right]}{\int d\,[U]\ e^{-S_G(U)}\ \underset{f}{\Pi}\,\det\left[\Delta_f(U)\right]} \tag{45}$$

The determinant of $\Delta_f(U)$ contains the effects of quark loops on the gluon Green functions as explained in Fig. 10. For a fixed gauge field configuration U , $\Delta_f^{-1}(U)$ is the quark propagator in presence of the external field U as shown in Fig. 11. Under suitable hypothesis on the fermion determinant we can write Eq. (45) as:

<u>FIG. 10</u> - Typical diagram contributing to det ($\Delta_f(U)$).

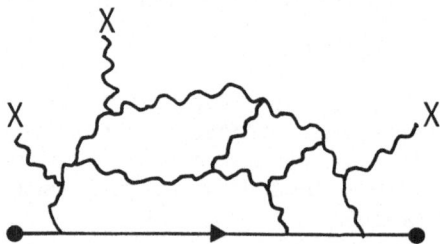

<u>FIG. 11</u> - Quark propagator in presence of the field $\{U\}$.

$$\langle \tilde{0} \rangle = \frac{\int d[U]\ e^{-S_{eff}(U)}\ \tilde{0}\left[U, \Delta_f^{-1}(U)\right]}{\int d[U]\ e^{-S_{eff}(U)}} \tag{46}$$

where $S_{eff}(U) = S_G(U) - \sum\ \text{tr}\left[\ln \Delta_f(U)\right]$.

384

In presence of quarks, in a Montecarlo procedure, the integral over the gluon fields will be replaced by the sum over the gauge field configurations generated to have an equilibrium distribution:

$$e^{-S_{eff}(U)} d\left[U\right] \tag{47}$$

The fermionic determinant is an highly non-local quantity and its computation turned out to be one of the most serious difficulties in Montecarlo simulations. Most of the results for hadron spectroscopy that will be discussed later where obtained in the so called "quenched" approximation[21]:

$$\det\left[\Delta_f(U)\right] = 1 \tag{48}$$

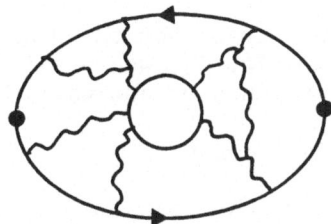

FIG. 12 - Diagram that enters in the computation of the width $\varrho \rightarrow 2\pi$.

Only the quark propagator in presence of an external field has to be computed and the gauge field configurations are generated as in the pure gauge case. The quenched approximation is expected to work reasonably well for several reasons: it is exact for $N_{colour} \rightarrow \infty$ at n_f fixed; Zweig rule is exactly satisfied and hadrons are made only by valence quarks and gluons, almost as much in the same way as it happens in real world; all the few known results for the spectroscopy which include the effects of fermion loops seem in good agreement (within $\sim 10 \div 15\%$) with the results of the quenched approximation. On the other side it is clear that there are quantities that cannot be computed in this approximation, like for example the width of the $\varrho \rightarrow 2\pi$ (see Fig. 12). Moreover, we expect that fermion loops are very important to establish the existence and the properties of the deconfinement phase transition: only recently the first complete Montecarlo simulations for this problem where started on large lattices (see next Sect.)

4. - THE HADRON SPECTRUM

4.1. - Prerequisites to a Realistic Computation of the Hadron Mass Spectrum

To trust the results for hadron spectroscopy in lattice QCD we must verify that three conditions are satisfied:

a)   We are deep enough in the weak coupling region so that the renormalization group behaviour of Eq. (18) is satisfied.

b)   Despite of the problem of fermion doubling we recover the chiral symmetry properties of the continuum theory.

c)   Different lattice actions give the same continuum physics (universality).

No definite answer can be given to the points a), .. c) and in some particular cases there are counter examples indicating that some of these conditions are violated. Still lot of work is needed to set definitively these points.

Several indications showed that there exists a region of the coupling constant, A+B in Fig. 13, where all dimensionless masses scale (for a given action) in a unique way:

$$m_i a = c_i \, f(g) \longrightarrow m_i/m_j = \text{const.} \tag{49}$$

$f(g)$ is common to all masses but it is not universal among

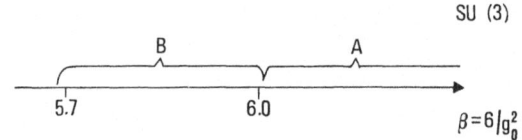

FIG. 13 - Regions A and B where asymptotic scaling and scaling hold.

different actions and it is different from the perturbative calculable $f(g)$ of Eq. (18). Only in a region of smaller g (A in Fig. 13), for which rather large lattices must be used, masses start to scale with the predicted law from perturbation theory. In region B universality seems to hold in a weaker sense suggested by the 1/N expansion. In SU(3), using the Wilson action for the gauge fields, we expect that region A starts at $\beta = 6/g^2 \simeq 6$.

However, because of Eq. (49), ratios of masses can be safely extracted also at slightly smaller values of β. For example the ratio of the deconfinement critical temperature T_c over the string tension $\sqrt{K}$ stays constant $\simeq 0.5 \div 0.6$ in the range $\beta = 5.7 - 6.0$.

Much less evidence for chiral symmetry breaking (with staggerend fermions) has been accumulated so far from Montecarlo simulations. The best data have been collected in SU(2) lattice gauge theory and I will briefly describe a detailed analysis made in Ref. (20). The signal for chiral symmetry breaking is that:

$$\lim_{m_f \to 0} < \overline{\psi}^f \psi^f > \neq 0 \qquad (50)$$

and correspondingly the pseudoscalar meson associated with the generator of continuum axial transformations becomes massless. Notice that, $< \overline{\psi}^f \psi^f >$ is zero for massless quarks to all orders in perturbation theory. On a finite lattice we have to compute $< \overline{\psi}^f \psi^f >$ at non zero values of the quark masses to avoid strong finite size effects coming from massless excitations. One can state the same by saying that there cannot be chiral symmetry breaking on a finite lattice. In Fig. 14 I report the data from a Montecarlo simulation for the SU(2) gauge group on $6^6 + 8^4$ lattice[20]. To obtain $< \overline{\psi}^f \psi^f >$ at $m_f = 0$ an extrapolation is

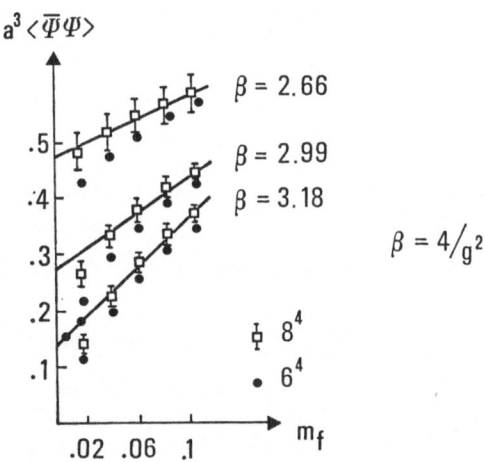

FIG. 14 – $< \overline{\psi}^f \psi^f >$ at several values of $\beta = 4/g^2$ for SU(2) on a 6^4 lattice (o) and on a 8^4 lattice (■). The lines are to guide the eye to extrapolate to $m_f = 0$.

needed; for volumes large enough and small quark masses the extrapolation will introduce a negligible systematic error in the evaluation of $< \overline{\psi}^f \psi^f >_{m_f=0}$. The measurement of $< \overline{\psi}^f \psi^f >_{m_f=0}$ as a function of β in the interesting scaling region allows the determination of its continuum value. This is shown in Fig. 15: at too large values of β the data deviate from the scaling behaviour as already discussed in the previous section and in Fig. 6.

387

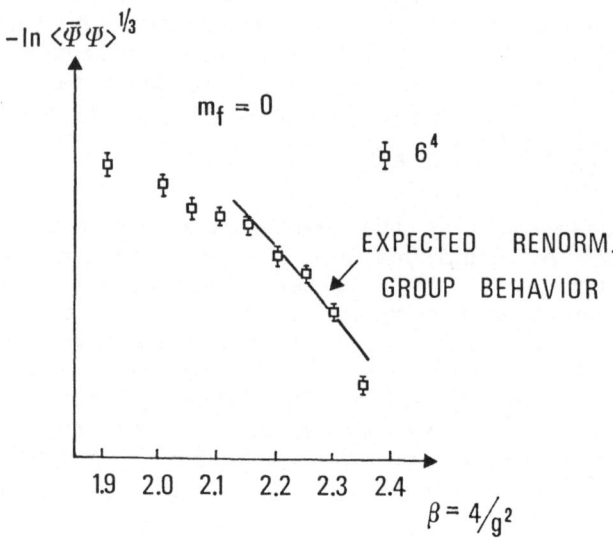

FIG. 15 - ln ($< \overline{\psi}^{\,f} \psi^{f} >_{m_f=0}^{-1/3}$) plotted against $\beta = (4/g^2)$ for SU(2) gauge theory on a 6^4 lattice.

Our conclusion is that Montecarlo results indicate beyond any reasonable doubt that $< \overline{\psi}^{\,f} \psi^{f} >_{m_f=0} \neq 0$ at any finite β and probably also for $\beta \rightarrow \infty$. No comparable results exist for SU(3).

4.2. - Strategy for the Computation of the Hadron Spectrum

To compute the hadron mass spectrum we start by defining operators carrying the same quantum numbers of the particles of which we want to measure the mass. For example for the pion, rho, proton and Δ^{++} particles we may use:

$$\pi^{+}(x) = \overline{u}^{A}(x)\, \gamma^5\, d^{A}(x),$$

$$\varrho_{\mu}^{+}(x) = \overline{u}^{A}(x)\, \gamma_{\mu}\, d^{A}(x),$$

$$P_{\delta}(x) = (u^{A}(x)\, C\, \gamma_5\, d^{B}(x))\, u_{\delta}^{C}(x) \in_{ABC} , \qquad (51)$$

$$\Delta_{\mu,\delta}^{++}(x) = (u^{A}(x)\, C\, \gamma_{\mu} u^{B}(x))\, u_{\delta}^{C}(x) \in_{ABC} ,$$

It is straightforward to compute the correlation function for these operators. For example for the pion:

388

$$G(x) = \langle \pi(x) \pi^+(o) \rangle = \qquad (52)$$

$$= \frac{\int d[U] \, e^{-S_G(U)} \, \text{tr} \left[S^u(x,0) \, \gamma^5 \, S^d(0,x) \, \gamma^5 \right]}{Z}$$

$S^f(x,0)$ is the propagator of a quark with flavour f between 0 and x.

Remember that det $\Delta_f(U)=1$. The r.h.s. of Eq. (52) is diagramatically represented in Fig. 16. In a Montecarlo simulation:

$$\int d[U] \, e^{-S_G(U)} \longrightarrow \underset{\substack{\text{link} \\ \text{configurations}}}{\Sigma} \qquad (53)$$

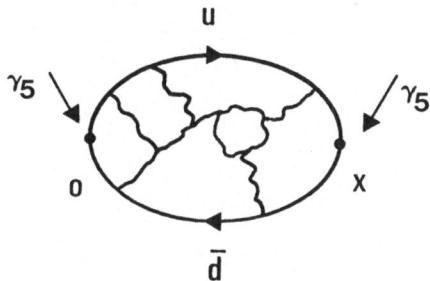

FIG. 16 - Typical diagram for the pion propagator in the quenched approximation.

We put the spatial components of the momentum equal to zero in Fourier space by summing over the spatial coordinates $\vec{x}$:

$$G(t) = \sum_{\vec{x}} G(x) \qquad (54)$$

G(t) will propagate all the possible intermediate states of mass m_n carrying the same quantum number of the pion: We expect:

$$G(t) = \sum_n Z_n \, e^{-m_n t} \qquad (55)$$

$m_n = m_\pi$, $m_{\pi'}$, $m_{\pi''}$ etc.

For large time distances the pole corresponding to the lowest lying state of mass m_π will steam out and we expect:

$$G(t) \xrightarrow[t \rightarrow \infty]{} Z_\pi \ e^{-m_\pi t} \ . \tag{56}$$

At finite t a systematic overestimation of the mass of the lowest
lying state is due to the presence of higher mass excitations. In
Fig. 17 I report the logarithm of the propagator plotted vs. t. At

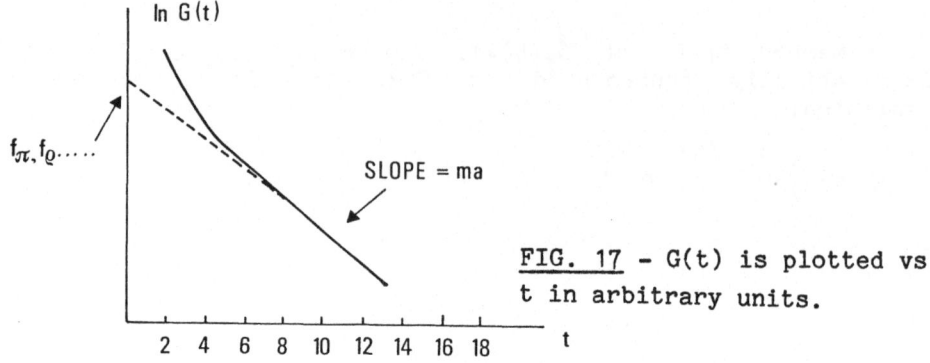

FIG. 17 - G(t) is plotted vs
t in arbitrary units.

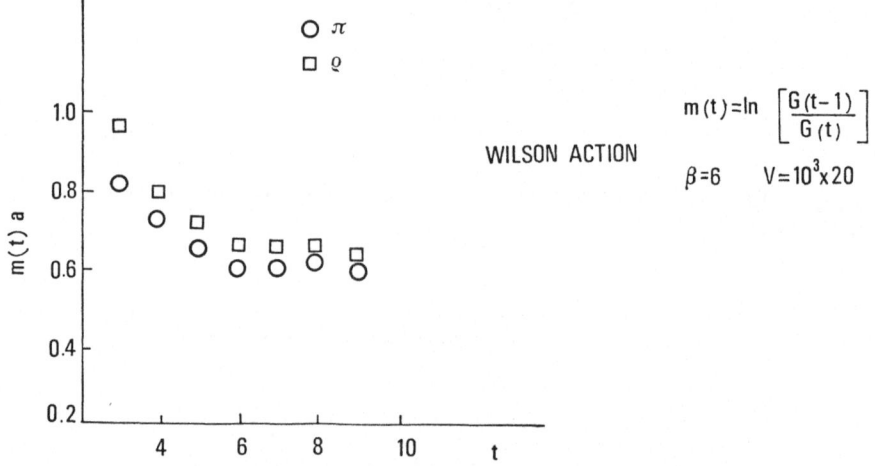

FIG. 18 - m(t)a for the ϱ and π mesons is plotted versus t. One
sees that they stabilize for t 6-8. The oscillations at large t
are due to statistical fluctuations.

large t is must behave as a straight line because only one particle
is propagating (Eq. 56). The slope of the straight line corresponds
to the particle mass in lattice units (ma) and for meson
propagators the intercept at t=0 allows the determination of the
meson decay constant f_π , f_ϱ ... etc. Also in Fig. 18 the "mass"
m(t) defined as m(t) = ln G(t)/G(t-1) for a lattice 10^3x20 at β =6

with periodic boundary conditions (max time distance $t_{max}=10$) is given. For $t \to \infty$ one expects

$$m(t) \to \text{constant} \qquad (57)$$

from above. We see that $m(t)$ stabilizes around $t=6-7$. A check of consistency on the results can be obtained by using some other operator with the same quantum numbers: different operators should lead to the same value of the mass. In the case of the pion a possible alternative operator is the four component of the axial current:

$$\pi'(x) = \overline{u}^A(x) \, \gamma^5 \, \gamma^o \, d^A(x) \qquad (58)$$

This kind of check have been succesfully used in actual Montecarlo experiments. Also variational methods can in principle be used. One defines an operator which is a linear combination of "pion" operators:

$$\Pi(x) = \sum_i C_i \, \pi_i(x) \qquad (59)$$

$$\Gamma(t) = \sum_{\overrightarrow{x}} \langle \Pi(x) \; \Pi(0) \rangle$$

The unknown variational parameters C_i are fixed by minimizing:

$$X = \min \frac{-1}{t} \ln \left(\frac{\Gamma(t)}{\Gamma(0)} \right) \qquad (60)$$

which gives the mass and the eigenfunction of the lowest lying state. This method has been used so far only in the computation of the glueball mass spectrum.

With actual lattices ($10^3 \times 20$) the evaluation of the baryon mass (proton and Δ^{++}) is much more difficult than the meson case. Contrary to what is seen in Fig. 18 the mass $m_B(t)$ never really stabilizes for $t \lesssim 10$ and in fact different operators lead to different results. The baryon mass has been evaluated with a systematic error $\sim 20-30\%$.

In order to give the masses of the particles in physical units one must fix a fundamental (strong interaction) scale and a mass parameter for each quark flavour. At fixed β we could set the scale from the mass difference between the ϱ and π (or from $\Lambda_{\overline{MS}}$ mesured in deep inelastic scattering or from the Regge slope) and then assign a mass to the quarks by fixing the π, K, ... meson masses.

$$m_\varrho^2 - m_\pi^2 = \frac{1}{a^2} \quad f(m_f a, \beta) \qquad m_\pi^2 = \frac{1}{a^2} g(m_f a, \beta) \tag{61}$$

f and g, which govern the behaviour of m_π^2 and m_ϱ^2 as a function of m_f, at fixed β, are measured in the Montecarlo experiment. For a given value of β, m_f and a are fixed by solving Eqs. (61). Once that m_π and m_ϱ are used as input all other particle masses, $\Lambda_{\overline{MS}}$, the Regge slope... are predictions of the theory. We expect all these quantities to scale in the perturbative predictable way with the cutoff a for $\beta \gtrsim 6 - 6.2$.

At $\beta \sim 6$ to really achieve this program we should have results in a range of parameters where the typical pion correlation length (for a pion of 140 MeV of mass) in lattice units

$$\xi_\pi = \frac{1}{m_\pi a} \simeq 10$$

implying a huge number of lattice points ($N \gtrsim 30^4$) to avoid boundary effects. If we try to obtain results in the small quark (pion) mass region with actual lattices ($10^3 \times 20$) we encounter the situation shown in Fig. 19. From a certain value of the quark mass

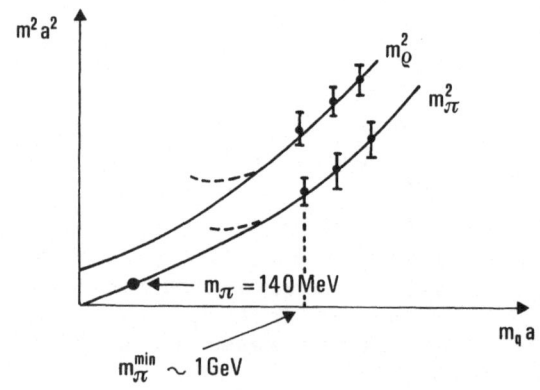

FIG. 19 - Schematic plot in arbitrary units of the Montecarlo data for m_π^2 and m_ϱ^2 as a function of the quark mass. The full line is the results one would obtain in the infinite volume limit. The dashed lines are the behaviour at small quark masses with a finite lattice. The arrow indicate the point corresponding to a pion with $m_\pi \sim 140$ MeV.

(depending on the lattice size) boundary effects deviate the Montecarlo results from the results in the infite volume limit. The extrapolation of the data to the region of m_f corresponding to the physical π and ϱ masses introduces a source of systematic errors. These have been estimated too be small for meson masses at $\beta = 6$ on a $10^3 \times 20$ lattice. However, the minimum measured pion mass ~ 1 GeV which is still too far from the physical region of interest for the study of chiral properties. Some selected references on hadron spectroscopy on the lattice are given in Ref. (21). Finally I would like to mention that techniques have been developed to evaluate by Montecarlo simulations other relevant hadronic quantities[22] like G_A/G_V the proton and neutron anomalous magnetic moment[23] and the matrix elements of operators entering in weak decays[24] (e.g. $< K^0 | H_W | \overline{K}_0 >$) with very promising results.

4.3. – Results for the Hadron Spectrum

In this subsection I will not write down numerical results obtained by different groups (these can be found in the literature)[21]. I will report the general conclusions that can be (or cannot be) understood out of the existing results.

Asymptotic scaling: contrary to the pure gauge sector no answer exist for the spectroscopy. No Montecarlo simulation explored the region $\beta > 6$. Results between $\beta = 5.7$ and $\beta = 6$ for Wilson fermions are compatible with asymptotic scaling within large uncertainties[21].

Scaling: for those results which have small systematic errors simple scaling (m_i/m_i = constant) seems at work. In this case, limited for the moment to the spectroscopy of pseudoscalar and vector mesons + some poorer result for the A_1, B and δ mesons, Montecarlo data are in good agreement with experiment. The proton over the ϱ mass ratio is significantly higher than the experimental result. The origin of this discrepancy is at present not fully understood. We remember once more that for the proton finite size effects where found rather large (20-30%) even on the largest volumes ($10^3 \times 20$). The quenched approximation could play a role for this ratio. Spin-spin effects, like the proton $- \Delta^{++}$ mass splitting, are generally found to have the correct sign, but too small in magnitude (this is also the case perturbation theory).

Finally universality seems to fail if one compares the results obtained with Wilson fermions and naïve fermions at $\beta = 5.7$. The situation could be better at $\beta = 6$, but the existing data did not clarify with the necessary accuracy the situation. Universalility under a changement of the pure gauge action was found to work well for the hadronic spectrum at $\beta \sim 5.7$ in the sense suggested by the 1/N expansion. Lot of work is needed. Still we can say that lattice techniques set a new standard in non perturbative strong interactions.

5. - QCD AT FINITE TEMPERATURE

5.1. - Generalities

The existence of a deconfinement phase transition in QCD as the physical temperature is increased was first conjectured by G. Parisi and N. Cabibbo in 1975[22]. It was subsequently rigorously proved that such a transition exists for pure gauge theories on the lattice[23]. However such rigorous result does not predict the order of the phase transition, the value of the critical temperature T_c and we cannot exclude that T_c moves to infinity in the continum limit. In recent years Montecarlo simulations provided a clear evidence for a phase transition in SU(2) and SU(3) on the lattice. For pure gauge theories the transition was recognized to be of first order in the SU(3) case and T_c and the latent heat were measured. Only more recently the first (ambiguous) results have been obtained in lattice QCD with fermions.

For the study QCD at finite temperature we will use its path integral formulation. Let us start from the partition function of a statistical system at a temperature $\beta = 1/T$:

$$Z = tr\,(e^{-\beta H}) \tag{62}$$

We can write Eq. (62) in the coordinate basis and use the usual tricks to arrive to the path integral expression of Z:

$$Z = \int dx < x \mid e^{-\beta H} \mid x > = \sum_{\substack{\text{periodic} \\ \text{path} \\ x(\beta)=x(o)}} e^{-\int_o^\beta dt \mathscr{L}(t)} \tag{63}$$

Finite temperature = finite time interval $\beta = (1/T)$ + periodicity in the time direction. For pure lattice gauge theories Eq. (63) becomes:

$$Z = \int d\left[U\right] e^{-S_\beta} \tag{64}$$

where $-S_\beta = (1/g^2) \sum_P tr \left[U(P) + U^+(P)\right]$. The lenght of lattice in the time direction L_t fixes the temperature:

$$L_t = N_\beta\, a = \frac{1}{T} \tag{65}$$

N_β is the number of lattice points in the time direction and as usual a is the lattice spacing. In Eq. (64) we must use periodic boundary conditions in time. On the contrary for fermions we have to use antiperiodic boundary conditions. In practical cases the spatial volume is finite. To obtain a good approximation of the continuum physics the following condition should be satisfied:

$$1 \ll N_\beta \leq N \tag{66}$$

The total number lattice points is

$$V = N_\beta \times N^3$$

where N is the number of lattice points in the space directions. Thermodynamical quantities can be obtained from the partition function by suitable derivatives with respect to the temperature or the volume:

$$\mathcal{E} = \text{energy density} = -\frac{1}{V}\frac{\partial}{\partial\beta}\ln Z\,(\beta\,,\,V) \tag{67}$$

For $T \rightarrow \infty$ we can exactly compute $\mathcal{E}$ because of asymptotic freedom $\mathcal{E}$ counts the number of degrees of freedom of the system (Stefan-Boltzmann law):

$$\mathcal{E}_{SB} = \frac{N^2-2}{15}\,\pi^2 T^4 \qquad \text{free gas limit} \tag{68}$$

At a large but finite T there are corrections due to gluon interactions which can be computed in perturbation theory. We can also find the expression of $\mathcal{E}$ on the lattice expressing it in terms of expectation values of Wilson plaquettes:

$$\mathcal{E} = \frac{6N}{g^2}\left[\langle P_S \rangle - \langle P_t \rangle + O(g^2)\right] \tag{69}$$

where $P_{s(t)}$ denotes the space (time) plaquette. Expression (69) holds at any value of the coupling constant and it is suitable for Montecarlo simulations.

5.2. – The Order Parameter of the Phase Transition

To recognize when the phase transition occurs we need to identify an order parameter. The pure gauge system posseses, beyond the usual gauge invariance, a global symmetry with respect to the (non-trivial) center of the group Z(N). Let us transform at fixed $t=t^*$) all the links pointing in the time direction according to:

$$U_t(x,t^*) \longrightarrow z\,U_t(x,\,t^*) \tag{70}$$

where z is an element of the center. It can be easily recognized that the transformation in Eq. (70) will leave unaltered the action:

$$\text{tr } U_t(n)\,U_\mu(n+t)\,U_t^+(n+\hat{\mu})\,U_\mu^+(n) \longrightarrow$$
$$\text{tr } z\,U_t(n)\,U_\mu(n+t)\,U_t^+(n+\hat{\mu})\,z^+\,U_\mu^+(n) = \tag{71}$$
$$= U_t(n)\,U_\mu(n+t)\,U_t^+(n+\hat{\mu})\,U_\mu^+(n)$$

Since by definition elements of the center commute with all the elements of the group. There are however gauge invariant quantities which are not invariant under the Z(N) transformation like for example the product of time-links along a line going from 0 to β :

$$L(x) = \text{tr} \left[\prod_{t=0}^{N_\beta} U(x,t) \right] \quad \text{Polyakov line} \tag{72}$$

As explained in Fig. 20 when the links are correlated on a distance $\xi/a \ll N_\beta$ then $\langle L(x) \rangle = 0$; for $g^2 \to 0$; which corresponds to $T = 1/N_\beta\, a \to \infty$, all the links are correlated on longer and longer distances so that the expectation value of $L(x)$ becomes different

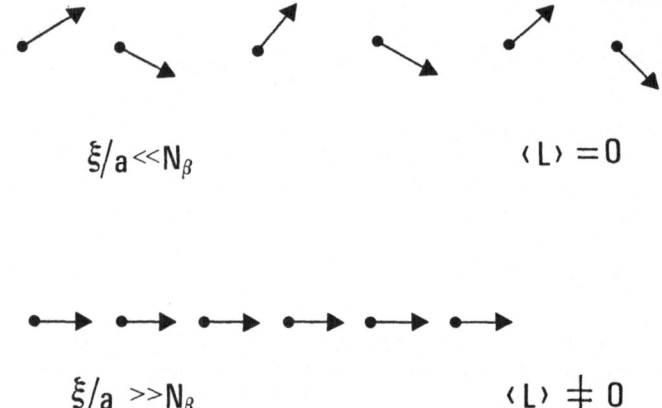

$$\xi/a \ll N_\beta \qquad\qquad \langle L \rangle = 0$$

$$\xi/a \gg N_\beta \qquad\qquad \langle L \rangle \neq 0$$

FIG. 20 - When $\xi/a \ll N_\beta$ the links fluctuate independently and $\langle L \rangle = 0$. When $\xi/a \gg N_\beta$ the links will be alligned along a common direction in group space.

from zero and the Z(N) symmetry is broken. This is precisely the signal of the phase transition. On the other hand the Polyakov line can be interpreted in terms of static colour sources:

$$\langle L(x) \rangle = \langle e^{\,i \int_o^\beta A_\mu dx_\mu} \rangle = e^{-F_q/T} \tag{73}$$

F_q is the free energy of an infinitely heavy quark. We have:

T small g large	$F_q \to \infty$	$\langle L \rangle = 0$	Confinement
T large g small	F_q finite	$\langle L \rangle \neq 0$	deconfined quarks and gluons

To measure T_c, the temperature at which the system undergoes the phase transition, the most commonly used method is to tune the coupling constant (i.e. the lattice spacing a) increasing the temperature (as g decreases) until $\langle L \rangle$ does acquire a non zero expectation value:

$$T = \frac{1}{N_\beta \, a(g)} \longrightarrow T_c \qquad\qquad (74)$$

$$T_c = \frac{1}{N_\beta \, a(g_c)}$$

Different values of N_β will corresponds to different values of g_c the value of the coupling constant at which $T = T_c$. By increasing N_β ($N \gg N_\beta$) g_c will decrease and we will approach the continuum limit. At g_c small enough we expect that T_c scales in the perturbative way (see Eq. 18):

$$T_c a \sim (\beta_0 g_c^2)^{-\beta_1/2\beta_0^2} \exp\left[-\frac{1}{2 \, \beta_0 g_c^2} \right] \qquad\qquad (75)$$

It can be shown that at high temperature a pure gauge theory SU(N) in (d+1) dimension is equivalent to a corresponding d-dimensional Z(N) spin theory. Universality then implies that the critical properties of the gauge theory should be equal to the critical properties of the corresponding spin system[24]:

$$SU(2)_{d=4} \longrightarrow Z(2)_{d=3} \equiv \text{3d Ising model}$$
$$\text{(2nd order phase transition)} \qquad\qquad (76)$$

$$SU(3)_{d=4} \longrightarrow Z(3)_{d=3} \quad 1^{st} \text{ order phase transition}$$
$$\text{(discontinuity in } \epsilon \text{ and latent heat)}$$

Finally we have to consider the effect of the quarks on the phase transition. As discussed in the previous section, the effect of virtual quarks (i.e. det $\Delta_f(U)$) is expected to have a small influence on hadron spectroscopy. This is not the case if quarks are introduced in thermodynamical problems. The fermionic action (fermions in the fundamental representation) destroys the global Z(N) symmetry of the pure gauge theory, and the expectation value of the Polyakov line is no more an order parameter. Universality arguments cannot be used to predict the fate of the deconfinement phase transition; the fermionic term can be seen as an external magnetic field in the corresponding Z(N)-spin theory[25]. If the transition was a second order (like for SU(2))even a small magnetic field is expected to suddenly destroy it; if the transition was first order (as for the more interesting case of SU(3)) then the transition disappears when the external field is strong enough.

Decreasing the quark mass increases the effect of quarks leading to a stronger magnetic field in the spin model: the real problem is to understand the fate of the phase transition for quarks (up, down and strange) which are light on the scale of the critical temperature $T_c \sim 200$ MeV. A realistic possibility is that there is no more a real phase transition but rather a rapid crossover region with a bump in the specific heat on a very small interval of temperature, related to the chiral symmetry restoring transition which was observed in the quenched case for $m_f = 0$:

$$\langle \overline{\psi}\psi \rangle \longrightarrow 0 \qquad \text{as } T \longrightarrow T_c^{chiral} \tag{77}$$

The rapid crossover between different dynamical regimes could still lead to observable physical effects.

5.3. - Montecarlo Results for QCD at finite Temperature

In this section I will report some of the more interesting results on Montecarlo simulations for QCD at finite temperature. Theoretical expectations for the order of the deconfinement phase transition based on the universality hypothesis imply that the phase transition is of second order for SU(2) and of first order for SU(3).

In Fig. 21 and Fig. 22 the expectation values of the Polyakof

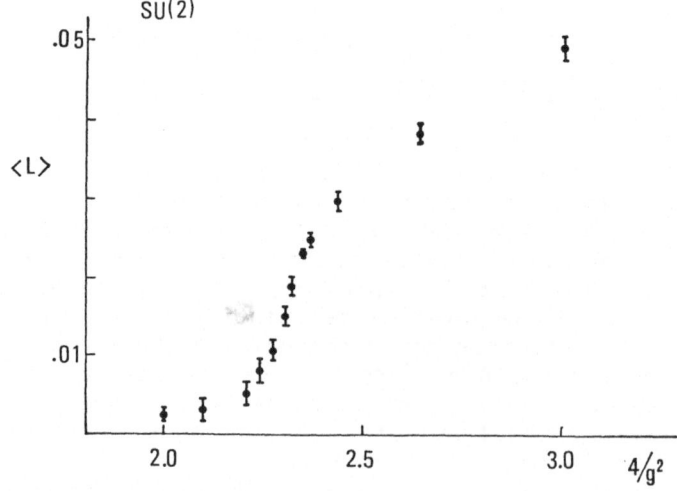

FIG. 21 - The expectation value of $\langle L \rangle$ for the SU(2) pure gauge theory as a function of the coupling constant $4/g^2$ on a $10^3 \times 4$ lattice from Ref. 26. Near T_c the data are well fitted by the expression in Eq. (79) of the text with $\gamma = 0.32 \pm 0.03$.

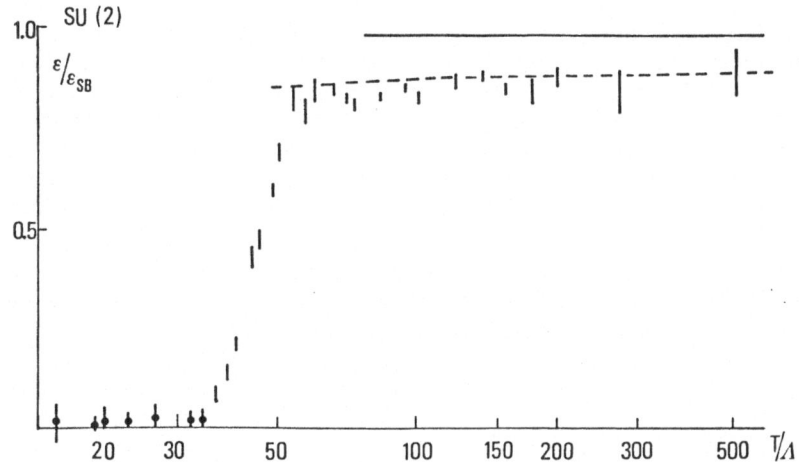

FIG. 22 - Energy density vs T on a 10^3x3 lattice normalized to the free gas energy density on a lattice of the same size (——). The dashed line include the $O(g^2)$ corrections.

line $\langle L \rangle$ and of the energy density[26] are given as a function of $4/g^2$ and of T respectively for SU(2)[26]. There is a rapid increase of $\langle L \rangle$, but not a discontinuity, at around $4/g^2 \simeq 2.3$. Correspondingly, in a very short range of T, $\in$ goes to its high temperature value. The critical temperature was estimated to be:

$$T_c \simeq 43 \, \Lambda_L \tag{78}$$

Because of universality, from the 3d-Ising model, one expects a well defined behaviour of $\langle L \rangle$ near the critical temperature:

$$\langle L \rangle \sim (T-T_c)^\gamma \qquad \text{for } T \longrightarrow T_c^+ \tag{79}$$

One knows that $\gamma_{ising} \sim 0.33$. A fit to the curve in Fig. 21 with the expression of Eq.(79) gives for SU(2):

$$\gamma_{SU(2)} = 0.32 \pm 0.03 \tag{80}$$

consistent with the scaling law for the Ising model and confirming the prediction from universality.

For SU(3), to identify the phase transition as a first order phase transition, it is commonly used the following method: one performs a Montecarlo simulation at a fixed temperature starting from a completely ordered (e.g. all the links U (x)=I where I is the identity matrix) and separately from a completely disordered (links chosen at random) configuration. After a certain number of configurations generated by some Montecarlo algorithm the two sequences will converge to a common stable thermodynamic equilibrium and averages of measurable quantities will coincide. At the critical temperature however the expectation values will attain two different values depending on whether we started with an ordered or disordered (cold or hot) configuration. On a finite lattice occasionally the system will flip from one state to the other. An explicit example of the application of this method to the determination of the critical temperature on a 8^3x3 lattice is reported in Fig. 23[27]. The critical coupling is g_c=5.55: at the critical coupling the two phase structure is stable over several thousands Montecarlo configurations. In Fig. (24) the energy density vs $6/g^2$ is also shown: a jump is observed at the critical

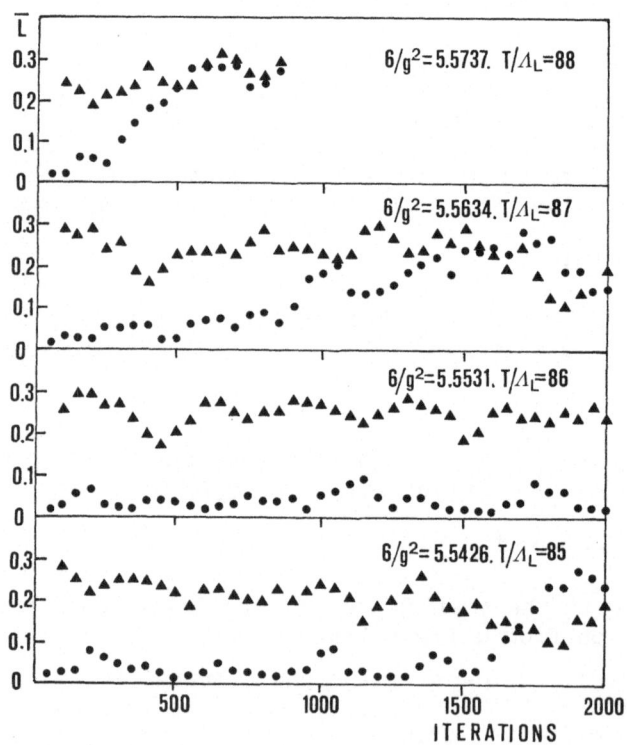

FIG. 23 - Lattice average $\langle L \rangle$ of the order parameter as a function of the number of configurations after ordered and disordered starts on a 8^3x3 lattice for various values of the coupling $6/g^2$ (SU(3)); also shown is the associated temperature using the renormalization group relation.

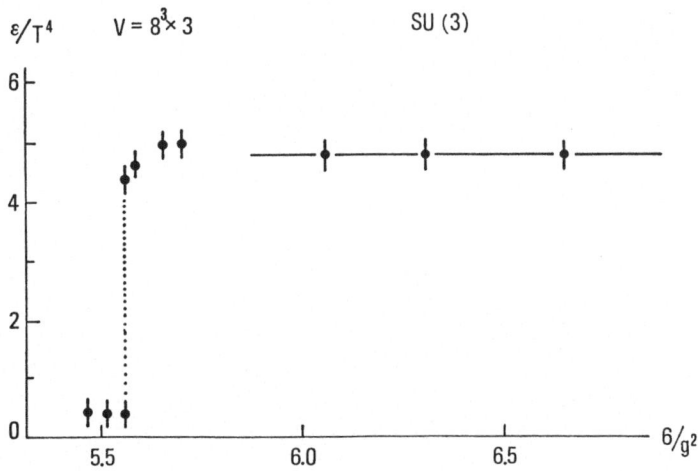

FIG. 24 - Energy density vs $6/g^2$ on a 8^3x3 lattice. The curve at high temperature is predicted by perturbation theory.

temperature from a value $\in \sim 0$ to the value predicted by perturbation theory. The sharpness of the variation of the value of $\in$ is an indication that the phase transition is a first order one. However, on a finite volume, the best method to determine if the transition is first order remains the two phase (hot and cold) start method discussed above. To our purpose it is relevant to find the values of T_c in physical units. If we measure T_c in a region of small coupling constants (large N_β) it should follows the renormalization group behaviour given in Eq. (75). A detailed analysis[28] which took into account finite size effects and extrapolated to the infinite volume limit gave the results reported in Fig. 25. As expected the asymptotic renormalization group behaviour set up at $6/g^2 \gtrsim 6$. (within large statistical errors). However the ratio of the critical temperature over the string tension is stable from smaller values of $6/g^2$:

$$
\begin{array}{cc}
6/g_c^2 & T_c/\sqrt{K} \\
5.70 \pm 0.01 & 0.56 \pm 0.03 \\
5.79 - 5.82 & 0.60 \pm 0.02 \\
5.877 & 0.56 \pm 0.01 \\
5.92 - 5.94 & 0.56 \pm 0.01 \\
6.0 \pm 0.02 & 0.50 \pm 0.02 \\
6.22 \pm 0.07 & 0.54 \pm 0.05
\end{array}
\qquad (81)
$$

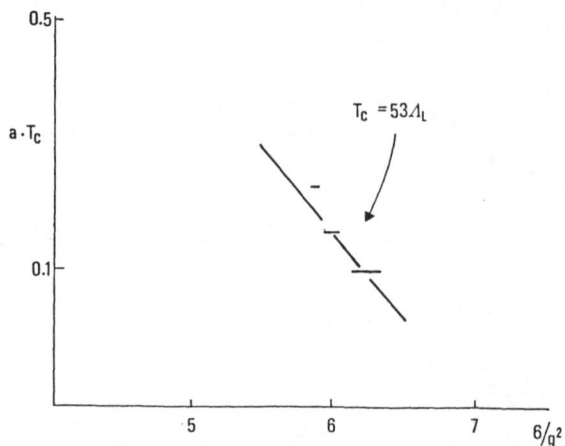

FIG. 25 - Critical temperature T_c versus $6/g^2$. The curve is the prediction of perturbation theory.

We conclude that:

$$T_c/\sqrt{K} \simeq 0.5 - 0.6 \qquad \text{for } 5.7 \lesssim 6/g^2 \lesssim 6.2$$

giving:

$$T_c \simeq (220 \pm 20) \text{ MeV} \tag{82}$$

The effect of quarks on the phase transition and the determination of the critical temperature in presence of quarks is still an open problem. Different results from several groups[29] lead to different conclusions. It is not clear if, for light quark, the transition will disappear: the results are not precise enough to distinguish rapid crossover from real phase transition. However, the transition connected to the chiral symmetry breaking still influences thermodynamical quantities and Montecarlo results seems to indicate that there is a rapid change of the dynamics in a rather small interval of temperature as it is shown in Fig. 26. We can try to give the value of the crossover (transition) temperature in physical units. In Ref. (29) they found:

$$\frac{T_c}{\Lambda_{\overline{MS}}} = 3.65 \pm 0.3 \qquad \text{4 flavour } (\frac{m_q}{T_c} = 0.4 - 0.32)$$

to be compared to:

$$\frac{T_c}{\Lambda_{\overline{MS}}} = 2.77 \pm 0.3 \qquad \text{"quenched"} \tag{83}$$

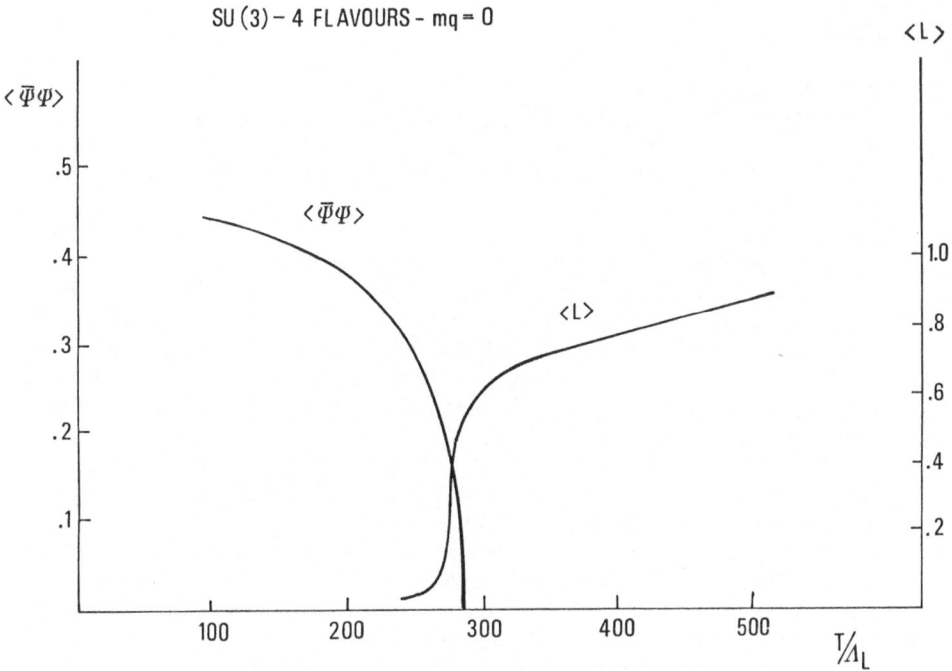

SU(3) – 4 FLAVOURS – $m_q = 0$

FIG. 26 – Expectation values of $< \bar{\psi}^f \psi^f >$ and $< L >$, (extrapolated to zero quark mass), vs T for SU(3) with 4 flavours on a lattice. We observe a rapid crossover behaviour on a small range of T.

Unfortunately no result is available with dynamical quarks for the determination of $\Lambda_{\overline{MS}}^{4\ flavours}$ from measurements of the string tension. If (but it is questionable) if we assume as in Ref. (29) $\Lambda_{\overline{MS}}^{quenched} = \Lambda_{\overline{MS}}^{4\ fermions}$ we obtain:

$$T_c \sim 300\ \text{MeV} \tag{83}$$

Still we are waiting for a study of the phase transition in presence of quarks at the same level of accuracy of what has been done for the pure gauge case in Ref. (28).

ACKNOWLEDGEMENTS

I acknowledge many enlighting discussions with F. Karsh on the present situation of QCD at finite temperature.

403

REFERENCES

1. K.G. Wilson, Phys. Rev. $D\underline{10}$, 2445 (1974).
2. C.N. Yang, R.L. Mills, Phys. Rev. $\underline{96}$, 191 (1954).
3. H.D. Politzer, Phys. Rev. Letters $\underline{30}$, 1346 (1973); D.J. Gross and F. Wilczek, Phys. Rev. Letters $\underline{30}$, 1343 (1973); Phys. Rev. $D\underline{8}$, 3633 (1973).
4. See for example E.S. Abers, B.W. Lee, Phys. Rep. $\underline{9}$, 1 (1973).
5. G.C. Wick, Phys. Rev. $\underline{96}$, 1124 (1954).
6. J.B. Kogut et al., Phys. Rev. Letters $\underline{43}$, 484 (1979); Phys. Rev. Letters $\underline{45}$, 410 (1980); Phys. Letters $\underline{98B}$, 63 (1981); G. Münster, Nucl. Phys. $B\underline{190}$ (FS3) 439 (1981); E $B\underline{205}$ (FS5) 648 (1982), G. Münster, P. Weitz, Phys. Letters $\underline{96B}$, 119 (1980).
7. M. Falcioni et al., Phys. Letters $\underline{102B}$, 270 (1981); Nucl. Phys. $B\underline{190}$ (FS3) 792 (1981).
8. A. Hasenfratz et al., Nucl. Phys. $B\underline{181}$, 353 (1981); C. Itzykson et al., Phys. Letters $B\underline{95}$, 259 (1980); M. Lüscher et al., Nucl. Phys. $B\underline{180}$, 1 (1980).
9. N. Metropolis, A.W. Rosenbluth, M.N. Rosenbluth, A.H. Teller and E. Teller, J. Chemical Phys. $\underline{21}$, 1087 (1953).
10. The method for SU(3) is explained in N. Cabibbo and E. Marinari, Phys. Letter $\underline{119B}$, 387 (1982).
11. G. Parisi Nucl. Phys. $B\underline{180}$ (FS2), 378 (1981).
12. D. Callaway and A. Rahman, Phys. Rev. Letters $\underline{49}$, 613 (1982).
13. For a recent discussion see: K.G. Bowler et al., Contributed paper to the XXII Intern. Conf. on High Energy Physics, Leipzig (1984).
14. J.D. Stack, Phys. Rev. $D\underline{29}$, 1213 (1984).
15. D. Barkai, K.J.M. Moriarty and C. Rebbi, BNL-34462 (1984).
16. For a recent review see: "Quarkonium Spectroscopy" W. Büchmüller CERN-TH 3938 (1984).
17. L.H. Karsten and J. Smit, Nucl. Phys. $B\underline{183}$, 103 (1981).
18. H.B. Nielsen and N. Ninomiya, Nucl. Phys. $B\underline{185}$, 20 (1981); $B\underline{193}$, 173 (1981).
19. E. Marinari, G. Parisi and C. Rebbi, Nucl. Phys $B\underline{190}$, 734 (1981).
20. J. Kogut, M. Stone and H.W. Wyld, Nucl. Phys. $B\underline{255}$ (FS9) 326 (1983).
21. The program to compute hadron masses by Montecarlo simulation started with the following papers: H. Hamber and G. Parisi, Phys. Rev. Lett. $\underline{47}$ 1792 (1981), E. Marinari, G. Parisi and C. Rebbi, Phys. Rev. Letter $\underline{47}$, 1795 (1981); D.H. Weintgarten, Phys. Letters $B\underline{109}$, 57 (1982). More recent results can be found in: H. Lipps, G. Martinelli, R. Petronzio and F. Rapuano, Phys. Letters $\underline{126B}$, 152 (1983). K.C. Bowler, D.L. Chalmers, A. Kenway, R.D. Kenway, G.S. Pawley and D.J. Wallace, Edimburg preprint no. 84/295 (1984); A. Billoire, E. Marinari and R. Petronzio, CERN-TH 3838 (1984). For a different approach based on the hopping parameter expansion

see for example P. Hasenfratz, I. Montway, Phys. Rev. Letters 50, 309 (1983) and references therein.

22. N. Cabibbo and G. Parisi, Phys. Letters 59B, 67 (1975).

23. C. Borgs and E. Seiler, Commun. Math. Phys. 91, 329 (1983).

24. B. Svetitsky, L.G. Yaffe, Nucl. Phys. B210 (FS6) 423 (1982).

25. T. Banks and A. Ukawa, Nucl. Phys. B225 (FS9), 145 (1983).

26. P.V. Gavai, H. Satz , Bielefeld preprint BT-TP84/14 (1984).

27. T. Celik, J. Engels and H. Satz, Phys. Letters 125B, 411 (1983).

28. A.D. Kennedy, J. Kuti, S. Meyer and B.J. Pendleton, S. Barbara preprint NSF-ITP-84-61 (1984).

29. P. Hasenfratz, F. Karsch and I.O. Stamatescu, Phys. Letters 133B 221 (1983); T. Celik, J. Engels and H. Satz, Phys. Letters 133B, 427 (1984); M. Fischler and R. Roskies, Pittsburgh preprint Pitt-84-18 (1984); R.V. Gavai, M. Lev and B. Petersson, Bielfeld preprint BI-TP-84/10 (1984); F. Fucito, C. Rebbi and S. Solomon, Caltech preprint CALT-68-1124 and 1127 (1984); J. Polonyi, H.W. Wild, J.B. Kogut, J. Shigemitsu and P.K. Sinclair, Phys. Rev. Letters 53, 644 (1984).

INDEX

Ablation-abrasion model 264, 265
Action functional 169
Anderson model 131-132
Asymptotic freedom 372

Backbending 103
Backward Heisenberg picture 167
Boltzmann equation 217, 223-225, 229
Boundary condition 168, 170-172, 180, 182, 186, 191-192, 198-199, 206

Central collisions 37, 42, 211, 230
Chaos 120, 128-129, 142, 145, 157
Characteristic functions 185-186, 195, 200, 206
Chiral bag 287
Chiral symmetry 285
Coalescence 239, 243
Collective flow analysis 251-262
Collective modes 72-79, 92
Collective variables 1, 2, 13
Colliders 309, 349-350
Color flux tubes 361-364
Complete fusion 37
Compound nucleus reactions 120, 123, 127, 131
Compound nucleus resonances 145, 153
Condensation theory 264, 267-272
Correlation width 124-125
Correlations 121, 187, 200
Cosmic rays 308

Delta production 216, 229, 249
Delta-3 statistics 121-122
Deuteron formation 239-247, 277
Disordered systems 120, 131-133, 142
Dissipation 1-3, 24-28, 70, 82, 90, 94
Drift coefficient 71

EMC effect 306
Elastic enhancement factor 140
Ensemble average 134
Entropy 1-3, 16-18, 25-28, 228, 233-239, 244, 270
Equation of states 210, 236-238, 266-278, 295
Equilibration 140
Ergodic system 156-158
Ericson fluctuations 139

Fermions on the lattice 380-385
Fire ball model 59, 211-214
Fission 69, 82-87, 90-96, 105
Fission fragments 42
Fluctuations 71, 120-123, 127, 130, 186, 193-200, 205, 235, 258, 275-277
Fokker-Planck equation 24, 29
Freeze out 234-244

Gamma spectroscopy 106-109, 111-113
Gauge invariance 370, 396
Gaussian orthogonal ensemble 121-122, 125-126, 129-131, 152-156
Gaussian unitary ensemble 152, 161
Generating function 133
Generating functional 369
Gluon fields 368
Goldstone modes 138
Grassman variables 134-135, 138-141

Hadron spectrum 367, 386-393
Hagedorn temperature 340
Hauser-Feschbach formula 140
Hedgehog 288, 300
High energy collisions 305
High spin states 99-110
Hubbard-Stratonovitch transformation 135-137
Hydrodynamics 235-239, 244, 251, 258-259, 276
Hyperbolic symmetry 137

Integrable system 156
Intranuclear cascade model 222-224, 230-235, 242-251, 258-262, 276
Irrelevant variables 18, 25
Irreversibility 16, 224
Isentropic expansion 234, 269-275

JACEE 330

LMG model 202
Landau equation 222-225, 235-236, 250
Lattice QCD 367-405
Lattice gauge theories 368
Level repulsion 151
Limiting temperature 48, 65
Linear momentum transfer 37, 44
Linear response 174-175
Link variables 370
Liouville equation 3-5, 221
Liouville space 173
Liquid-gas transition 211, 264-276

MIT bag model 284
Mass transfer 204
Massive modes 141
Master equation 24
Mean field approximation 180, 194
Mean free path 48, 213-218, 234, 236, 276
Multifragmentation 36, 48, 212, 265-276
Multiplicities 54, 245, 248-251, 265, 277, 315-319

Navier-Stokes equation 236-238
Nearest neighbour spacing 121-122
Nuclear temperature 39, 41, 62, 73, 79, 83, 107-108, 113

Pair correlations 101-102, 104
Participant-spectator model 212-216, 230
Particle emission 53-54
Particle production 361-364
Peripheral collisions 44, 211
Pion production 215, 248-252, 262
Plaquette 370

Poisson spectrum 150-151, 155-156, 160
Projection method 1-4, 10, 17, 20
Pseudo Hamiltonian 169, 178, 190
Pseudo rapidity 321

QCD 367-405
QCD at finite temperature 394-403
QCD fondamental scale 372
Quark deconfinement 210, 307, 340, 367, 394
Quark fields 368
Quark matter 297
Quark-antiquark potential 378-379
Quark-gluon plasma 306, 333, 337, 343

Random matrix model 128, 130, 133, 142, 146
Random phase approximation 184
Rapidity 319-336
Regular motion 119, 127
Relaxation time 223, 226-228
Replica trick 134, 137, 139-142
Residual nucleus 37
Rotational motion 101, 104, 111-112
Roughening transition 375

S-matrix 125-126, 133, 138
Saddle-point method 137-139
Shape transition 101-102
Sinai's billiard 129, 150, 155, 158
Single particle observable 180
Skyrmion-skyrmion interaction 291, 300-301
Skyrmions 283, 288-292, 299
Soliton model 283
Spectator physics 262-264, 267
Spectral fluctuations 145, 149-158
Spectral rigidity 151
Spin alignment 101-104, 113
Spin-spin correlations 69, 84, 90, 93, 97
Stationary value 171
Statistical equilibrium 72-75, 79
Stochasticity 120, 123, 127-130, 142

Stopping power 330, 333
Strangeness production 343-349
Strong coupling expansion 374
Superdeformed states 102, 110

Thermal model 241, 243, 249
Thermodynamic limit 140-141
Time Dependent Hartree-Fock 2,
    70, 180-206, 221, 264-266,
    268-275
Topological charge 286
Topological current 287
Transition probalities 185, 187,
    193
Transmission coefficients 138
Transport equations 4,   23-25,
    70-71, 276-277

Transverse energy 337-343

Uehling-Uhlenbeck equation 222,
    228
Universality of fluctuations
    152-156

Variational principle 169
Viscosity 235-237
Vlassov equation 221

Wave packet spreading 201
Wigner representation 7, 219-221
Wilson action 370
Wilson loop 375

Yrast states 74, 101-104, 107